# PRACTICAL HANDBOOK

## *of*

# Spatial Statistics

EDITOR-IN-CHIEF
## Sandra Lach Arlinghaus, Ph.D.
Institute of Mathematical Geography
and The University of Michigan

SPECIALIST ASSOCIATE EDITOR
**Daniel A. Griffith, Ph.D.**
Syracuse University

ASSOCIATE EDITORS
**William C. Arlinghaus, Ph.D.**
Lawrence Technological University

**William D. Drake, Ph.D.**
The University of Michigan

**John D. Nystuen, Ph.D.**
The University of Michigan

CRC Press
Taylor & Francis Group
Boca Raton London New York

CRC Press is an imprint of the
Taylor & Francis Group, an **informa** business

CRC Press
Taylor & Francis Group
6000 Broken Sound Parkway NW, Suite 300
Boca Raton, FL 33487-2742

First issued in paperback 2019

No claim to original U.S. Government works

ISBN 13: 978-0-367-44889-9 (pbk)
ISBN 13: 978-0-8493-0132-2 (hbk)

**Visit the Taylor & Francis Web site at**
**http://www.taylorandfrancis.com**

**and the CRC Press Web site at**
**http://www.crcpress.com**

## Library of Congress Cataloging-in-Publication Data

Practical handbook of spatial statistics / editor-in-chief, Sandra
    Lach Arlinghaus ; specialist associate editor, Daniel A. Griffith ;
    chapter contributors other than editors, Daniel G. Brown ... [et
    a;].
        p. cm.
    Includes bibliographical references (p. - ) and index.
    ISBN 0-8493-0132-7 (alk. paper)
    1. Geography—Statistical methods. 2. Spatial analysis
(Statistics) I. Arlinghaus, Sandra L. (Sandra Lach). II. Griffith,
Daniel A.
G70.3..P73 1995
910'.021                                                    95-24710
                                                            CIP

Library of Congress Card Number 95-24710

## Foreword

## THE IMPORTANCE OF POSITION IN SPACE

### Sandra L. Arlinghaus and John D. Nystuen

*"The rain is raining all around,*
*It rains on land and sea..."*

unknown children's song.

Most of us have watched the rain fill potholes in our streets or the snow pile up in the bird baths in our yards as weather commentators report that these are "scattered showers" or "snow flurries." The reports do not tally with our own observations. The problem is that the official forecasts of widespread phenomena, such as rain or snow, are based on a small set of observations at separated locations; the measurements derived from these locations are then attributed to an entire region. But precipitation is spatially and temporally heterogeneous, perhaps even fractal in nature. Finer and finer meshes of observations yield higher and higher variances, yet the pattern is not random in space or time. Nearby locations experience similar intensities of precipitation as rainfall sweeps over a region leaving moist rain tracks. After many repeated rain events, amounts of rain recorded at each location begin to converge. This type of phenomenon calls for a measurement strategy that focuses on the spatial character of widespread phenomena. [Nystuen, McGlothin, and Cook, 1993] Spatial statistics does just that.

On April 29, 1986, U. S. scientists detected the nuclear incident at Chernobyl prior to official Soviet acknowledgment that any event of note had occurred (see Sadowski and Covington, 1987 for a technical discussion). Unusually high energy emissions, well above normal reflected energy levels, were detected by multispectral scanners operating with 30 to 80 meter resolutions from a commercial satellite orbiting the Earth. Immediate reference to data from another satellite with ten meter resolution confirmed that the hotspot was confined to a very small region implying an even higher, very localized, energy release. Calculations revealed that the temperature of the energy release was that of burning graphite. A meltdown was occurring. Spatial, spectral, and temporal precision made this detection possible. A strategy to detect rare events in space and time was employed. Positional accuracy in data recording is a prerequisite for most spatial statistics applications whether the phenomena are widespread or located at unique points.

# I. DEPENDENCE OF OBSERVATION ON SPATIAL POSITION

As Griffith notes in the next chapter,

*Spatial statistics differs from classical statistics in that*
*the observations analyzed are not independent...*

In the case of the rainfall problem, the weather stations scattered across a region are generally viewed, for the purposes of forecasting, as independent. However, we all know that for the most part, they are not independent--the distance between them, and the distance between local residents and rain gauges, matters. The fundamental assumption of spatial statistics acknowledges this difficulty, involving dependence on spatial position, at the outset.

This handbook displays various techniques from spatial statistics and illustrates them using real world data. It also illustrates how techniques for spatial statistics differ from, or are similar to, corresponding techniques for classical statistics.

## II. A CLASSICAL VIEWPOINT

The roots of useful ideas often have multiple branches that penetrate different disciplinary horizons. The notion of distance decay is one that can be traced to Isaac Newton; in geographical interpretations, it later rested in the hands of Tobler [1961, 1992], Warntz [1965], and others. Currently, note that Griffith's discussion of spatial weights matrices, and Griffith's and Can's consideration of land use patterns around the central city, refer to distance decay in their chapters in this handbook.

Another powerful notion---that of a transformation---has served as the backbone of twentieth century mathematics. It has altered disciplinary focus from the study of individual mathematical systems to the study of relations between mathematical systems. D'Arcy Thompson explored this transformational approach to biology [1917]; Tobler has employed transformations and some of Thompson's work in various aspects of cartographic analysis [1961, 1992] as has Bookstein in the measurement of biological shape [1978]. Clarke (1995) presents a transformational view of cartography in his text on analytical and computer cartography. Thirty years ago Michael Dacey and others contributed to the development of spatial statistics in highly original ways. Dacey used the idea of a dimensional transformation to permit evaluations of the spatial association of point and area phenomena. Indeed, one of the figures in Vasiliev's chapter calls to mind some of Dacey's earlier efforts [1964]. Today, note that Wong employs the transformation of spatial aggregation to move across a hierarchy of scale changes; Feng sees the fertility transition as one transformation in the broader transition theory framework (for a complementary current perspective on transition theory see Drake in Ness, Drake, and Brechin, 1993); Long uses the Jacobian as a normalizing factor to transform a correlated mathematical space into an uncorrelated one; Brown considers the role of the transformation in a Geographic Information Systems (GIS) context; and, Li adopts a transformational viewpoint in both content and written format in his essay on parallel processing. The concept of transformation is critical in these, and in other, chapters of this handbook.

Ideas capable of widespread application have deep and far-flung roots. It is the aim of this handbook to show the clearly visible, tangible branching structure of the spatial statistics 'tree;" we hint here at its roots in the hope that the diligent reader will be encouraged to dig into the rich and fertile minds from which this tree was raised.

## REFERENCES

**Bookstein, Fred L.** *The Measurement of Biological Shape and Shape Change.* Berlin, Springer-Verlag, 1978.

**Clarke, Keith C.** *Analytical and Computer Cartography,* 2nd edition. Prentice-Hall, Englewood Cliffs NJ, 1995.

**Dacey, Michael F.** Imperfections in the uniform plane. *Michigan Inter-University Community of Mathematical Geographers,* John D. Nystuen, Editor, May, 1964. Reprinted, *Solstice: An Electronic Journal of Geography and Mathematics,* Summer, 1994.

**Ness, Gayl D.; Drake, William D.; Brechin, Steven R., eds.** *Population-Environment Dynamics: Ideas and Observations.* Ann Arbor, University of Michigan Press, 1993.

**Nystuen, Jeffrey A.; McGlothin, Charles C.; and, Cook, Michael S.** The underwater sound generated by heavy rainfall. *Journal of the Acoustical Society of America,* 93 (6), 3169-3177, 1993.

**Sadowski, Frank G. and Covington, Stephen J.** *Processing and Analysis of Commercial Satellite Image Data of the Nuclear Accident near Chernobyl, U.S.S.R.* U. S. Geological Survey, Bulletin 1785. U.S. Government Printing Office, Washington, D. C., 1987.

**Thompson, D'Arcy W.** *On Growth and Form.* Cambridge University Press, 1917.

**Tobler, Waldo R.** *Map Transformations of Geographic Space.* Ph.D. Thesis, University of Washington, 1961.

**Tobler, Waldo R.** Preliminary representation of world population by spherical harmonics. *Proceedings of the National Academy of Sciences of the United States of America,* 89:6262, 4 Jul. 15 1992

**Warntz, William.** *Macrogeography and Income Fronts.* Philadelphia, Regional Science Research Institute, Monograph #3, 1965.

# Preface

*Anyone who already uses statistics and maps will benefit from using spatial statistics.*

This handbook is a reference work that illustrates the differences between classical statistics, and spatial statistics--those techniques which account, in some manner, for geographical position. It does so at both the abstract level and the real world level.

Useful features in this handbook include:
- 1. Comparisons of classical and spatial statistical techniques;
- 2. Rules-of-thumb capturing the essence of selected techniques;
- 3. Real-world data used to illustrate abstract concepts;
- 4. Real-world locales of timely current nature;
- 5. Cutting-edge topics in spatial statistics;
- 6. Spatial index that maps relative locations of terms by chapter;
- 7. Reference lists grouped by chapter and for the book as a whole.

Editors and authors alike have worked hard to create a useful and uniform document; however, as is always the case, there is no perfect document. If you find something that you wish to share with us for the next edition--new material, corrections, or whatever--please communicate with:

William C. Arlinghaus
ATTN: CRC Handbooks
Department of Mathematics and Computer Science
Lawrence Technological University
21000 West Ten Mile Road
Southfield, MI 48075
arlinghaus@LTU.edu

Thank you in advance; we have tried to be careful and to contribute something that is useful and different; in the end, despite all the care of the many who have generously offered time, effort, and advice, the blame for errors, omissions, or poor judgment must rest with the Editor-In-Chief, alone.

Sandra Lach Arlinghaus, Ann Arbor, MI                    June 7, 1995.

## Affiliations of Editors and Authors

*Editor in Chief:*
Sandra Lach Arlinghaus, Ph.D. Geography.
Founding Director, Institute of Mathematical Geography (independent),
2790 Briarcliff, Ann Arbor, MI 48105; and,
Adjunct Professor of Mathematical Geography and Population-
    Environment Dynamics
School of Natural Resources and Environment
The University of Michigan
Ann Arbor, MI 48109

*Associate Editors:*
William C. Arlinghaus, Ph.D. Mathematics.
Professor and Chairman,
Department of Mathematics and Computer Science,
Lawrence Technological University
21000 W. Ten Mile Rd.
Southfield, MI 48075

William D. Drake, Ph.D. Operations Research.
Professor,
School of Natural Resources and Environment,
School of Public Health, and,
College of Architecture and Urban Planning,
The University of Michigan
Ann Arbor, MI 48109

Daniel A. Griffith, Ph.D. Geography.
Professor and Chairman,
Department of Geography and Program in Interdisciplinary Statistics
144 Eggers Hall
Syracuse University
Syracuse, New York 13244-1160

John D. Nystuen, Ph.D. Geography.
Professor,
Geography and Urban Planning,
College of Architecture and Urban Planning,
The University of Michigan
Ann Arbor, MI 48109

*Chapter contributors other than editors:*

Dr. Daniel G. Brown
Department of Geography
Michigan State University
East Lansing, Michigan 48824

Dr. Ayse Can
Fannie Mae
Office of Housing Research
3900 Wisconsin Avenue, NW
Washington, D.C. 20016-2819

Mr. Michael Feng
Department of Geography
144 Eggers Hall
Syracuse University
Syracuse, New York 13244-1160

Dr. Bin Li
Department of Geography
P.O. Box 8067
University of Miami
Coral Gables, Florida 33124-2060

Dr. Daniel Long
Northern Agricultural Research Center
Star Route 36, Box 43
Montana State University
Havre, Montana 59501

Dr. W. Scott Overton
Department of Statistics
44 Kidder Hall
Oregon State University
Corvallis, Oregon 97331-4606

Dr. Stephen V. Stehman
Faculty of Forestry
322 Bray Hall
SUNY/CES&F
Syracuse, New York 13210

Ms. Irina Ren Vasiliev
Department of Geography
State University College
Geneseo, New York 14454

Dr. David Wong
Department of Geography and Earth Systems Science
George Mason University
Fairfax, Virginia 22030

## Acknowledgments

One of the nice parts of writing a book with many co-editors and co-authors is having the opportunity to get to know an academically diverse set of interesting individuals. The editors extend their heartiest, and primary, thanks to all of the authors and hope one day to have the good fortune to meet each of them in person. They have all been enjoyable to work with and hopefully that constructive spirit of interaction is evident in the fruit it has borne. The editors in Ann Arbor also wish to extend particular congratulations to Daniel A. Griffith; Dan has worked endlessly and painstakingly to make this work come to life!

In addition to the authors, we thank, bearing in mind the unfortunate possibility that we might inadvertently have omitted someone, the following individuals and organizations.

Community Systems Foundation (CSF) of Ann Arbor has generously provided access to computing resources for this project. Their continuing support of handbook projects has been indispensable. William E. Arlinghaus provided complimentary professional copyright procurement assistance when needed. His efforts in these matters is greatly appreciated. Syracuse University generously extended copyright permission to include segments of computer programs written by Griffith.

Wayne Yuhasz at CRC Press was our original editor. He was delightful to work with. Nora Konopka, also of CRC Press, has served tirelessly as a friendly liaison during the transition in change in editorship at CRC. How fortunate we are, now, to have Robert Stern as the new editor to help see us through the final stages of manuscript preparation. His assistance with, and concern for, a variety of matters is greatly appreciated. We also salute the many others at CRC who have been so helpful.

# TABLE OF CONTENTS

Estimating an Autoregressive Response Model
Comparison of AR-based ANOVA and Conventional ANOVA
Conclusions
Acknowledgments
References

# Chapter 1

## INTRODUCTION: THE NEED FOR SPATIAL STATISTICS

### Daniel A. Griffith

Geographic information and analysis (GIA) is a critical, emerging scientific discipline. [Cook *et al.*, 1994] The establishment in the late 1980s of the National Center for Geographic Information and Analysis (NCGIA), with funds from the National Science Foundation, attests to its importance. Data that are tied to position on the Earth's surface, that are spatial or geo-referenced data, often serve as the empirical backbone of much of the research that is presently done in this general context. The statistical analysis of spatial data forms the subject matter of "spatial statistics." Indeed, in writing about his Cornell Theory Center supercomputer project, Durrett notes [1994, p. 4] that

> [f]or a half century, the literature ... has been dominated by models in which spatial location is ignored and each individual [site] is assumed to interact equally with all the others. Such models provide an acceptable approximation in many contexts, but there is a growing list of examples of phenomena that must be treated by models that are spatially explicit ... .

Others echo this need for models that are spatially explicit: the *Chorley Report* [1987] released in Great Britain; reports of the National Research Council in the United States entitled Spatial Data Needs and Renewing U.S. Mathematics [1990a, b]; Warnecke's survey of state activities [1990, 1991]; and, International Business Machine's (IBM's) feature article in 1991.

## I. COMPONENTS OF GEOGRAPHIC INFORMATION AND ANALYSIS

The academic subject matter of Geographic Information and Analysis comprises three principal components: geographic information systems (GISs), spatial statistics, and classical spatial analysis. From a broad perspective, geographic information systems are a form of applied computer science, spatial statistics is a form of specialized applied multivariate statistics, and classical spatial analysis is a form of quantitative geography.

GISs constitute a powerful new technology that can address many information needs of decision makers working with geographically

0-8493-0132-7/95/$0.00+$.50

(geo-) referenced data--data that are tagged, or identified, by locational coordinates. Often this tagging is for coordinates on the Earth's surface; today many tags are created with the aid of the satellite-based technology of the global positioning system (GPS). GISs are unique combinations of computer hardware and software--including high-resolution graphic displays, large-capacity electronic storage devices, efficient strategies for data organization, high-volume communication channels, specialized algorithms for data integration and reliability analysis, and specialized query computer languages. These components, together with massive amounts of highly complex geo-referenced data, are organized efficiently (through a sequence of electronic interfaces) to store, inventory, manage, search, manipulate, display (instantaneously), and analyze information contained in a geo-referenced database. The goal is to combine tabular attribute data with computerized maps in an enlightening way, achieving this goal by having a large storage capacity, a rapid response time, and a wide repertoire of analytical functions. Together, these support a dramatic mode of scientific visualization.

Generally speaking, spatial statistics is concerned with the statistical analysis of geo-referenced data. In 1991 a National Research Council report characterized this subject area as (p. vii)

> *one of the most rapidly growing areas of statistics, rife*
> *with fascinating research opportunities.*

Yet, despite these opportunities, many statisticians remain unaware of them and most students in the United States are never exposed to course work in spatial statistics--this handbook attempts to bridge that gap. *Spatial statistics differs from classical statistics in that the observations analyzed are not independent; this single assumption violation is the crux of the difference.* Cressie [1991, p. 3] characterizes this problem area as follows:

> *Independence is a very convenient assumption that*
> *makes    much    of    mathematical-statistical    theory*
> *tractable.    However, models that involve statistical*
> *dependence are often more realistic.*

Moreover, observations are correlated strictly due to their relative locational positions (referred to as spatial autocorrelation), resulting in spill-over of information from one location to another (locational information). This spill-over causes redundant information to be present in data values. The redundancy increases as the degree of locational dependence increases. This duplication of information produces complications in the statistical analysis of geo-referenced data that remain dormant in the statistical analysis of traditional data composed of independent observations. That is, invoking an assumption of independent observations suppresses potential data complexities. The net result is that classical statistics applied to geo-referenced data fail to capture locational information, raising questions of estimator sufficiency, bias, efficiency, and consistency. These four cardinal statistical properties might also be coupled with two others: robustness and minimum variance. Geo-referenced data are highly complex, with spatial dependence introducing further complications. Examples of studies that demonstrate changes in statistical inferences when a traditional ordinary least squares (OLS) regression model estimation is replaced with a spatial statistical one include explaining the Huk rebellion [Cliff and Ord, 1981, p. 237], predicting county wage rates [Anselin, 1988, pp. 191, 193], and estimating mean density of coffee production [Griffith, 1989].

Classical spatial analysis has been treated conceptually for about a century, and algebraically for several decades. It has played a central role in the quantitative scientific tradition in geography. Spatial analysis involves spatial operations research (such as minimum route selection), logical overlaying (identifying areal units possessing joint categorical attributes), triangulated irregular networking (TIN; which in a sense forms the basis of spatial statistics), and buffering (distance bands around points or lines), among others. Most of these procedures currently are available as functions in GIS tool kits; they automate what once were tedious manual tasks.

## II. BACKGROUND:
## THE IMPORTANCE OF LOCATIONAL INFORMATION

Scholarly awareness of complications attributable to locational information latent in spatial data, especially in terms of their impact on the validity of traditional statistical analyses, has emerged recently among scientists, catapulting spatial statistics into the forefront of much data analysis discussion. In fact, the analysis of spatial data has become a major preoccupation of numerous statisticians only rather recently. Accordingly, increasing attention has begun to focus on the general field of spatial and geo-statistics (and spatial econometrics). For instance, the announced goals in the solicitation of proposals for the NCGIA [NSF, 1987] included the objective of promoting advances in spatial statistics within the context of GISs. And, the Board of Mathematical Sciences of the National Research Council [1990b] has targeted spatial statistics as one of twenty-seven topics of national concern in mathematics (its rank is 17). Similar evidence has been made available by the British scientific community, particularly through that country's Regional Research Laboratories initiative. (For example, as a cooperative effort, the Department of Mathematics at Lancaster University and the North West Regional Research Laboratory in Great Britain initiated a project to integrate statistical and GIS software, while the U.K. Economic and Social Research Council (ESRC) funded an "experts" workshop, held at the University of Sheffield in March of 1991, on this same theme. [see Goodchild, Haining, and Wise, 1992]

In 1989 a symposium entitled "Spatial statistics: past, present, and future" was hosted by the Department of Geography, Syracuse University. Reported findings of this symposium [Griffith, 1990] include

- (1) there is a need for a MINITAB or SAS for spatial statistics [Ripley, p. 56; Haining, p. 101; Doreian, p. 105; Wartenberg, p. 153; Upton, p. 158];
- (2) there is a need for many more additional relevant empirical applications of spatial statistical techniques [Martin, p. 27; Richardson, p. 130; Upton, p. 354; Wartenberg, p. 393]; and,
- (3) as the issue of computational intensity subsides, and GIS software becomes increasingly user-friendly, more ubiquitously available, and a source for implementing spatial statistical techniques, the danger of malpractice by the non-specialist practitioner grows [Anselin, p. 73; Martin, p. 124].

Similar sentiments are echoed in Cressie [1991; pp. 657, 699], while a review of the literature demonstrates that little spatial statistics and spatial econometrics technology has been adopted in scientific research [see Anselin and Griffith, 1987; Anselin and Hudak, 1992], highlighting the existence of a dissemination problem. In response to this first point, Griffith [1989, 1993c] has developed both MINITAB and SAS macros for undertaking spatial statistical analyses, whereas Anselin [1992] has developed SPACESTAT for executing spatial econometrics.

Therefore, although a scholarly awareness of spatial statistics currently exists in various fields (especially those in the geosciences), important research needs have been defined by leading researchers in the field, and although both curricular developments and dissemination endeavors are underway, insufficient synthesis and consulting materials are available. The literature is piecemeal, specialized, and diverse. Its content consists of books that tend to be introductory [Goodchild, 1986; Griffith, 1987; Odland, 1988] or advanced [Ripley, 1988]; little exists at the intermediate level. One book covers pattern models [Ahuja and Schachter, 1983], another surface partitionings [Okabe, Boots, and Sugihara, 1992]. Some books treat mostly point pattern analysis [Cliff and Ord, 1981], others geo-statistics [Cressie, 1991], spatial autoregressive models [Griffith, 1988], or spatial econometrics [Paelinck and Klaasen, 1979; Anselin, 1988]. Some books are extremely theoretical [Bartlett, 1975; Matérn, 1986], while others are very applications-oriented [Upton and Fingleton, 1985; Isaaks and Srivastave, 1989; Haining, 1990]. This handbook fills the gap identified here, and in so doing explicitly addresses the two points of furnishing additional relevant empirical applications of spatial statistical techniques, and providing guidance to help non-specialist practitioners avoid improper spatial statistical practice. The style, nature, and scope of this volume is designed to pique the curiosity of graduate students and spatial scientists alike.

## III. BACKGROUND: STATISTICAL ESTIMATOR PROPERTIES

Complications mentioned in the preceding discussion can be referenced to the statistical properties of estimator sufficiency, unbiasedness, efficiency, and consistency. An estimator is sufficient if, when reducing the sample data to its corresponding summary statistic(s), it does not foster a loss of information pertinent to the population to which an inference is to be drawn. Exactly all of the information relevant to a population that is contained in a sample is condensed into a sufficient statistic. This definition means that all the knowledge about a parameter than can be gleaned from both the individual sample values and their ordering *must* be gained from the value of the estimator alone. Griffith [1988] uses Neyman's factorization theorem, and the theorem on completeness for the exponential family [Lindgren, 1976] to show that in the presence of locational information the mean and standard deviation are not sufficient statistics. Rather, a four-dimensional statistic is needed that incorporates both the geographic arrangement of observations and the nature and degree of their spatial dependence. *These are the estimators found in spatial statistics, the ones that should be employed in the statistical analysis of geo-referenced data.*

An estimator is unbiased if the mean of the sampling distribution generated by it equals the parameter that it is supposed to estimate. On average, the estimator value is equivalent to its corresponding population parameter value. Many common statistics involving simple linear combinations of geo-referenced data, such as the arithmetic mean and regression coefficients, are unbiased. But ones involving more complicated arithmetic operations, such as the variance and correlation coefficients, tend not to be. For example, Cordy and Griffith [1993] discuss how, in some cases, the OLS regression coefficient estimators provide a reasonable alternative to their spatial statistical counterparts, while the usual variance estimators are severely biased when regression errors are spatially autocorrelated. The main problem with using OLS in the presence of spatial autocorrelation is that the usual standard error estimator tends to underestimate the true standard error. This result indicates that the geographic arrangement of observations and the nature and degree of prevailing spatial dependence affects levels of statistical significance, and hence the precision of any single set of sample estimates, as well as prediction error. This consequence raises questions about OLS results reported in many existing studies involving the analysis of geo-referenced data.

One of two candidate estimators is relatively efficient if its sampling distribution has the smaller variance of the two, both of which are unbiased, making it the more reliable measure. An unbiased estimator that attains the lower bound established in the Cramér-Rao inequality is an efficient estimator; efficiency, then, may be defined as the ratio of the Cramér-Rao lower bound to the actual variance of an unbiased estimator (in general this lower bound is not attainable). This statistical trait coupled with the Cramér-Rao inequality implies that the more efficient an estimator is, the more information it provides about a target population. In the limit, then, a population constant would have zero variance, and its companion estimator would yield perfect information about this parameter. Cordy and Griffith [1993] found that, in general, the need to take spatial autocorrelation into account in variance estimation for geo-referenced data tends to negate advantages due to computational simplicity affiliated with the use of OLS estimation. For the common case where the spatial autocorrelation parameter needs to be estimated, some of the potential gains in efficiency by employing spatial statistical estimators are not realized. Sometimes if either this spatial autocorrelation parameter is negligible or the sample size is small, traditional statistical estimators can be more efficient that their spatial statistical counterparts (likely due to edge effects). Frequently, however, the spatial autocorrelation parameter value is positive, and moderate in magnitude (roughly indicating juxtaposed correlations falling into the range of 0.3-0.5).

An estimator is consistent, which is a large sample or limiting property, if for large $n$ it takes on values that are very close to the value of its corresponding parameter. This will occur only if both the variance of and the bias of an estimator tend to zero as n tends to infinity. It suggests that when the sample size is sufficiently large, there is near certainty that the error made with a consistent estimator will be less than any small preassigned positive constant. The larger the sample size the better the inference one could expect to make.

Moreover, if an estimator is unbiased and its variance goes to zero as $n$ increases to infinity, then it is a consistent estimator of its respective parameter. Consistency is a neglected topic with regard to spatial statistics. Ord [1975] broached the topic, noting that least squares and one popular spatial statistical estimator are inconsistent, and that at least certain maximum likelihood estimators are consistent. Mardia and Marshall [1984] stress that when geo-referenced data involve a single map being observed, which often is the case, consistency of estimators is not obvious. They maintain that only weak consistency can be established in general. Traditional estimators tend to display statistical consistency when the size of a geographic region is increased without limit, although their performance deteriorates as the degree of spatial dependence in, and areal unit articulation of, a geo-referenced data set increase. In contrast, when more and more geo-referenced samples are drawn on a continuous variable in a region of fixed size (i.e., the sampling density increases), many of the commonly used estimators tend to become inconsistent. This more realistic latter situation is pertinent to investigations of almost any geo-referenced data set, as well as changing geographic scales (resolution) of analysis, where the nature of large samples causes the average distance between sampled points to decrease, and the degree of observed spatial dependence to increase. One complicating factor is that a lower limit exists for impacts of scale change due to the discreteness of punctiform geographic distributions, which probably prevents perfect positive spatial autocorrelation from being attainable for many geographically distributed phenomenon.

## IV. OVERVIEW OF THE TOPICS

The principal objective of this *Handbook* is to illustrate how to implement spatial statistical techniques, and to show clearly basic differences between spatial and conventional statistical analyses using real-world data sets. Its contents are organized into ten chapters, outlined in the Table of Contents, illustrating the differences between spatial and classical statistics. In the discussion that follows, a view of the topics of the chapters is taken from a different vantage point to enhance overall perspective. (The reference database is STARS from the World Bank; however, a variety of additional databases are employed, too.)

The critical research frontier issues in spatial statistics, particularly with regard to implementation of its techniques, are represented by the ten topics that compose these chapters. Moreover, interlaced with methodological recommendations, cautions, and guidance, and coupled with exemplary illustrations, are treatments of pressing contemporary spatial statistical issues.

## A.  DEFINING SPATIAL WEIGHTS MATRICES

Stetzer [1982] was one of the first researchers to systematically address the issue of how a spatial weights matrix should be specified. Often a spatial scientist needs to know the answer to questions asking

- (1) which spatial autoregressive model and accompanying spatial weights matrix performs best, irrespective of sample size and mis-specification;
- (2) about the consequences of over-specifying the weights matrix (i.e., including false spatial dependence adjacencies); and,
- (3) about the consequences of under-specifying the weights matrix (i.e., omitting true spatial dependence adjacencies).

Griffith [1993b], and Griffith and Csillag [1993] indicate that a comprehensive treatment of the spatial weights matrix specification is warranted as part of any quantitative geographic research project. This treatment should deal with edge effects, questions of internal partitioning of a geographic landscape, regional shape, and the nature and degree of the prevailing spatial dependence.

## B.  IMPLEMENTING SPATIAL STATISTICS WITH SUPERCOMPUTERS

Massively large geo-referenced data sets are becoming the norm, rather than the exception, in today's world of high resolution imagery and high performance computing. The volume of data in a standard GIS database, accompanied by the numerically intensive requirements of many spatial statistics calculations, suggests that an integration of sophisticated spatial analysis and GIS requires the current capabilities of supercomputers in order to circumvent computational bottlenecks. Implementation of spatial statistical models in these situations frequently requires that supercomputing play a central role. Li [1993] has found that two-dimensional geographic distributions map nicely onto the connection machine, with its substantial number of CPUs. In addition, Griffith [1990], and Griffith and Sone [1992, 1993], have found that spatial statistical procedures can be easily implemented with vector supercomputers (the Cray 2 and IBM 3090). The sum total of these outcomes is that numerically intensive spatial statistical calculations can be performed in a very reasonable time, with the time increase accompanying increasing sample size remaining quiet manageable. This finding has particular relevance to the newest novelty to appear in the desktop computing environment, namely workstations like the IBM RISC/System 6000 machine.

## C. SPATIAL STATISTICS FOR REGULARLY SPACED DATA

Classical experimental design is based on the three concepts of randomization, blocking, and replication. Randomization strives to neutralize the effects of spatial correlation and renders valid tests for the hypothesis of equal treatment effects. Grondona and Cressie [1991] have shown that resorting to a spatial statistical model can yield more efficient estimators of the treatment contrasts than classical statistical approaches. In so doing, such a spatial statistical analysis gives a more complete understanding of the phenomena influencing crop yield. For example, by focusing on the influence of one variable (using partial derivatives) on the effects of spatial autocorrelation on variance estimates, detection of treatment differences is enhanced.

Meanwhile, Griffith [1993c] has found that the special nature of the uniform structuring of a regular square tessellation surface partitioning allows measures of spatial autocorrelation to be computed, and spatial statistical model parameters to be estimated, without having to explicitly construct a geographic weights matrix. The regular two-dimensional arrangement of the geo-referenced data allows standard time series LAG functions to be used in order to identify pixels lying immediately to the east and south of a given areal unit. Invoking this function necessitates the addition of a missing values column to the map, since the extreme eastern pixels have no values immediately to their east. By including a sequential numbering scheme the data can then be reordered by sorting on this numbering variable. The consequence is that a LAG function can then be used to identify values within the boundary that are adjacent to each pixel. Regular square tessellation data require far simpler computer code for undertaking spatial statistical analyses. This simplicity is less demanding of computer memory, allowing substantially larger problems to be analyzed.

## D. AGGREGATION EFFECTS IN GEO-REFERENCED DATA

Geographic scale can vary between something that is quite coarse, a category heading that many surface partitions fall under, to something that is quite fine (high resolution). Customarily (as in conventional univariate statistics), some aggregation of geo-referenced data is necessary in order to unmask pattern from detail, although too much aggregation also obscures pattern; aggregation into a single large areal unit blends regional distinctions, camouflaging them into oblivion. One theme in the quantitative geographic literature acknowledging complications attributable to changes in scale is known as the modifiable areal unit problem (MAUP; see Amrhein, 1994; and, Chapter 5 in this book).

A spatial scientist needs to properly analyze the variablility of data over space at an appropriate scale. Thus, knowing which linear statistics are, and which are not, sensitive to variations in geographic scale and zoning (surface partitioning) systems is exceedingly important. [See, for example, Green, 1993] This variation arises from three major sources: natural variability (stochastic error), measurement error, and sampling error. Each source of variability adds uncertainty to analysis results and confounds understanding of the nature and degree of relationships among geographic distributions. This uncertainty can be complicated by geographic aggregation, which propagates and convolutes these various sources of error. In the end, a spatial scientist routinely seeks methods to disentangle aggregation effects that constrain the drawing of inferences based upon geo-referenced data from one geographic scale to others.

## E.  SPATIAL SAMPLING

GISs have revolutionized geo-referenced data handling, in general, and the visualization of spatial data, in particular. But for this technological advance to be informative, data entered into a GIS database must be collected in a meaningful fashion. One way to ensure meaningfulness is to implement a proper sampling design for collecting the geo-referenced data. Another is to comply with a proper data analysis. And a third is to extract appropriate interpretations from the data analysis.

Published theoretical and applied works concerning spatial sampling designs span more than fifty years and embrace many disciplines. Initial applications of statistical sampling theory to problems involving spatially autocorrelated variables appeared in the late 1930s. Meanwhile, discussions addressing the relative efficiency of random sampling versus systematic sampling of data distributed in two dimensions appeared nearly fifty years ago. The general conclusion reached was that systematic sampling is superior to random sampling of geo-referenced data, with the optimal sampling network design being a superimposed equilateral triangular grid. To date, almost without exception, research on sampling network design has focused on improved estimation of averages for predefined regions; the goal has been to obtain a minimum variance estimate of this mean from a fixed number of sample points.

The problem of spatial sampling designs has been revisited by Stehman and Overton (see, for example, 1989; and, Chapter 3 in this book). To begin, they emphasize that systematic sampling has many advantages over unrestricted random sampling. One practical advantage is ease of implementation. One theoretical advantage is increased precision of estimators. One geo-referenced data context advantage is the ability to furnish information on spatial or temporal patterns in a target population. In fact, geo-referenced data properties of estimators based upon a systematic sample are determined by the natural geographic ordering of a response variable. But, although systematic sampling is widely accepted as a practical sampling design, obtaining an unbiased estimation of variance with it has continued to be problematic. In practice, though, a standard recommendation is to treat the systematic sample as an unrestricted random sample, assuming that the response variable occurs randomly along the underlying natural ordering. In this context the standard formula unfortunately overestimates the true variance in most circumstances. Evaluation results obtained by Overton and Stehman [1990], using the criteria of precision and suitability of variance estimation, argues for use of the triangular network tessellation-stratified sampling design.

## F.  SPATIAL STATISTICS AND GIS

Presently there is both evidence and prevailing expert opinion indicating that spatial statistical techniques need to be converted into GIS functions. [See, for example, Griffith, 1993a] Access to standard GIS functions for managing, transforming, and displaying spatial input data and visualizing model output and residuals enhances a spatial statistical analysis. Candidate techniques for inclusion in a GIS toolbox include indices of spatial autocorrelation and selected spatial autoregressive models, which, for instance, can be used to enhance satellite digital data classification procedures. [Brown and Walsh, 1993] These procedures should be augmented with an elementary multiple linear regression function, a spatial weights generator, and a Moran Coefficient function to test for spatial autocorrelation in regression residuals, spatial correlogram and semivariogram functions. The descriptive functions can be used to examine data for spatial patterns, which may suggest causal processes or, in some cases, reveal systematic errors in the data. [Brown and Bara, 1994] An extremely useful geo-statistical function is the semivariogram, which can serve as a tool for suggesting optimum cell sizes in process models linked with raster GIS databases [Brown *et al.*, 1993] and for the correction of systematic errors in digital elevation models. [Brown and Bara, 1993] Moreover, spatial statistical procedures constitute an essential element of a complete battery of functions that should be available to the quantitative spatial scientist. When they are not embedded in GIS software, the scientist may well have no other recourse than to export geo-referenced data from the GIS package into a statistical or custom-designed software package.

## G.  VISUALIZATION OF SPATIAL DEPENDENCE

One of the hallmarks of GIS software is its ability to support and foster scientific visualization and computer mapping (cartographic representation, and display and analysis of geo-referenced data). Dealing with the relationship between spatial autocorrelation and scientific visualization dates back, at least, to Olsen's [1975] seminal piece.  One important element in the overall look of a choropleth map, for instance, is the relationship between neighboring values as they appear on the map.  Hence maps and spatial statistics should go hand-in-hand to help a spatial scientist understand how much of an effect neighboring attribute values have on each other.  By so doing, the scientist will begin to accumulate the knowledge necessary for eventually acquiring an intuitive understanding of the effects of locational information simply from visual inspection.

## H.  EMPIRICAL APPLICATIONS

Applied work frequently is guided by example.  The case dealing with dependence in geo-referenced forestry data endeavors to provide guidance for the use of spatial statistics in the analysis of forestry data, much of which is geo-referenced, and to demonstrate the value and utility of spatial analysis for natural resources problems (Chapter 7). Spatial scientists need to gain a better understanding of the geographic and attribute complexities latent in forestry data.

The example dealing with spatial dependence in geo-referenced urban data seeks to examine the geographic distribution of population in an urban area, which is arranged in such a way that it displays conspicuous patterns of spatial autocorrelation (Chapter 9).  Anselin and Can [1986] already have investigated a number of different specifications of the negative exponential component of population density gradients, deciding upon the simultaneous autoregressive errors model to account for the presence of spatial autocorrelation.

Finally, spatial dependence in geo-referenced population data ventures to explore reaction and interaction  processes in demographic fertility transition with spatial statistical procedures.  Feng's study in Chapter 8 deals with these concerns in the context of China's current population policy.

## V. SUMMARY

In summary, geo-referenced data are highly complex with spatial dependence introducing further complications. These complications are similar to those found in time series analysis. They are exacerbated by the multi-directional, two-dimensional nature of spatial dependence (time series entails dependencies that are unidirectional along a single dimension), and the far more complex geometric infrastructure involved (classical time series entails a regular linear geometry). The cost of such oversights can be considerable. Incorporation of such features can be achieved by following the guidance and prescriptions offered in this book.

## REFERENCES

**Ahuja, N., and Schachter, B.** *Pattern Models.* Wiley, New York,1983.

**Amrhein, C.** Searching for the elusive aggregation effect: evidence from statistical simulations. *Environment and Planning A*, 26, 1994.

**Anselin, L.** *Spatial Econometrics.* Kluwer, Dordrecht, 1988.

**Anselin, L.** SPACESTAT TUTORIAL: A workbook for using SpaceStat in the analysis of spatial data. *Technical Software Series S-92-1*, NCGIA, Santa Barbara CA, 1992.

**Anselin, L., and Can, A.** Model comparison and model validation issues in empirical work on urban density functions. *Geographical Analysis*, 18, 179-197, 1986.

**Anselin, L., and Griffith, D.** Do spatial effects really matter in regression analysis? *Papers of the Regional Science Association*, 65, 11-34, 1987.

**Anselin, L., and Hudak.** Spatial econometrics in practice: a review of software options. *Regional Science and Urban Economics*, 22, 509-536, 1992.

**Bartlett, M.** *Statistical Analysis of Spatial Pattern.* Chapman and Hall, London, 1975.

**Brown, D., Bian, L., and Walsh, S.** Response of a distributed watershed erosion model to variations in input data aggregation levels. *Computers and Geosciences*, 19(4), 499- 509, 1993.

**Brown, D. and Walsh, S.** Spatial autocorrelation in remotely sensed and GIS data. *Proceedings* of the ACSM/ASPRS Annual Convention, New Orleans, LA, Vol. 3, 13-39, 1993.

**Brown, D. and Bara, T.** Recognition and reduction of systematic error in elevation and derivative surfaces from 7 1/2-minute DEMs. *Photogrammetric Engineering and Remote Sensing*, 60, 189-194, 1994.

**Chorley, R.** [chairman]. *Handling Geographic Information.* Her Majesty's Stationary Office, Report to the Select Committee on GIS, London, 1987.

**Cliff, A., and Ord, K.** *Spatial Processes.* Pion, London, 1981.

**Cordy, C., and Griffith, D.** Efficiency of least squares estimators in the presence of spatial autocorrelation. *Communications in Statistics--Simulation and Computation*, 22, 1161-1179, 1993.

**Cressie, N.** *Statistics for Spatial Data.* Wiley, New York, 1991.

**Durrett, R.** Stochastic spatial models, *Forefronts* (newsletter of the Cornell Theory Center), 9 (#4, Spring), 4-6, 1994.

**Goodchild, M.** *Spatial Autocorrelation.* CATMOG, Norwich, England, 1986.

**Goodchild, M., Haining, R., and Wise, S.** Integrating GIS and spatial data analysis: problems and possibilities. *International Journal of Geographical Information Systems,* 6, 407-423, 1992.

**Green, M.** Ecological fallacies and the modifiable areal unit problem. *Research Report No. 27,* North West Regional Research Laboratory, Lancaster University, UK, 1993.

**Griffith, D.** *Spatial Autocorrelation.* Association of American Geographers, Washington, D. C., 1987.

**Griffith, D.** *Advanced Spatial Statistics.* Kluwer, Dordrecht, 1988.

**Griffith, D.** Spatial regression Analysis on the PC: spatial statistics using MINITAB. *Discussion Paper #1,* Institute of Mathematical Geography, Ann Arbor, MI, 1989.

**Griffith, D.** (ed.). *Spatial Statistics: Past, Present, and Future.* Institute of Mathematical Geography, Ann Arbor, MI, 1990.

**Griffith, D.** Which spatial statistics techniques should be converted to GIS functions? in *Geographic Information Systems, Spatial Modelling and Policy Evaluation,* edited by M. Fischer and P. Nijkamp. Springer-Verlag, 103-114, Berlin, 1993a.

**Griffith, D.** Advanced spatial statistics for analysing and visualizing geo-referenced data. *International Journal of Geographical Information Systems,* 7, 107-123, 1993b.

**Griffith, D.** *Spatial Regression Analysis on the PC: Spatial Statistics Using SAS.* Association of American Geographers, Washington, D.C., 1993c.

**Griffith, D., and Csillag, F.** Exploring relationships between semi-variogram and spatial autoregressive models. *Papers in Regional Science,* 72, 283-295, 1993.

**Griffith, D., and Sone, A.** Trade-offs associated with computational simplifications for estimating spatial statistical models. *Working Paper,* l'Institut de Mathématiques Economiques, Université de Bourgogne, Dijon, France (with French Resumé), 1992.

**Griffith, D., and Sone, A.** Some trade-offs associated with computational simplifications for estimating spatial statistical/econometric models: preliminary results. *Discussion Paper* No. 103, Department of Geography, Syracuse University, 1993.

**Grondona, M., and Cressie, N.** Using spatial considerations in the analysis of experiments. *Technometrics,* 33, 381-392, 1991.

**Haining, R.** *Spatial Data Analysis in the Social and Environmental Sciences.* Cambridge University Press, Cambridge, England, 1990.

**IBM** Exploring new worlds with GIS. *Directions,* Summer/Fall, 12-19, 1991.

**Isaaks, E. and Srivastava, R.** *An Introduction to Applied Geostatistics.* Oxford University Press, Oxford, England, 1989.

**Lindgren, B.** *Statistical Theory,* 3rd ed. Macmillan, New York,1976.

**Mardia, K., and Marshall, R.** Maximum likelihood estimation of models for residual covariance in spatial regression. *Biometrika,* 71, 135-146, 1984.

**Matérn, B.** *Spatial Variation,* 2nd ed. Springer-Verlag, Berlin,1986.

**National Research Council (Mapping Science Committee; Commission on Physical Sciences, Mathematics, and Resources).** *Spatial Data Needs: The Future of the National Mapping Program,* National Academy Press, Washington, D.C., 1990a.

**National Research Council (Board on Mathematical Sciences).** *Renewing U.S. Mathematics: A Plan for the 1990s.* National Academy Press, Washington, D.C., 1990b.

**National Research Council (Panel on Spatial Statistics and Image Processing).** *Spatial Statistics and Digital Image Analysis.* National Academy Press, Washington, D.C., 1991.

**National Science Foundation.** *Solicitation: National Center for Geographic Information and Analysis.* Biological, Behavioral, and Social Sciences Directorate, Washington, D.C., 1987.

**Odland, J.** *Spatial Autocorrelation.* Sage, Beverly Hills, CA, 1988.
**Okabe, A., Boots, B., and Sugihara, K.** *Spatial Tessellations: Concepts and Applications of Voronoi Diagrams.* Wiley, New York, 1992.
**Olsen, J.** Autocorrelation and visual map complexity. *Annals,* Association of American Geographers, 65, 189-204, 1975.
**Ord, K.** Estimation methods for models of spatial interaction. *Journal of the American Statistical Association,* 70, 120-126, 1975.
**Overton S., and Stehman, S.** Statistical properties of designs for sampling continuous functions in two dimensions using a triangular grid. *Technical Report* No. 143, Department of Statistics, Oregon State University, 1990.
**Paelinck, J. and Klaassen, L.** *Spatial Econometrics.* Saxon House, Farnborough, England, 1979.
**Ripley, B.** *Statistical Inference for Spatial Processes.* Cambridge University Press, Cambridge, England, 1988.
**Stehman, S., and Overton, W.** Variance estimation for fixed-configuration, systematic sampling. *Technical Report* No. 134, Department of Statistics, Oregon State University, 1989.
**Stetzer, F.** Specifying weights in spatial forecasting models: the results of some experiments. *Environment and Planning A,* 14, 571-584, 1982.
**Upton, G., and Fingleton, B.** *Spatial Data Analysis by Example,* vol. 1. Wiley, New York, 1985.
**Warnecke, L.** GIS in the states: applications abound. *GIS World,* 3, (# 3), 54-58, 1990.
**Warnecke, L.** *State Geographic Information Activities Compendium.* Council of State Governments, Lexington, KY, 1991.

# Chapter 2

## VISUALIZATION OF SPATIAL DEPENDENCE: AN ELEMENTARY VIEW OF SPATIAL AUTOCORRELATION

Irina Ren Vasiliev

## I. EDITORIAL NOTE

In 1977, Anthony C. Gatrell wrote a paper summarizing elements of spatial visualization and information theory; in his Abstract [Gatrell, 1977], he notes:

*The complexity of binary maps that is provided by the areal arrangement of colors is considered [in his paper], and measured using information theory. In addition, information theory provides other measures that have an interpretation in a map context. One of these, redundancy, is examined and found to bear a striking empirical relationship to a spatial autocorrelation statistic. It is argued that spatial autocorrelation is, conceptually as well as empirically, the two-dimensional equivalent of redundancy. It too measures the extent to which the occurrence of an event (color) in an areal unit constrains, or makes more probable, the occurrence of an event in a neighboring areal unit.*

So that readers first approaching the idea of spatial statistics might have at least one conceptual key to unlock the vast array of many fine books and articles from the past, such as Gatrell's, we thought it prudent to include a chapter at the elementary level to illustrate the need for visualization in spatial statistics. The example chosen here, in Vasiliev's interesting case studies, is the ubiquitous concept of spatial autocorrelation. Readers interested in further motivational material might enjoy Griffith's fine treatment of polyomino games and spatial autocorrelation. [Griffith, 1993] Readers interested in a provocative summary of visualization might enjoy Moellering's essay on "real maps, virtual maps and interactive cartography" [1984].

0-8493-0132-7/95/$0.00+$.50
© 1996 by CRC Press, Inc.

## II. INTRODUCTION

Much of spatial statistics is devoted to measuring neighborly influence. Maps and spatial statistics can work together to offer a visual understanding of the extent to which neighboring values of quantified information influence each other. When coupled with a positional understanding of where these values are in geographic space, visualization, the use of one's imagination to try to understand information in a pictorial or graphic manner, occurs. When the information is spatial, the results can be portrayed as a map--of abstract or of actual locations.

An example of the need for both maps and spatial statistics is found in Griffith [1987:46; 1988:8] in which the classical mean is calculated for values in a partitioned abstract region (a square; see Figure 1). In each of the twelve simple square maps there are four subregions or "states." The classical mean of each region containing four states is 3; from the viewpoint of classical statistics, these regions cannot be distinguished from each other. All values are the same but are distributed according to different spatial patterns. In situations such as this one, the geographic information concerning the states themselves is lost if there is no map to accompany the classical statistic.

## III. THE SPATIAL MEAN AND OTHER BASIC CONCEPTS

One approach to overcoming situations such as the one above, in addition to using maps, is to build some of the geographic information into the spatial measure: hence, a spatial statistic. To calculate the spatial mean of the regions in Figure 1, proceed as follows. In the region in the upper left-hand corner of Figure 1, the top row has states containing values of 1 and 3 (from left to right) and the bottom row has states containing values of 4.

- Assign each of these values to the central point in each state, as a weight.
- Superimpose a rectangular coordinate system on each region with the origin at the lower left-hand corner of each large square.
- Label the x- and y-axes in the natural manner with a unit representing the width of a state.
- Label the four central points of each state with coordinates: (1/2, 1/2), (3/2, 1/2), (1/2, 3/2), and (3/2, 3/2).

The spatial mean is obtained as an ordered pair that takes into account the position of the centroid of each state and the weight associated with the centroid.

- To calculate the first coordinate of the spatial mean add the weights in a column of states and multiply by the first coordinate of the centroid for that column. Repeat this procedure for the second (and further) column of states. Add the results from the calculations and divide by the sum of the weights for the entire region. In the case of the upper left-hand corner of Figure 1, therefore, the first coordinate of the spatial mean is: $((1+4)\times1/2 + (3+4)\times3/2)/12 = 1.08$.

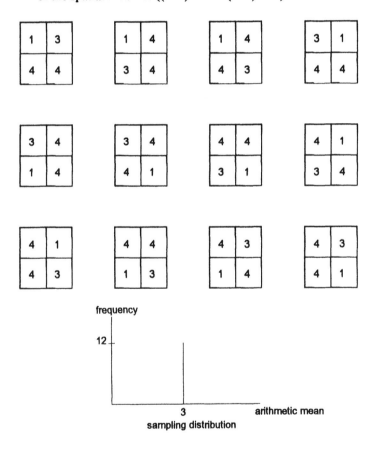

**Figure 1.** The classical mean. An example of different spatial distributions of the same set of values. In each case the classical mean is 3–obtained by adding the values of each of the four cells, yielding a sum of 12 in each case, and then dividing by the number of cells, 4 in each case. This mean does not distinguish among the different spatial patterns, for its standard error here is zero. (Adapted from Griffith 1988:9.)

•   To calculate the second coordinate of the spatial mean add the weights in a row of states and multiply by the second coordinate of the centroid for that row. Repeat this procedure for the second (and further) row of states. Add the results from the calculations and divide by the sum of the weights for the entire region. In the case of the upper left-hand corner of Figure 1, therefore, the second coordinate of the spatial mean is: $((4+4)\times1/2 + (1+3)\times3/2)/12 = 0.83$.

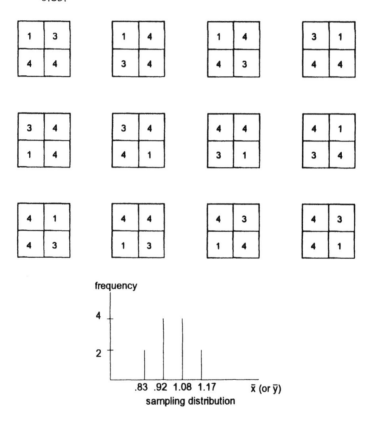

**Figure 2.** The spatial mean. This weighted mean is an ordered pair based on a rectangular coordinate system with the origin at the lower left hand corner of each of the 12 figures. The unit distance on x and y axes corresponds to the width of a "state" so that each of the twelve regions is 2 units by 2 units. The values from Figure 1 are assigned as weights for the centroids of each state. The spatial mean (x,y) is calculated for the upper left example as: $x=((1+4)\times0.5+(3+4)\times1.5)/12=1.08$; $y=((4+4)\times0.5+(1+3)\times1.5)/12=0.83$. The spatial means ( $x$ , $y$ ) for each of the twelve regions in Figure 1, reading from left to right across each row are:  (1.08, 0.83), (1.17, 0.92), (1.08, 0.92), (0.92, 0.83), (1.17, 1.08), (0.92, 1.08), (0.92, 1.17), (0.92, 0.92), (0.83, 0.92), (1.08, 1.17), (1.08, 1.08), (0.83, 1.08). The symbol on each figure indicates the approximate location of the spatial mean.

For the example of the upper left region of Figure 1, the spatial mean is the ordered pair (1.08, 0.83). The point (1,1) is in the center of the square region at the central intersection point common to the four states. The spatial mean is located below and to the right of (1,1); it serves as a balance point for the surface weighted with the given values. In Figure 2, the spatial mean is indicated for each of the square regions from Figure 1. The spatial mean is different for each of the 12 different figures and is located as an intuitive balance point for the weighted surface. As the locations of the weights are shifted, so too is the spatial mean: the patterns of symmetry in the ordered pairs mirrors the symmetry in the placement of the weights in each region. The spatial mean is indeed a measure which includes locational elements.

In real-world examples more complex than the simple regions in Figures 1 and 2, it is often difficult to visualize the relationship between areal units and variable weights without a map. It is also difficult to calculate a spatial mean without the use of a map on which to overlay the Cartesian coordinate grid. A map that shows the changing spatial mean of a set of information--the centroid of the United States population (areal units are counties)--is useful in understanding a variety of associated spatial and temporal trends (Figure 3). The population center of the United States has been moving slowly to the west/southwest over the decades; an observation that is made clear by the map.

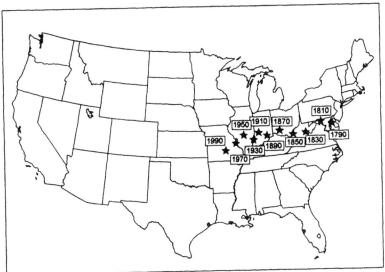

**Figure 3.** The spatial mean – or population centroid, in this case – has been shifting to the west in the USA with each Census. Here, only the odd decade Census years are mapped [Adapted from Plane and Rogerson 1994:35]

The spatial mean has the useful property of being independent of the choice of coordinate system; the same point relative to the location of all points will emerge as the spatial mean regardless of the choice of axes. If the classical idea of median were recast as a spatial median--an ordered pair with half the x-values on one side of the first coordinate and half the y-values on one side of the second coordinate--the parallel situation would not hold. A spatial median depends on the choice of axes and so the direct parallel is not particularly useful; instead, related concepts, such as the point of minimum aggregate travel are often employed to capture this sort of idea in a spatial context. The classical concept of a statistical mode, the most frequently occurring value in a list, has a natural extension in space--the cell with the most observations in it. The chapter on spatial sampling will focus on extensions of this classical idea.

Beyond the traditional initial concepts of mean, median, and mode, one might naturally wonder how other basic statistical notions, such as randomness, can be cast in a spatial setting. One way to consider randomness is based on the nearest neighbor statistic; there are many others, some of which will be considered elsewhere in this handbook. The literature on geographical analysis is rich in published examples (Haggett, Cliff, and Frey, 1977; Dacey, 1962). As does the spatial mean, the simplest form of the nearest neighbor statistic also depends both on the map itself and on the database from which the map is built. To calculate the nearest neighbor statistic, R, calculate the distance between all neighbors within a region and divide the sum of these values by the total number of neighbors in the region. A map of the region is required to see which points are neighbors and which are not; determining the neighborhood relationship often proves problematic. When the calculations are executed, certain relationships between the points are evident from the statistic alone. When R is near zero, the points are clustered and distances between them are small. When R is near 1, the distribution is close to random (Figure 4). When R is nearly 2.15, spacing between points is maximal: the distribution is uniform--a hexagonal lattice. It is for this reason that a number of the chapters in this handbook invoke hexagonal lattices in examples.

At a general level, the notion of neighbor is an interesting one. Neighbors are defined by adjacency patterns, which in turn must also be defined. Generally, areal units sharing a polygonal boundary are considered to be neighbors; they are not considered to be neighbors when this boundary shrinks to a single point (Figure 5). If it appears that there is sufficient influence between areal units whose common boundary is but a single point, then the definition of neighbor must be clearly evaluated. Indeed, if actual local boundaries are ill-defined, even though the map boundaries appear clear, the pattern of real boundary intersection will be a zone rather than a point and the diagonal regions become neighbors.  As is often the case, new definitions can be suggested by the analysis of real-world problems:  the careful researcher needs to be sensitive to this sort of interplay.

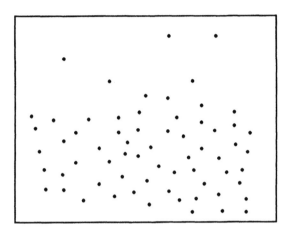

**Figure 4.**  Is this distribution of neighboring points random or clustered?

**Figure 5.**  In the Four Corners area of the United States, Utah and Colorado are neighbors, but Utah and New Mexico are not.

## IV. SPATIAL AUTOCORRELATION

A more sophisticated summary measure of the influences that neighbors have on each other in geographic space is that of spatial autocorrelation; generally, neighbors are assumed to be those areal units that share a non-zero length of polygonal boundary (only). There are a number of ways to calculate this statistic: in all cases, however,

- positive spatial autocorrelation indicates that similar values for the variable in question are clustered together in space;
- negative spatial autocorrelation indicates that similar values are separated by intervening dissimilar value--that is, dissimilar values are clustered in space; and,
- no spatial autocorrelation indicates that the pattern of values is random across space.

The map of Population in Buffalo, 1981, provides an example of positive spatial autocorrelation (Figure 6): more densely populated tracts are nearer each other. A clustering of population is obvious in the northern parts of the city (north is at the top of the map in Figure 6). The map of Arson Arrests in Buffalo, 1981, furnishes an example of negative spatial autocorrelation: the percentage of arrests is scattered--adjacent regions do not necessarily have related arrest rates; indeed, adjacent regions appear for the most part to have dissimilar arrest rates (Figure 7).

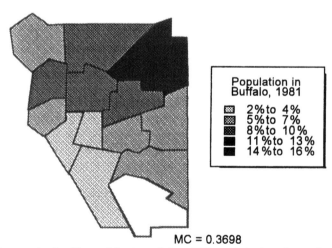

**Figure 6.** An example of positive spatial autocorrelation: 1981 population by police precinct in Buffalo, NY.

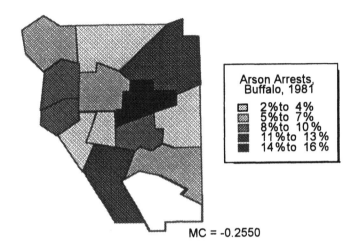

**Figure 7.** An example of negative spatial autocorrelation: 1981 arson arrests by police precinct in Buffalo, NY.

More generally, adjacency patterns of areal units on a map may be easily captured in a matrix: enter a 1 in the (i,j) position of the matrix if areal unit i is adjacent to areal unit j; enter a 0 otherwise. Figure 8 shows a matrix representing an abstract configuration; the matrix aids in quantifying locational information. The join-count statistic is a binary classification measure consisting of the number of 1s that are next to each other on the map, the number of 0s that are next to each other on the map, and the number of 1s and 0s (corresponding areal units) that are next to each other on the map. The word join is used in the set-theoretic sense: adjacent regions of like value are joined. These combinations are often written, using W to play the role of 1 and B to play the role of 0, as WW, BB, and BW, respectively. They are counted by pre-multiplying the binary attribute vector by this matrix and compared to the values expected for the three statistics under conditions of no spatial autocorrelation (Figure 8).

In terms of very simple counting, one might expect the combination WW 0.25% of the time, BB 0.25% of the time, BW 0.25% of the time, and WB 0.25% of the time--as in counting possibile permutations for hybrids in simple genetics experiments. Of course, when map adjacency is also considered, values this large are not generally possible from a topological standpoint. In the simplest case, though, BW is twice either BB or WW (not both). As the values for WW or BB increase, a corresponding decrease results for BW. When this happens, similar values are clustering, and spatial autocorrelation is on the rise.  If the value for BW is increasing, as those for BB or WW are decreasing, then dissimilar values are clustering and spatial autocorrelation is becoming increasingly negative (Figure 8).  If the values for BB, WW, and BW are close to the expected values, then the variable is randomly distributed in geographic space.   To visualize the clustering or randomness of the distribution, return to the original map and trace the pattern of the areal units.   Such a visual image can offer perspective on each attribute value's correlation with its neighbors.

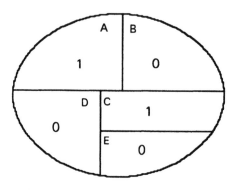

|   | A | B | C | D | E | W=1: urban; B=0: rural |
|---|---|---|---|---|---|---|
| A | 0 | 1 | 1 | 1 | 0 | |
| B | 1 | 0 | 1 | 0 | 0 | join-count statistic: |
| C | 1 | 1 | 0 | 1 | 1 | WW = 2; AC, CA |
| D | 1 | 0 | 1 | 0 | 1 | BB = 2; DE, ED |
| E | 0 | 0 | 1 | 1 | 0 | BW = 10; AD, DA, AB, BA, BC, CB, CD, DC, CE, EC. |

Expected under conditions of no spatial autocorrelation and equally likely chance of a region being urban or rural.
    BW=BB+WW

Simplest case:
    There are 25 entries in the matrix; 20 not on the main diagonal.
Regions DB and BD are not adjacent; regions BE and EB are not adjacent; regions AE and EA are not adjacent.  The remaining 16 region pairs are accounted for in the join-count statistic.

BW>BB+WW; dissimilar values appear more clustered than do similar ones; 7 is the expected value for BW--the actual value is 10 for this example..

**Figure 8.**  A configuration matrix representing areas that are neighbors with the map used to generate it.

There are other indices of spatial autocorrelation, in addition to the join-count statistic.  The Moran Coefficient (MC) and the Geary Ratio (GR) are often used to index spatial autocorrelation:  to understand the mathematical developement of these ratios, see Li's chapter in this handbook.  In this chapter, a summary illustrating how to interpret these indices, once they have been calculated, is shown for simple maps.  For the Moran Coefficient:

- when MC approaches +1, there is strong positive autocorrelation;
- when MC approaches -1, there is strong negative autocorrelation;
- when MC approaches $-1/(n-1)$ there is a random distribution of values.

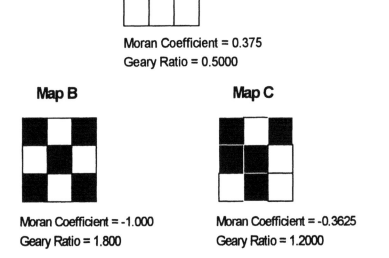

**Map A**

Moran Coefficient = 0.375
Geary Ratio = 0.5000

**Map B**

Moran Coefficient = -1.000
Geary Ratio = 1.800

**Map C**

Moran Coefficient = -0.3625
Geary Ratio = 1.2000

**Figure 9.** Maps A, B, and C--hypothetical maps to show positive, negative, and random spatial autocorrelation for a binary attribute.

For the Geary Ratio:

* when GR approaches 0, there is strong positive autocorrelation;
* when GR approaches 2, there is strong negative autocorrelation;
* when GR approaches 1, there is a random distribution of values.

A more detailed discussion of these measures can be found in Cliff and Ord [1981] as well as in Li's chapter in this handbook. Maps A, B, and C in Figure 9 illustrate visual differences in these measures. Map A illustrates a situation of positive autocorrelation using either index; Map B illustrates negative autocorrelation. Map C approaches a random distribution on both indices although this fact is somewhat difficult to discern visually. For these simple maps, the indices measure, fairly well, what we actually see.

## V. MAP COMPLEXITY

A map of values distributed over space often reveals clustering of similar values. Mapping techniques depend on our ability to see the varied use of pattern and color to locate information in space. The larger the number of categories of values, however, the more difficult visualization can become. Too many ranges of values destroy information as to where the clusters really are, let alone understand their significance. Detail can overwhelm and mask pattern. Olson addresses the issue of map complexity in relation to the autocorrelation exhibited on choropleth maps (1975). She has found that it was difficult for individuals to determine the degree of autocorrelation as the maps moved away from exhibiting strong positive and negative autocorrelation measures towards exhibiting more random distributions of values.

In cases of visually complex maps spatial statistics are of particular value. In addition, as Dykes (1994) notes, the visual impression of latent spatial autocorrelation can be altered by varying class delimitation on choropleth maps. Both the Moran Coefficient and the Geary Ratio distinguish maps based on different distributions whose nature and degree are not visually discernible.

Multivariate maps are even more visually complex than are their single variable counterparts. Spatial statistics for each variable, singly, can help disentangle visual complexity on a multivariate map, without producing separate univariate maps. Individual univariate spatial statistics do not, however, show the locational information that is common among the variables. We need the map to see this information.

## VI. MAP REPRESENTATIONS OF CHANGES IN SPACE AND TIME

Often it is useful to consider how spatial dependence changes over time. To date, the best method of representing such changes has been with a series of maps: each map showing the geographic distribution of the variable of interest at a certain point in time. Using paired maps, placed side by side in chronological order, changes that occur in space can be visualized in a pairwise fashion. Recent advances in the computing world, particularly in the realms of geographic information systems and of animation software, permit views of map series that appear far more promising than any arrangement of sequences of paper maps. Animation is an exciting cutting-edge idea that can enable one to move back and forth through both time and space.

## VII. SUMMARY: RULES OF THUMB FOR SPATIAL AUTOCORRELATION

The visualization of spatial dependence is an extension of both cartography and spatial statistics. Spatial statistics help us to understand the nature and the extent of the effects that neighbors have on each other. Maps show us where those effects are occurring. To see the statistics and the maps together, it is critical to have a good grasp of the visual nature of spatial autocorrelation. The following rules of thumb offer a summary of how one might go about doing this.

- The larger the areas of similar or identical information, the more positive is the spatial autocorrelation.
    Abstract map example: Map A in Figure 8 exhibits strong positive spatial autocorrelation.
    Real-world map example: population map of Buffalo (Figure 5).
- The more dispersed and fragmented the areas with similar values, the more negative the spatial autocorrelation.
    Abstract map example: Map B in Figure 8 exhibits strong negative spatial autocorrelation.
    Real-world map example: arson arrests map of Buffalo (Figure 6).
- Random spatial autocorrelation is difficult to detect visually because patches of areas display similar information while patches of areas contain fragmented locational information. It is difficult to tell at a glance whether Map C in Figure 8 displays negative or random

spatial autocorrelation.   The spatial statistics (Moran Coefficient and Geary Ratio) suggest that the distribution is random.

Spatial statistics and maps, taken together, offer us the capability to make better maps of the phenomena in which we are interested and to understand better the information from the spatial statistics.   It is this combination that is the heart of visualization.

## REFERENCES

**Boots, B., and Dufouraud, D.**   A programming approach to minimizing and maximizing spatial autocorrelation statistics.   *Geographical Analysis* 26(1):54-66,  1994.

**Cliff, A. and Ord, J**.   *Spatial Processes, Models, and Applications*.   Pion, London, 1981.

**Dacey, Michael F.**   Analysis of central place and point patterns by a nearest neighbour method.   Lund Studies in Geography, B, Human Geography, 24, 55-75, 1962.

**Dykes, J.**   Area-value data:  new visual emphases and representations, in *Visualization in Geographical Information Systems*, edited by H. Hearnshaw and D. Unwin.   Wiley, New York, pp. 103-114, 1994.

**Gatrell, Anthony C.**   Complexity and redundancy in binary maps.   *Geographical Analysis*, Vol. IX, No. 1, 29-41, 1977.

**Getis, A. and Boots, B.**   *Models of Spatial Processes:  An Approach to the  Study of Point, Line, and Area Patterns*.   Cambridge University Press, Cambridge, 1978.

**Griffith, Daniel A.**   *Spatial Regression Analysis on the PC:  Spatial Statistics Using SAS*.   Association of American Geographers, Washington, D.C., 1993.

**Griffith, Daniel A**. *Spatial Autocorrelation. A Primer*.   Washington, D.C.:  Association of American Geographers, 1987.

**Griffith, Daniel A**.   *Advanced Spatial Statistics*.   Kluwer, Dordrecht, 1988.

**Haggett, Peter; Cliff, Andrew D.; and Frey, Allan**.   *Locational Analysis in Human Geography* (in two volumes), 2nd edition.   Wiley, New York, 1977.

**Moellering, Harold**.   Real maps, virtual maps and interactive cartography, pp. 109- 132 in Gaile and Willmott (ed.), *Spatial Statistics and Models*,  Reidel, 1984.

**Olson, Judith M**.   Autocorrelation and Visual Map Complexity.   *Annals of the Association of American Geographers* 65(2):189 - 204, 1975.

**Plane, David A. and Rogerson, Peter A**.   *The Geographical Analysis of Population*.   Wiley, New York, 1994.

# Chapter 3

## SPATIAL SAMPLING

### Stephen V. Stehman and W. Scott Overton

## I. INTRODUCTION

Spatial sampling covers the broad topic of how to collect data for geo-referenced populations. Because most real-world objects can be associated with a spatial location, spatial considerations are potentially important to any sampling problem. But in classical sampling texts [Cochran 1977, Deming 1950, Jessen 1978, Kish 1965, Yates 1981] spatial concerns are not usually given detailed treatment because the spatial information is not used in the analysis, and because until recently, survey objectives did not usually require the techniques of spatial statistics. Similarly, most treatises on spatial sampling give no formal attention to issues of probability sampling. With the recent increased attention and accessibility to spatial analyses, the interaction between sampling design and spatial statistics is becoming more critical. Advantages accrue to the melding of the two fields by applying probability sampling methods of selection and analysis to the special problems of characterizing spatially distributed populations. Spatial sampling techniques may be applied to a wide variety of objectives and problems. In most real world problems it is impossible or impractical to obtain a complete census of all objects or an entire region of interest, so sampling a subset of the objects or region becomes a necessity. Further, sampling is advantageous if information is needed in a timely manner, or when well-trained personnel are needed to collect accurate data. Lastly, the precision of a full census is seldom needed, even if it can be conducted accurately and in a timely fashion, so that the precision of a sample usually is adequate for the objectives.

A common objective motivating a sampling approach is estimation of population parameters, such as means, totals, or proportions. Parameters such as these can be estimated with known precision from a probability sample, using what is commonly called "design-based" methodology. When the population is distributed in space, the spatial coordinates of the units in the sample may be useful in analyzing the sample, as may other variables whose spatial patterns are known, such as topographic elevation. Incorporating such information into sample analysis is the subject of a general treatment of the use of models to "assist" in the analysis of sample data [Särndal *et al.* 1992]. Greater dependence on models is sometimes made, even to the extent of not utilizing a probability sample at all. Many common spatial approaches are of this type ("model dependent"), but it should be emphasized that some samplers [cf. Royall 1970] rely on models to this extent even when no spatial element is present. In yet a third circumstance, spatial information may be used in developing a more efficient sampling design, but enters the analysis only through design. For example, geographic stratification or systematic spatial sampling are both designs that take advantage of spatial information and may improve efficiency of estimators.

Here we will confine attention to spatial considerations in one or two dimensions. Most concepts and techniques translate easily from one- to two-dimensional space. Many sampling problems focus interest on a continuous two-dimensional region such as a forest or estuary. For a response variable that is assumed continuous across such a region for example soil pH in the forest or dissolved oxygen in the estuary a descriptive objective might be the mean soil pH, or the distributionfunction of dissolved oxygen. Again, spatial information may be used explicitly in the estimator, or it can be used via stratification or systematic sampling to improve efficiency.

Sometimes spatial analyses are a critical component of the objectives. Haining [1990] describes one type of sampling problem as a problem in which "the spatial variation of some variable is specifically required, either in the form of a map or in the form of a summary measure (such as a variogram or correlogram) to highlight important scales of variation." For instance, in addition to parameters of a population of households, objectives may also identify the spatial pattern of household income. This objective could be satisfied by a contour map of income using a technique such as kriging, or by estimating mean income for specified subregions.

Spatial sampling considerations are relevant to the design of monitoring networks. For example, where should stations be located to monitor the concentrations and distribution of ash from a trash incinerator, or to monitor ozone levels in a city or National Park? This issue is often addressed in the context of spatial statistics without considering the potential contribution of probability sampling. Estimating the area or proportion of area in various land cover and land use types is another typical problem in which spatial sampling plays a major role; this is more often approached from a sampling perspective than from a spatial one, but the methods of spatial statistics can contribute to the answer. Evaluating the accuracy of maps is itself a problem in spatial sampling. For example, estimating attribute error for information contained in a GIS coverage requires spatial sampling of the actual region represented by the coverage to determine if the mapped attributes correspond to the true attributes of the region. Spatial sampling may be of interest when objectives specify description, prediction, or spatial analyses. The focus of this chapter is on spatial sampling for description.

## II. SPATIAL UNIVERSES AND POPULATIONS

The universe, **U**, is the spatial extent or region of interest in a sampling problem. The elements of this universe are the entities thatare measured. A discrete universe consists of a finite set of discrete objects, $U = \{u_1, u_2,...,u_N\}$, such as households, business establishments, people, lakes, or trees. Often these objects possess a spatial location, which we will usually identify as a unique point, but the objects themselves are self-contained, separable entities, and it is possible to sample these objects without reference to their spatial location (see Figure 1). Other objects, such as people, are not stationary in space; still, it is often possible to identify them with a spatial location, as by their residence. Many animals are highly mobile and difficult to pin down to a permanent spatial location, but they, too, may be surveyed, as by the number that are in a random quadrat at a particular instant in time. Mobile populations often pose formidable problems in sampling.

A continuous universe is a continuous area or region, such as a particular county, park, or marsh. That is, a continuous universe may be defined by all possible sample locations within a well-defined, bounded region $U = \{u: u \in R^d , d = 1, 2\}$. Even if discrete objects are available, we may still choose the continuous view of the universe and sample the space containing these objects rather than sampling the objects directly. By partitioning a continuous universe, we can convert it into a finite set of universe segments (see Figure 2). Such a structure imposed on a universe is often designated a frame, but it may also be appropriate to think of this as an alternate universe. The ability to make such a conversion illustrates that the view taken of the universe is a matter of choice and it is not strictly a characteristic determined by the entities being sampled.

The population is the response or attribute defined on the universe; real-world objects typically have many attributes, of which only a small set will be designated the population for any particular survey. For the discrete universe, we will denote the population values by $y_u$, where u (to simplify notation, we drop the subscript on $u_i$ when used with the population value $y_u$). For the continuous universe, the population value at point u will be denoted as $y(u)$, where u represents the spatial location in two-dimensional space of the point. Populations defined on the discrete universe, such as income of a household, gross sales of a business establishment, surface area of a lake, height of a tree, or political preference of a person, are examples of typical finite population attributes. The populations defined on a continuous universe often vary continuously over the universe). This type of population may be thought of as a smooth surface with a high degree of regularity generated by some biological, geological, economic, or other process. Deviations or irregularities in the surface may be present, with smooth surfaces still existing between these irregularities. Description of the irregularity features may be of interest as well as description of the smooth surfaces in the domains between the irregularities. Whether populations are continuous or discrete influences the choice of sampling strategy.

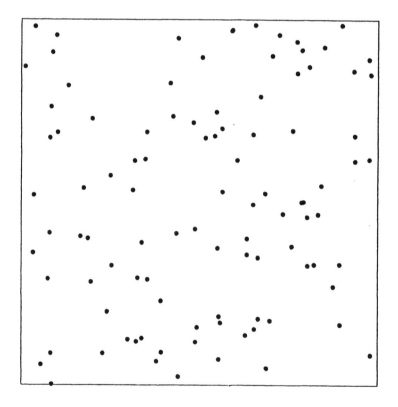

**Figure 1.** Locations of residences in a continuous spatial universe. The universe of residences could be sampled by constructing a list frame, or by sampling the spatial region itself and devising a protocol by which residences are selected into the sample by their spatial locations.

Given that the target universe and populations of interest have been defined, construction of a frame that unambiguously represents the universe in a manner consistent with stated objectives is necessary to provide the basis for sample selection.  A frame is "any material or device used to obtain observational access" to the universe of interest. [Särndal *et al.* 1992, p. 9]  The two main types of frame used are a list frame and a map frame.  List frames consist of a list of elements of the universe and are commonly used in traditional survey work.  The entities of a list frame are the sampling units.  These sampling units may be the actual elements of U, in which case selection of the sampling unit is equivalent to selection of a unit of U.  Alternatively, the sampling unit may consist of a set of elements of U, such as the set of households in a city block, all of which could be selected as a cluster.  Spatial information may be incorporated into this frame by clustering or stratification, or by constructing a frame population representing the spatial coordinates of the location of the units to be sampled.

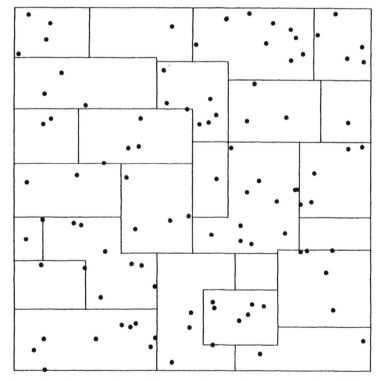

**Figure 2.** Arbitrary spatial partitioning converting a continuous universe into a finite set of universe segments.  Locations of residences are represented by bullets.  Residences may be sampled by selecting a sample of the areal sampling units, then sampling those residences within the selected areal units.

A map frame consists of a map of the spatial universe of interest. The map frame may display spatial locations of the objects to be sampled, such as lakes, houses, or parcels, or it may simply be a representation of well-defined boundaries within with little additional detail. Examples of this latter type of map frame are the boundary of a county, national forest or park. A map frame could be converted to a list frame to sample a discrete universe. For example, a map displaying all the lakes in a state could be used to construct a list frame of such lakes. The conversion process may be costly, and it is difficult to construct a list frame that retains the spatial pattern of the objects in the universe. Retaining this spatial information is one advantage of a map frame. Even if a list frame is available, spatially distributed populations usually should be sampled in a manner to ensure a representative spatial distribution of the sample, so some spatial device is desirable. Geographic stratification is one such device, and systematic sampling directly from the map frame is another.

Recognition of distinctions between a target universe and a frame universe is important. [Kish 1987] The target universe is the universe of interest specified by a study's objectives. The frame universe is that universe represented by the frame that is used for sampling. The statistical inferences available from the sampling exercise apply to the frame universe. If the frame universe fails to exactly correspond to the target universe, it is necessary either to modify the objectives to correspond with the frame, or to redefine sampling units to cover the missing elements of the universe. This exact issue arose in the United States Environmental Protection Agency's (U.S. EPA) National Lake Survey. [Linthurst *et al.* 1986; Overton *et al.* 1986] The list frame of lakes was constructed from USGS 1:250,000 scale maps. Lakes represented on these maps usually had surface area greater than 4 hectares, with smaller lakes not indicated. This well defined frame population was surveyed, but the survey was criticized because smaller lakes were excluded from the frame universe. To remedy this problem, a supplementary sample was defined via the Thiessen polygons of the lakes represented in the initial frame. The sample of polygons identified with the sample of lakes provided a sample of small lakes (surface area smaller than 4 hectares), with these small lakes located from higher resolution maps covering the sample polygons. The sample of small lakes could readily be subsampled to provide comparable data on small lakes. Such a solution resolves the discrepancy between the frame and target universes, but requires greater dependence on the map frame than was made in the original sample. This particular solution also required a simple redefinition of the sample unit, from the selected lake, to the Thiessen polygon defined for that lake.

Frame "errors" have essentially the same nature, and can usually be corrected, or otherwise compensated for, at the level of the sampling unit. As an example, a map showing spatial locations of the objects of interest may be out of date so that if used as a map frame it is missing some units of the target universe (added since construction of the map) and contains extra units no longer extant in the target universe. Jessen [1977, Chapter 6], Kish [1965, Sections 2.7, 9.6, Chapter 11], Särndal *et al.* [1992, Section 14.7], and Yates [1981, Sections 4.8-4.21] discuss additional details of frames.

## III. SAMPLING FUNDAMENTALS

The basic concepts of sampling are typically presented in the context of a finite, discrete population. These same fundamentals apply to spatial sampling, so we will present the basic ideas following the traditional approach, then extend the ideas to the spatial setting in the following section. Recall that the discrete universe is defined as a finite set, $U = \{u_1, u_2,...,u_N\}$, where the elements $u_i$ are the objects composing the universe. These individual objects always have individual identity. In the spatial context, an explicit identity is provided by the spatial location of the object, but this explicit identity is always there, even without a spatial context. The value $y_u$ is the population value for element u. Parameters of the population are defined by operations on $U$. For example, $T_y = \sum_U y_u$ is the population total and $T_y / N$ is the population mean. Similarly, the population variance is defined as,

$$V_y = \sum_U (y_u T_y / N)^2 / (N - 1).$$

If we have another response variable, $x_u$, defined for $u \in U$, we can specify other parameters of interest relating these two response variables. For example, the ratio $R = T_y / T_x$ may be of interest (e.g., the ratio of total income to total expenditures), or we could define the finite population correlation and regression coefficients,

$$RHO = \frac{\sum_U (y_u - T_y / N)(x_u - T_x / N)}{(N - 1)\sqrt{V_x V_y}},$$

and

$$B = \frac{\sum_U (y_u - T_y / N)(x_u - T_x / N)}{(N - 1)V_x}.$$

Another important parameter in many sampling problems is the population distribution function, $F_y(t)$ = proportion$\{u \in U: y_u \leq t\}$. To distinguish the parameters of a real, finite population from those of a hypothetical model, we use notation such as $T_y / N$, $V_y$, RHO, and B instead of the mathematical statistics notation of $\mu$, $\sigma^2$, $\rho$, and $\beta$. The general pattern is to define a finite population parameter simply by applying the usual formula for a sample statistic to the entire universe. These parameters, which are properties of the finite set, are characteristics of real populations and exist irrespective of the validity of a model [Kish 1987, p. 12]. The parameters of random variables are conceptually much different.

The objective of a traditional sampling problem is to estimate parameters like these. A sampling strategy consists of a sampling design and an estimator. The sample consists of a subset of the universe, and the sampling design is the protocol or set of rules by which the sample elements are selected from the universe. This protocol determines the sample space, **S**, the set of all possible samples, and a function specifying the probability of obtaining a particular sample, $p(S)$. The first-order inclusion probability, denoted $\pi_u$, is the probability that element u is included in the sample, and is calculated by the following formula:

$$\pi_u = \sum_{\{S:u \in S\}} p(S)$$

(sum over all samples in **S** containing element u).

The pairwise, or second-order, inclusion probability, $\pi_{uv}$, is the probability that elements u and v are both included in the sample, and is calculated as:

$$\pi_{uv} = \sum_{\{S:u, v \in S\}} p(S)$$

(sum over all samples in **S** containing both elements u and v).

Inclusion probabilities are a characteristic of the sampling design, not of a single sample. For a given sampling design, we can specify these inclusion probabilities, and they are relatively simple for most of the common designs. For example, the inclusion probabilities for simple random, stratified random, systematic, and one- and two-stage cluster sampling are listed in Särndal *et al.* [1992] and Overton and Stehman [1995]. The first- and second-order inclusion probabilities characterize a sampling design. The first-order inclusion probabilities appear in the estimators of parameters of interest, and, in addition, the second-order inclusion probabilities are required to determine the variance and standard error estimators for a particular estimator.

The inferential framework of classical finite population sampling relies on the method of probability sampling. A probability sample requires that $\pi_u > 0$ for all $u \in U$ (each element of the target universe has a positive probability of being included in the sample), and $\pi_u$ must be known for the elementsin the sample. [Overton and Stehman 1995] Probability sampling does not require that every element of the universe have the same probability of being sampled, only that the inclusion probability be known. Unequal inclusion probabilities may arise in spatial sampling, and this feature must be recognized. If inclusion probabilities are unequal and unknown, the sample is "selection biased" [Stuart 1962], or the phrase "biased sample" may be correctly applied. Such samples are not probability samples, because of the unknown inclusion probabilities. Unequal inclusion probabilities are readily accommodated in sampling practice by incorporating them into the estimator, as developed below.

An estimator is a formula for estimating a parameter of interest. A general unbiased estimator of the population total, the Horvitz-Thompson estimator, $\hat{T}_y = \sum_{u \in S} y_u / \pi_u$ , provides the basis for a general strategy of estimation in sampling. Estimators of other parameters of interest are readily obtained via the Horvitz-Thompson estimator if the parameter can be formulated as a population total, or as a function of population totals. For example, N is itself a total, $N = \sum_{u \in U} 1$, and thus $\hat{N} = \sum_{u \in S} 1 / \pi_u$ is an unbiased estimator of N. If N is known, an unbiased estimator of the population mean is simply $\hat{T}_y / N$. If the sample size, n, is a random variable, the estimator $\hat{T}_y / \hat{N}$ has smaller mean square error than $\hat{T}_y / N$ (even if N is known). These general estimator forms can be used with any sampling design, and they reduce to familiar special case versions for common designs such as simple random, stratified random, systematic, and cluster sampling. The formulæ are presented as special cases in texts such as Cochran [1977], and the more general treatment of estimation using the

Horvitz-Thompson estimator as the unifying framework is provided by Overton and Stehman [1995] and Särndal *et al.* [1992].

For parameters that are functions of population totals, such as R = T$_y$ /T$_x$ (ratio of population totals), RHO, and B, we need a strategy for obtaining consistent estimators, where consistent in the context of finite population sampling is defined by Overton and Stehman [1995] and Särndal *et al.* [1992, Section 5.3]. The practical meaning of consistency is that the estimator estimates the desired parameter of the population of interest, not some other parameter. The Horvitz-Thompson estimator is an example of a consistent estimator. Functions of Horvitz-Thompson estimators are also consistent, and this is the basis for a general strategy of consistent estimation. To estimate a parameter that is a function of population totals, simply replace those totals by their appropriate sample-based Horvitz-Thompson estimators. For example, the population regression coefficient is estimated by

$$\hat{B} = \frac{\sum_s x_u y_u / \pi_u - (\sum_s x_u / \pi_u)(\sum_s y_u / \pi_u) / \hat{N}}{\sum_s x_u^2 / \pi_u - (\sum_s x_u / \pi_u)^2 / \hat{N}}.$$

Formulæ for the variance and variance estimators of estimated totals and means for standard designs may be found in Cochran [1977]. Wolter [1985] and Särndal *et al.* [1992, Section 5.5] review approaches for estimating the variance of an estimator that is a function of population totals (e.g., $\hat{B}$ above). Because of the special importance of systematic sampling in spatial problems, we briefly review some variance estimation issues for this design. In practice, a standard approach to variance estimation for systematic sampling, in the equal probability case, is to use the formula for the estimated variance from simple random sampling to approximate the variance from the systematic sample. This approximation overestimates variance if systematic sampling has resulted in a reduction in variance relative to simple random sampling. If systematic sampling has higher variance than simple random sampling, this approximate estimated variance will be an underestimate of the true systematic sampling variance. This phenomenon extends to variable probability versions of systematic sampling. Much confusion surrounds the issue of bias in systematic sampling, and several texts present misleading statements on this point. Unbiased estimators of totals, means, and proportions are readily available for systematic sampling, regardless of any periodicity or other characteristics of the population. It is unbiased estimation of variance that is problematic, and this distinction needs to be kept clear.

We noted earlier that spatial information may be incorporated into the sampling design in several different ways. To obtain a sample that is spatially well distributed, either stratified or systematic sampling may be used. That is, the population can be stratified geographically, and a sample taken within each of the spatial strata defined. Systematic sampling achieves a similar effect; systematic sampling in the spatial context is treated further in the next section. Spatial information also may be used in forming clusters. Objects close together may be grouped to form clusters, or the objects may be clustered naturally in space, such as houses within a city block. Cluster sampling usually allows for larger sample sizes by reducing transportation costs, but variances per observation are often increased because of intracluster correlations. These intracluster correlations may be viewed as positive spatial autocorrelation among the objects. Two-stage cluster sampling often is more efficient than one-stage cluster sampling when the intracluster correlation is high. In two-stage cluster sampling, the primary sampling units (clusters) are subsampled. By observing fewer elements in each cluster, the variance per observation is reduced relative to one-stage cluster sampling because the effect of intracluster correlation is reduced. Different sampling designs result in different $\pi_{uv}$s and therefore different variances, so any spatial influence on the precision of estimators for a design are incorporated via the $\pi_{uv}$s.

## IV. SAMPLING A CONTINUOUS UNIVERSE

The basic sampling concepts described in Section III apply to sampling continuous universes. Sampling patterns for continuous universes are most often either sets of points or sets of spatially distributed "plots." Particularly when there are spatial components stated in a study's objectives, regular patterns of points or plots are usually favored. Two regular point grids are commonly used: those having the points at the intersections of two sets of equally spaced parallel lines, with the sets at right angles to each other (a square grid), and those having the points at the intersections of two similar sets of lines at an angle of 60 degrees to each other (a triangular grid) (see Figure 3). There are also two commonly used sets of plots: that set formed by the tessellation of a square grid (a square tessellation) and that formed by the tessellation of a triangular grid (a hexagonal tessellation).

A randomly selected point in a continuous universe will have an inclusion density analogous to the inclusion probability of a sample unit from a discrete universe, where the inclusion density is the expected number of sample points in a unit area. Cordy [1993] provides a continuous universe point sample extension of the Horvitz-Thompson Theorem, and presents formulæ for the variance of the estimator that are analogous to those used in the discrete case. This theory can be applied to an unrestricted random sample of points or to a systematic grid.

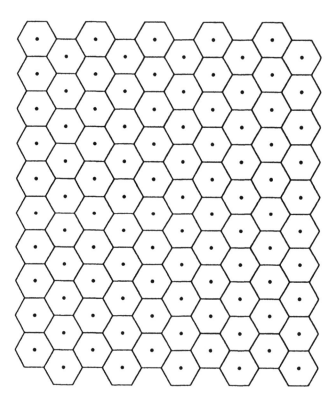

**Figure 3**. Triangular grid with hexagonal tessellation.

The sampling grid is often placed arbitrarily, particularly when it is to be used with spatial methods like kriging, this then resulting in a model-dependent methodology.   However, if the grid is randomly placed, then design-based methods are available, sometimes to considerable advantage. Randomization of the grid must be carefully implemented.  The simplest way is to designate one grid point as the "pointer," rigidly fix the orientation of the grid, and outline a tessellation cell in a fixed position (see Figure 4).  Then randomly locate the pointer in the fixed cell, holding the grid in its rigidly fixed orientation.  This guarantees that the inclusion density is uniform in all parts of the universe.  It is also possible to randomize the orientation of the grid, but this extra step contributes nothing to the randomness of the sample.

A continuous population is defined as a continuous function on a spatial (continuous) universe.   The sampling objective may be estimation of the population mean or total, description of spatial pattern (e.g., a response surface), or model prediction.   Recall that the spatial universe is bounded and has area A, where A is the "size" of U, analogous to N of the discrete universe.   The forms of population parameters for the continuous universe are defined differently than the parameters of a population on a discrete universe, but the analogy is clear.  The population total is now the integral of the surface over the two-dimensional Universe, $\int_U y(u)du$, analogous to $\sum_U y_u$, and the population mean is $\int_U y(u)du/A$, analogous to $T_y / N$.  For a finite universe, the population distribution function is the proportion of objects in which the response variable is less than a specified value.   For a continuous U, we would be interested in the proportion of area less than a specified value, which alternatively can be interpreted as the proportion of randomly selected points.  The inferential framework is again design-based.

Traditional geo-statistics has similar objectives but the methods and terminology are different.   Usually the "population" is modeled by a spatial stochastic process.   Kriging methods are then employed to interpolate between sample locations and to construct contour plots.  A kriging approach also may be used to estimate the population mean. Brus and De Gruijter (1993) compare kriging (model-dependent) estimators with design-based estimators of a spatial mean and find that meaningful practical differences in the estimators may occur. They state that model-dependent strategies do not always produce unbiased estimates of the spatial mean, and that the quality of the variance estimator depends on the quality of the model of the spatial structure.

Olea [1984] evaluates several sampling designs on the basis of precision for estimating spatial functions by universal kriging, and Cressie [1991, Section 5.6.1] compares different sampling designs based on the criterion of minimizing kriging variance. Haining [1990] provides a good overview of some spatial sampling issues pertinent to kriging.

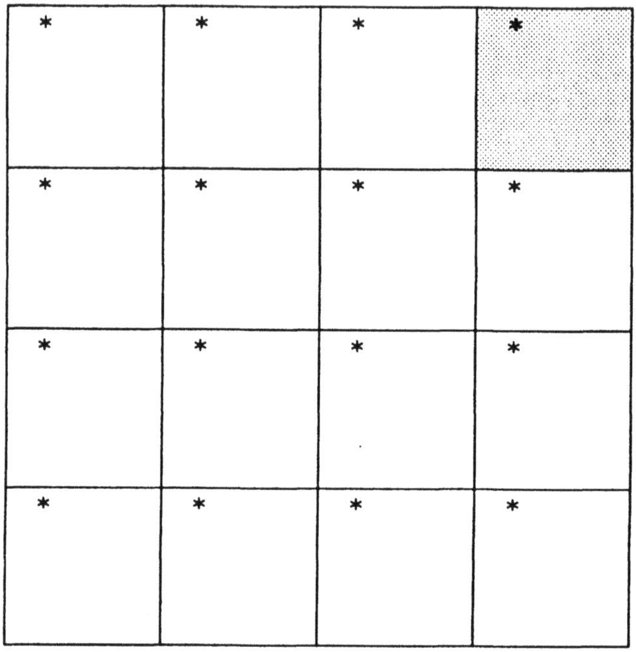

**Figure 4.** Systematic sample with a randomized start. The tessellation cells of a fixed starting grid (grid points in the centers of the tessellation cells) are shown. The shaded cell contains the "pointer" grid point that was randomly located within the shaded tessellation cell. The starting grid is then shifted to correspond to this pointer grid point to yield the systematic pattern shown.

A. **POINT SAMPLING OF A CONTINUOUS POPULATION**

Several spatial sampling designs based on point samples may be considered. We will discuss three of these designs, namely unrestricted random sampling (URS), systematic sampling (SYS), and tessellation-stratified sampling (TSTR); descriptions of other designs may be found in Olea [1984]. The three designs mentioned above are equal probability sampling designs, so an unbiased estimator for the population mean is simply the sample mean, $\bar{y} = \sum_{u \in S} y(u) / n$. An important criterion for choosing a sampling design is the variance of the estimated mean, $V(\bar{y})$. The design with the smallest variance for fixed sample size is preferable.

Unrestricted random sampling is the continuous spatial analog of a simple random sample. Sample points are obtained by selecting n independent, random locations in the universe. Systematic sampling is implemented by a randomized, regular grid of points. The third design of interest is tessellation-stratified sampling (see Figure 5). In this design, the strata are squares or hexagons formed by the tessellation of a regular point grid; the grid may be purposively placed or randomly placed. Sample points are then randomly selected within each stratum, and y(u) is obtained at these sample locations. Selecting one random point per stratum has an advantage from the view of precision, but if unbiased estimation of variance is a critical concern, two or more random points may be selected per stratum. Clearly, a less dense sampling grid forming fewer tessellation cells is needed in the case of two or more observations per stratum. Both the systematic and tessellation-stratified designs provide some control over the spatial distribution of sampling locations.

Precision comparisons in two-dimensional sampling may be approached from two perspectives. [Bellhouse 1988] Quenouille [1949] and Das [1950] assume that the population consists of MN units, either points or quadrats, arranged in M rows and N columns. Zubrzycki [1958] assumes that "the population consists of a number of non-overlapping domains which are congruent by translation." This latter approach does not "discretize" the universe as does the former approach. Precision comparisons of the three designs for estimating the mean in one-dimensional systematic sampling provide insight into the expected performance of the designs in the two-dimensional case. For populations with a linear trend, the order of preference is TSTR, SYS, and URS. For quadratic or sinusoidal surfaces, the preference between SYS and TSTR depends on the specific surface chosen, and neither design is categorically better than the other for sinusoidal (periodic) populations. [Bellhouse 1988, Cochran 1977] Cochran [1946] shows that SYS is preferable to TSTR and URS for an autocorrelation model in which the correlogram is concave upward.

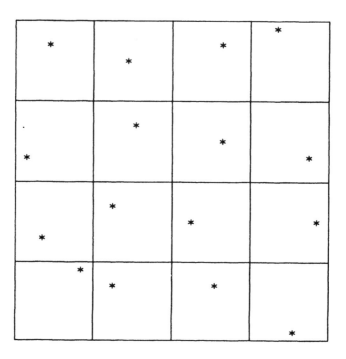

**Figure 5.** Tessellation-stratified sampling from a square grid, with one sample observation per stratum selected at random.

Results for two-dimensional sampling generally follow those for one-dimensional sampling. For Quenouille's [1949] approach and a linear trend model, TSTR has smaller variance than SYS, and SYS has better precision than URS (Bellhouse 1981). Neither TSTR nor SYS is uniformly better than the other for periodic populations, and no simple relationship between SYS and TSTR has been found when the Zubrzycki [1958] approach is applied to an autocorrelation model. [Bellhouse 1985] For a two-dimensional spatial autocorrelation model, systematic sampling is more efficient than stratified or random sampling. [Bellhouse 1977]  Matérn's [1986] order of preference is

SYS, TSTR, and URS, based on his analysis of a population model employing a decreasing correlation function. Overton and Stehman [1993] present some numerical results illustrating the magnitude of differences in precision of the three designs under various population models. In general, TSTR is the most precise design, but SYS sometimes has smaller variance, depending on the population model.

Variance estimation for both the systematic and tessellation-stratified (one sample point per cell) designs is problematic. For the systematic design, a common approach to variance estimation is again to treat the systematic sample as a URS and to estimate variance by

$$\hat{V}(\bar{y}) = \sum_{u \in S} (y(u) - \bar{y})^2 / (n - 1). \tag{1}$$

This formula usually overestimates variance (Ripley 1981). Several alternative variance estimators for systematic sampling from a square grid have been described. [Yates 1981; Matérn 1986] A variance estimator based on the mean-square-successive difference adapted to a triangular grid provides an estimator with small bias for the tessellation-stratified design, and performs better than equation (1) for the strict systematic design. [Overton and Stehman 1993] Wolter [1985, Chapter 7] provides simulation results illustrating the performance of several variance estimators under one-dimensional systematic sampling.

## B. AREAL SAMPLING OF A CONTINUOUS UNIVERSE

Areal samples also may be used to sample continuous populations, where an areal sampling unit is a plot or quadrat or other spatial unit on which the populations of interest will be described. For areal samples, the observations, $y_u$, used in estimation are typically integrals of the population value over the areal sampling unit; this might be expressed as the mean value over the sampling unit, together with the unit size. Then if all units are the same size, the unit mean is unambiguous. Similarly, mean values and areas over portions of the areal sampling unit, with those portions representing qualitative classes of response, are feasible observations for areal samples.

Relative to point samples, much additional information is available from the areal sampling unit. Information on spatial patterns within the sampling unit may be recorded, as for example, the portion of the unit occupied by a particular class (e.g., ecological community type, or age-class of forest stand). Areal samples are also useful for populations that do not quite fit the definition of continuous and also are not appropriate for treatment as discrete. Marshes and grasslands are examples. We have used the term "extensive" for these populations, and their treatment should often be closer to continuous than discrete. This treatment is a matter of scale and resolution. For instance, a very small sampling unit in a sward of forbs and grasses could be described in terms of individual plants, whereas a large plot in the same sward would necessarily be described in more continuous terms. At whatever scale one is sampling, some populations will be discrete, some continuous, and some will fall in between.

In areal sampling, the spatial universe is partitioned into a large number of cells of varying size. For a single sample point positioned at random in the entire spatial universe, the areal unit into which the point falls is chosen. The inclusion probability is $\pi_u a_u/A$, where $a_u$ is the area of the selected unit and A is the area of U. The estimator of a population total from that single sample point is then $z_u = y_u/\pi_u$, which is a Horvitz-Thompson estimator, and where $y_u$ is the appropriate value for plot u. A useful and simple way to take more than one sample is to employ URS (Section IV.A), by which n sample points are chosen independently. This generates n independent realizations of the random variable Z, and we use standard results for iid samples to obtain the overall estimate $\hat{T}_y = \bar{z} = \sum_{u \in S} z_u / n$, and $V(\bar{z}) = V(Z)/n$. An unbiased

estimator of V(Z) is the sample variance, $s^2 = \sum_{u \in S} (z_u - \bar{z})^2 / (n-1)$,

and $V(\bar{z})$ is estimated by $s^2/n$. Different ways of obtaining the partition on U and of selecting a sample of areal units will be treated in the remainder of the section.

## 1. Frames for areal sampling

### a. Traditional areal sampling

Areal (area) sampling has a long history of use. Traditionally, samples for agriculture and households were often cluster designs, allowing the surveyor to spend considerable time in a selected neighborhood before moving to another. The clusters were areal units containing many farms or households. The frames for areal sampling received great developmental effort, and were used year after year, usually with considerable additional effort spent in keeping them current. Jessen [1987, Section 6.7], Kish [1965, Chapter 9], and Yates [1981, Sections 4.14, 4.17] provide additional background on areal sampling.

This sampling design is easily implemented in spatial sampling by partitioning the continuous spatial universe into areal sampling units. A sample of these areal units is an areal sample. Given any variable observed on the sampled areal units, the general design-based methodology appropriate for the selection scheme is available (see Section III). Several options are available for constructing a partition on the continuous spatial universe for the purpose of areal sampling. This partition can be completely arbitrary, constructed simply for the purpose of sampling. Additionally, there are many natural partitions, such as townships, city blocks, watersheds, or crop fields, that will provide more natural areal units for sample description. The areal units need to be moderately uniform in size, but not necessarily equal.

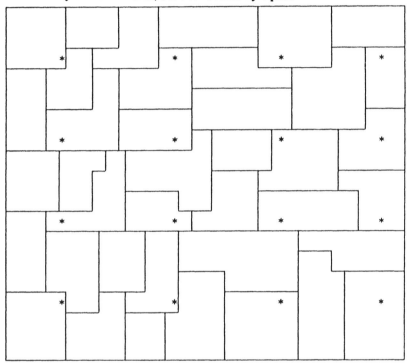

**Figure 6.** Arbitrary spatial partitioning of a continuous universe with a systematic grid (*s) superimposed. Areal sampling units are selected with probability proportional to the area of the unit.

On account of the spatial context, we prefer to sample these areal units via a regular point grid. Let the resolution of the grid be coarse enough that there is little or no chance that two or more points of the grid can fall in a single unit, randomize the position of the grid, and select in the sample all the units that contain a grid point (see Figure 6). Because the areal units are generally of uneven size, they will be selected with variable probability by this selection scheme. Each unit in the sample occupies a known area, $a_u$, and the inclusion probability for that unit is $\pi_u = a_u D$, where D is the density of the systematic sampling grid. A two-dimensional systematic sample of the sampling units provides a spatially well-distributed sample of areal units. The estimated population total is

$$\hat{T}_y = \sum_{u \in S} y_u / \pi_u = \frac{1}{D} \sum_{u \in S} y_u / a_u,$$

allowing for unequal sized areal sampling units, and the approximate variance of $\hat{T}_y$ is estimated (with $u \neq v$) by

$$\hat{V}(\hat{T}_y) = \sum_{u \in S} (y_u^2 / \pi_u^2)(1 - \pi_u) + \sum_{u \in S} \sum_{v \in S} [(\pi_{uv} - \pi_u \pi_v)/\pi_v](y_u y_v / \pi_u \pi_v),$$

the Horvitz-Thompson estimator of variance, with $\pi_{uv}$ approximated by $(2(n - 1)\pi_u \pi_v) / (2n - \pi_u - \pi_v)$. [Stehman and Overton 1994] This formula provides an approximation analogous to the use of the variance for a simple random sample to the variance for an equiprobable systematic sample. Several authors have investigated alternate formulae to account for efficiency of systematic sampling. [Cordy and Thompson 1995, Dunn and Harrison 1993]

*b. A rigorous equal-probability areal sample*

The tessellation dual of a high resolution regular point grid provides an excellent frame for a probability areal sample of an extensive population; the cells of the tessellation partition the universe into a collection of sampling units, all of the same size and shape. So structured, this can be considered an alternate universe of spatial units, each with the response function of the parent population. Parameters of this "little" population defined on an areal sampling unit then become the variables to be observed for selected units.

A sample of these areal units can be selected by any finite sampling design, but in the spatial context, it is convenient to make the selection without generating a list frame. This is easily implemented via a regular point grid of somewhat lower resolution than the tessellation. If this grid is randomized on the spatial universe, in the manner prescribed in Section IV, then the cells of the areal frame are selected with equal probability; each cell will have the same probability of containing a sample point after randomization. Equal probability is insured by the equal sizes of the cells, the regularity of the grid, and the randomization protocol (see Figure 7).

In general, although this design is quite satisfying, it may be somewhat inconvenient, as the cell selected by a particular point will usually have a different spatial relation to the point than will the cell selected by another point. That is, the position of the random point in one selected cell will be different than the position in another selected cell. As a consequence, it is necessary in general to draw the entire tessellation in order to use the method. However, if the resolution of the tessellation and the grid are "compatible," then the grid can be randomized on the continuous universe, and the selected cells constructed as centered on the grid points. "Compatible" means simply that when the sampling grid is laid over the frame grid, all points of the sampling grid will coincide with points of the frame grid. Having to construct only the selected cells is a great saving of effort.

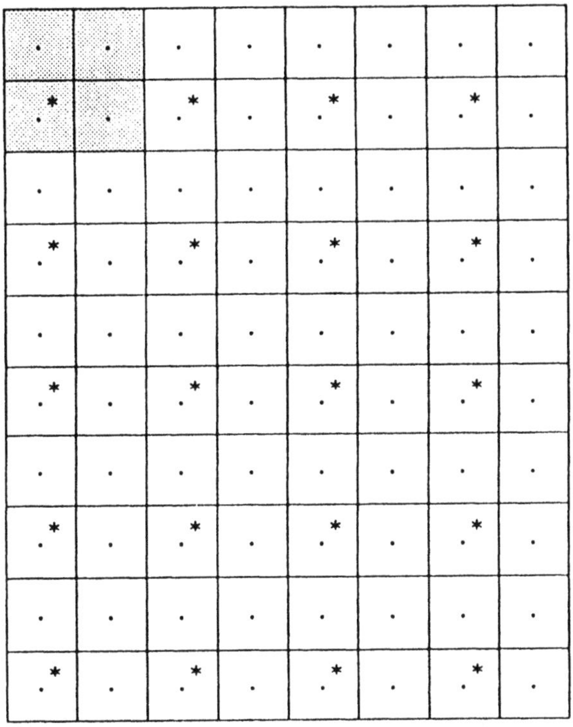

**Figure 7.** A rigorous equal-probability areal sample based on a square grid. The grid points shown represent a high resolution, regular point grid, and the tessellation cells may be considered a universe of spatial units. A sample of these areal units would be obtained by selecting a random starting position (the pointer grid point) in the shaded tessellation cell, and then selecting those areal units intersected by the grid points (*s) of the lower resolution grid superimposed. In this figure, the ratio of grid densities is 1:4 (k=4), so $\pi_u = \frac{1}{4}$.

This design is being considered for EMAP [Overton *et al.* 1991, pp. 3-4, 4-2], in which the density of the triangular sampling grid is approximately 1:640 ha$^2$ and the density of the triangular frame grid is approximately 1:40 ha$^2$, with the sampling ratio exactly 1/k, where k=16 in this case. Interpreting the systematic sample of 40 ha$^2$ hexagons as a sample from the universe of all such 40 ha$^2$ hexagons tiling the U.S. allows simple and direct estimators and variance approximations. This scheme selects a sample of tessellation cells from a randomly positioned tessellation.    It is interpreted as a conventional areal sample by conditioning on the position of the sampling grid.    This fixes the position of the tessellation, and selects the sample as a $1/k^{th}$ sample of the cells from that position. A total of k-1 other positions of the sampling grid would yield the same position of the tessellation.  As a result of this conditioning of the position of the tessellation, it is possible to treat the sample as a traditional areal sample.  Thus, conditionally, $\pi_u = 1/k$, and the estimator of the population total is $\hat{T}_y = k \sum_{u \in S} y_u$.

We can again invoke the *iid* model for random variables to construct a variance approximation, treating the individual areal samples as independent.  That is,

$$\hat{V}(\hat{T}_y) = k^2 m s_y^2, \tag{2}$$

where $s_y^2$ is the sample variance of the total of the ys for the m areal units in the sample (De Vries 1986, p. 218).  Unless k is very large, it would be appropriate to include a finite population correction factor $((k - 1) / k)$ in equation (2) to obtain

$$\hat{V}(\hat{T}_y) = k(k-1)m s_y^2. \tag{3}$$

Because the actual sample is systematic, equation (2) employs the unrestricted random sampling model for estimating the variance of the systematic sample.  As we have noted before, this variance estimator will overestimate the variance if the systematic sample achieved a gain in precision over that of an unrestricted random sample.  A useful improvement on either equation (2) or (3) when the sample is systematic is to replace $s_y^2$ by $\delta^2$, the mean square successive difference,

$$\delta^2 = \frac{1}{2(m-1)} \sum (y_u - y_v)^2,$$

where summation is over all m adjacent pairs.  Overton and Stehman [1994], Stehman and Overton [1989], and Wolter [1985] provide extensions to variable probability.

## 2. Support

For a point sample it is often appropriate to actually make the measurements over an areal sample rather than precisely at a point.  Such an areal sample is commonly called the support of the point.  This is necessary simply because some measurements cannot be made at a point; common examples are species diversity and biomass per hectare.

The areal units used for support may be circles, squares, or hexagons. It might be appropriate to make both supported point estimates and areal sampling estimates from the same areal samples, but this would seem uncommon. Support samples should be as small as possible, and the support area should vary from variable to variable. Fundamentally, the two designs have different objectives; for a point sample, the aim is to characterize the points, whereas for an areal sample the aim is to characterize an areal fragment.

## V. SAMPLING SPATIALLY DISTRIBUTED OBJECTS VIA AREAL SAMPLES OF THE CONTINUOUS UNIVERSE

Spatially distributed objects are readily sampled via application of one of several options for sampling the continuous universe. These all can be perceived as areal sampling, with the discrete units associated with the areal sampling units; point sampling may be explicitly involved, but always turns out to be a device for sampling specified areas associated with the objects. These methods often require that each discrete object be identified by a unique spatial location point. For example, lakes can be located by their centroids and stream reaches by their upper nodes; any position will suffice so long as it is unambiguous and objectively identified.

For an areal sample of the continuous universe, a sample plot will contain 0, 1, or more objects whose position point is within the plot (see Figure 8); that is, this areal sample also provides a sample of objects from the finite universe. The inclusion probability for each object in the areal unit is identical to that of the areal unit, but will be modified if subsampling is employed (a topic that will not be treated here). Pairwise inclusion probabilities for objects in the same areal unit will be equal to the first order inclusion probabilities. Pairwise inclusion probabilities for objects in different areal units will be identical to the pairwise inclusion probabilities for those units. Estimation of the number of objects in the finite universe, N, is a special case of the formula for $\hat{T}_y$,

$$\hat{N} = k \sum_{u \in S} n_u, \text{ for equal sized areal samples, and}$$

$$\hat{N} = \frac{1}{D} \sum_{u \in S} n_u / a_u, \text{ for variable sized areal samples,}$$

where $n_u$ is the number of objects in areal sampling unit u and D is the density of the systematic grid.

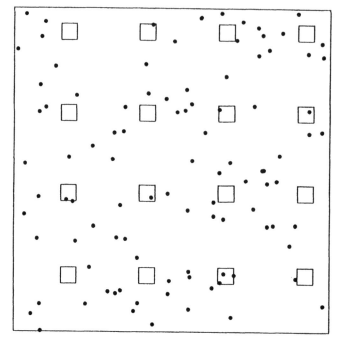

**Figure 8.** Areal sample of discrete objects as identified by square areal units associated with a randomly positioned systematic grid. The *s represent the unique spatial location identifiers of the objects sampled. Note that the areal sampling units may contain 0, 1, or more of the objects of interest.

Another option for selecting discrete objects that is practically convenient to implement is to select the closest object to a random point. The Thiessen polygon [cf. Ripley 1981, p. 38] surrounding an object represents all points located closer to that object than to any of the other reference objects; therefore this polygon represents the area within which a random point results in selection of this particular object. A useful modification of this selection rule is that no object be selected if it is greater than c measurement units from the random point. This effectively truncates the polygon by a circle of radius c, centered on the object. This approach also can be thought of as just another way of constructing a frame for areal sampling via the Thiessen polygons of the universe of objects, so that this topic could have been treated in Section IV. It is necessary to actually establish the polygons only for those objects selected in the sample, which provides a great savings in labor.

## VI. INFERENCE IN SPATIAL SAMPLING

The mode of inference used with traditional sampling methods differs from that of most spatial analyses as well as from that of conventional statistics. In design-based inference, the classical mode of inference in sampling, the population is considered fixed. That is, if we observed (without measurement error) all elements in a finite universe or all points in a continuous universe, no uncertainty in the parameter would remain. The basic uncertainty in design-based inference is due to use of the sample, rather than complete enumeration and is generated by the sample selection method. Thus properties of estimators depend on the design employed, and the same estimator (for example, the sample mean) may have different properties for different designs.

The formulæ for the expected value and variance of an estimator in the design-based framework help clarify the inferential approach. Let $\hat{\theta}$ be an estimator of the finite population parameter $\theta$, and $\hat{\theta}_s$ be the value of $\hat{\theta}$ for sample $s$ (e.g., $\hat{\theta} = \bar{y}$, and for sample $s$, $\hat{\theta}_s = \bar{y}_s$). Then the expected value of $\hat{\theta}$ is $E(\hat{\theta}) = \sum_S p(s)\hat{\theta}_s$, where $p(s)$ is the probability of obtaining sample $s$, and the summation is over all possible samples, $S$. The variance of $\hat{\theta}$ is $V(\hat{\theta}) = \sum_S p(s)[\hat{\theta}_s - E(\hat{\theta})]^2$. $V(\hat{\theta})$ is a parameter that depends on the estimator and the sampling design. It is important to recognize the distinction between variance and estimated variance. $V(\hat{\theta})$ is the quantity of interest when comparing different sampling strategies, so a hypothetical assessment of $V(\hat{\theta})$ may be important at the planning stage of a sampling project. In practice, $V(\hat{\theta})$ is unknown, and we use the estimated variance, a statistic based on the sample data, to estimate $V(\hat{\theta})$. The properties of an estimator do not depend on a model of the observed variables, but the influence of the sampling design is apparent in the dependence of the properties on $p(s)$. The basic concepts of design-based inference are illustrated well by the many examples in Stuart [1962].

Another approach to inference sometimes applied in sampling problems is called model-based inference. Others have called this type of inference model-dependent, to indicate complete reliance on a model, and Brus and De Gruijter [1993] use the term geo-statistical inference to indicate dependence on a geo-statistical model. In this mode of inference, the population observed is regarded as one realization of a process generating the universe and population. The observation y(u) obtained at location u is viewed as a random variable generated by a model. The quantity of interest may be the spatial mean, $\int_U y(u)du/A$, where A is the area of the region, or the model mean, $\mu = E_t[y(u)]$, where $E_t$ denotes expectation over the model [De Gruijter and Ter Braak 1990]. In the terminology of finite population sampling, the spatial mean would be analogous to a finite population parameter, and the model mean would be a superpopulation parameter. In a model-dependent approach, properties of estimators, including variance, are determined by the model specification.

While the stationary, spatial stochastic model commonly employed in geo-statistics may be an appropriate representation for some spatial physical phenomena, other model representations may be more appropriate for other responses. For example, for most ecological phenomena, a non-stationary spatial stochastic process provides a more realistic modeling perspective for studying and understanding the spatial organization of the underlying ecological systems. That is, we could define $\eta(u) = E_t[y(u)]$ to represent the fixed realization of an underlying process that has produced the current, observed pattern of the physical environment. If the current patterns arose as the result of a stochastic process, that process was (nearly) finished eons ago, and what we are observing is a "frozen accident." The spatial mean, $\int_U (u)du/A$, might then be a parameter of interest for describing the spatial universe with respect to the physical environmental variables. However, this structure would be fixed for successive stochastic realizations of the ecological processes that are operating to produce the ecological responses. These higher resolution ecological processes may impart additional spatial pattern to the strong underlying pattern of the physical environment, but these processes overlay the geophysical processes and should be distinguished from them.

The modeling perspective that is taken is important because it determines the analysis of spatial data obtained for the purpose of studying specific processes. The impact of the perspective on analysis is illustrated by Jager and Overton [1993] in which focus on ecological processes led to use of response surfaces to explain patterns of ecological variables (acid neutralizing capacity of lakes in the Northeast United States) in terms of recognizable environmental variables (elevation and pH of precipitation). The response surface analysis was followed by applying geo-statistical techniques to residuals to search for evidence of spatial pattern not accounted for by the explanatory variables.

Design-based inference does not depend on the validity of a model, and the descriptive inferences provided in this framework therefore are appealing in situations in which model-dependent results are subject to criticism, such as for public policy decisions. Further, the objectivity of a probability sample may be a useful design characteristic in matters of debate in which a purposive sample may be questioned. Probability sampling does not preclude model-dependent analyses, although certain complex probability sampling designs (e.g., variable probability sampling) may complicate these analyses. Model-dependent inference is often necessary for spatial prediction, particularly when the data were not obtained via a probability sample, and it is sometimes suggested to accommodate missing data and measurement error in sampling applications.

The approach to inference also has implications with respect to the issue of "independence." De Gruijter and Ter Braak [1990] and Brus and De Gruijter [1993] discuss at length the meaning of independence in the design-based and model-dependent frameworks, and point out that many authors erroneously claim that classical sampling methods are not appropriate for spatially autocorrelated data. For example, Arbia [1993, p. 346] states that when common sampling techniques such as simple random, systematic, and stratified are used, "the selection of areal units is made as if the generating process which underlies the observed values is constituted by a sequence of random variables $X_1$, $X_2,...,X_n$ which are independently and identically distributed (i.i.d.)." This is simply not true with respect to any design-based inference. That it similarly is not true with respect to traditional spatial analyses, such as kriging, is evidenced by the fact that these methods are considered legitimate for any set of points, whether chosen haphazardly, purposively, or by a probability design; the model supplies the relevance of the predictions.

"Independence" is a standard assumption of most conventional statistical procedures. Haining [1990] classifies this as one of his three spatial sampling problems, the need to "obtain observations that are independent or nearly independent so that classical statistical procedures can be used to classify data or establish relationships between variables." Within the design-based framework, methods were discussed in Section III for estimating regression and correlation coefficients in which the lack of independence among sample units was accounted for by the pairwise inclusion probabilities employed in the estimators. Skinner, Holt, and Smith [1989] and Särndal *et al.* [1992] describe analyses of contingency tables for complex sampling designs. Again, lack of independence of the sampling units is accounted for by the pairwise inclusion probabilities employed in these analysis. Others have approached conventional statistical techniques from a model-dependent perspective, and have derived analyses that take into account the spatial autocorrelation of the observed data [Cressie 1991; Griffith 1978; Upton and Fingleton 1985, 1989].

The term independence has different meanings depending on the field of application, so the potential for confusion exists, especially when crossing fields. Independence, or its lack, could refer to the manner of selecting sample units, sometimes resulting in greatly diverse values of $\pi_{uv}$, and where for near-independence, $\pi_{uv}$, would be close to $\pi_u \pi_v$. As we have stated earlier, lack of independence of sample selection in design-based inference is taken into account by the variance estimator structure, although some designs pose severe difficulties for some uses. In the model context, independence may also be discussed in reference to the manner in which the population values arose, as (possible) independence of observations generated from a stationary, stochastic process. This view of independence may be taken by geo-statisticians. Still another use of independence arises in the statistician's concern about independence of the residuals, as is appropriate in a regression analysis. In this context, the residuals themselves are dependent on the specific model used.

## VII.  APPLICATIONS OF SPATIAL SAMPLING

Special sampling techniques have been developed for some applications of sampling spatially distributed populations. Line-intercept sampling, a special case of line transect sampling, is often used in plant ecology to estimate the area or proportion of area covered by a particular type of vegetation [cf. Buckland *et al.* 1993], or the linear extent of roads or down timber. [DeVries 1986]   Gilbert [1987, Chapter 10] discusses sampling techniques to detect "hot spots" of radioactivity or other environmental pollutants and provides a general treatment of sampling to estimate spatial patterns or totals of environmental contaminants.   Dunn and Harrison [1993] compare various sampling methods for estimating area under different land uses, and report some results on variance estimators for systematic sampling. In forestry, variable plot sampling (also called angle-gauge or horizontal point sampling) is used to obtain a sample of trees based on selecting random or systematic points in a forest. [De Vries, 1986]  This design is particularly efficient for estimating total basal area of trees without constructing a list frame.   Webster and Oliver [1990] provide a good review of spatial sampling applications in soil science.   Mandallaz [1993] takes a geo-statistical approach to incorporate auxiliary information into forest inventory estimators for small areas.

## VIII.  EMPIRICAL EVALUATION OF SAMPLING STRATEGIES

An important task in an investigation requiring spatial sampling is choosing an adequate sampling design.   This task is often aided by empirical evaluation of various design options, and given the improvements in modern day computing, such empirical analyses are now practical to implement for most situations.   This section outlines a procedure for a Monte Carlo or simulation approach for assessing various properties of a sampling strategy.   The empirical assessment may examine artificial populations generated according to specified models and statistical distributions, or natural populations selected because of their similarity to the populations expected to be encountered in an actual survey. Generally precision of estimators of the population mean, total, or distribution function is the primary criterion of interest, but empirical assessment may be used to evaluate possible bias of estimators or variance estimators, and properties of confidence intervals constructed using the proposed estimators. We illustrate the general approach using an example from the U.S. EPA Environmental Monitoring and Assessment Program. The specific problem was to estimate the mean concentration of dissolved oxygen in the Chesapeake Bay. The response variable was represented as a continuous variable over the two-dimensional region, and the sampling problem was to identify which of several design options would provide the most efficient estimate of the mean concentration. The problem is simplified to two dimensions by considering dissolved oxygen at the surface and we ignoring variation of the response with depth.

For continuous response variables, populations are usually generated using one of the following models. A surface model may be used to generate a hypothetical surface, $z(x,y)$, where the observation generated at any point is specified by the model

$$z(x,y) = g(x,y) + \epsilon,$$

and $\epsilon$ may represent either local high-resolution irregularities in the surface or errors of measurement, and is modeled as independent, homogeneous noise. Choice of the error distribution ($\epsilon$) may be used to vary the signal-to-noise ratio of the surface. Spatial autocorrelation models also may be used to generate the surface. This approach is described in more detail by Matérn [1986] and Bellhouse [1977]. For example, a one-dimensional superpopulation model with autocorrelated errors may stated as follows: $E(y_u) = \mu$, $E(y_u - \mu)^2$, and $E(y_{u+v} - \mu)(y_u - \mu) = \sigma^2 \rho(v)$, where $\rho(v)$ is the correlation between units $v$ distance apart.

In the Chesapeake Bay study, three general surface models have been evaluated, namely a planar surface, a quadratic surface, and a sinusoidal surface. [Overton and Stehman 1993] These models were generated using several versions as follows:

- Planar:      $g(x, y) = g_0 + ax + by$
- Quadratic:     $g(x, y) = g_0 + ax + by + cax^2 + cby^2$
- Sinusoidal:    $g(x,y) = g_0 + \sin(2\pi x/ad) + \sin(2\pi y/bd)$, where d is the distance between points in the grid, a and b are the periods of the sine function in the x- and y-dimension, respectively, with a and b measured in units equal to the distance between points in the systematic grid.

Overton and Stehman [1993] provide a rationale for using response models in environmental science applications. The focus of the investigation was estimation of the spatial mean, $\overline{Z}$, but the same general method could be used to evaluate other properties of the sampling strategy such as the variance estimator proposed for use with a systematic or tessellation-stratified design.

The simulation approach follows the conceptual foundation of design-based inference (Section VI). For a given population, the sampling strategy is repeated M times, and the properties of the sampling strategy are calculated from these M replications. For example, the bias of $\overline{z}$ is calculated as $\sum_{k=1}^{M} \overline{z}_k / M - \overline{Z}$, where $\overline{z}_k$ is the mean of the $k^{th}$ replication, and the variance of the strategy is calculated as $\sum_{k=1}^{M} (\overline{z}_k - \overline{Z})^2 / M$. The bias of various proposed variance estimators can be calculated in a manner similar to that for the bias of $\overline{z}$. The primary use of variance estimators is in constructing confidence bounds, so empirical evaluation of variance estimators should focus on coverage of confidence bounds obtained using these estimators. Observed coverage is simply the proportion of samples in the simulation in which the true parameter is contained within the confidence bounds based on a given variance estimator. Overton and Stehman [1993] and Wolter [1985] provide illustrations of how to evaluate variance estimators using a simulation approach.

## IX. SUMMARY

Spatial sampling presents some challenges to classical sampling theory and practice, but understanding classical techniques provides necessary fundamental concepts to apply to sampling spatially identified populations. In those cases in which the spatial sampling problem can be recast into sampling a discrete universe of objects, classical finite sampling methods readily apply. We have summarized several methods for sampling from a discrete universe. Constructing a list frame from map data would be the most direct way to translate a spatially located population into a discrete universe, but this approach is rarely practical. Other methods rely on a spatial partitioning, either by fixed size polygons such as squares or hexagons, or by areal sampling units of possibly unequal size constructed from existing fields, or using roads, streams, and other natural boundaries. Selecting the closest object to a randomly located point sets up another partitioning of the spatial universe that permits use of methods for sampling from a discrete universe.

Sampling a continuous population defined on a spatial universe fits a structure analogous to the classical finite sampling framework as demonstrated by Cordy's [1993] work, which provides a connection between the continuous and finite population sampling approaches. We have discussed three designs for sampling a continuous population, namely unrestricted random sampling, systematic sampling, and tessellation-stratified sampling. In all three a point sample is used, and observations are made at the selected points. In general, tessellation-stratified sampling is more efficient than unrestricted random sampling for most surfaces likely to be found in practice, and this design avoids some of the problems characteristic of systematic sampling (lack of a reliable variance estimator and poor precision in strongly periodic populations if the sampling interval is in phase with this periodicity). Both systematic and tessellation-stratified (one unit per stratum) sampling require approximations when estimating variance, and various methods have been presented.

On account of advances in modern computing, it is now practical to empirically investigate various sampling strategies. Simulation techniques may be used to compare empirically precision of various sampling designs under consideration for a particular application. We recommend that empirical validation of any non-standard methodology become a routine part of sampling protocol.

We conclude by reviewing some practical issues of spatial sampling. Most real sampling problems have several objectives, so optimizing the design with respect to a single parameter (or response variable) usually will result in poorer precision for estimating other parameters. While it may be instructive to know the spatial variation of the response variables when planning a survey, tailoring the design to optimize efficiency on the basis of this information is possible only if a single variable is of interest. The multiobjective nature of most surveys makes simplicity a desirable design characteristic, both to ensure that the design has reasonable efficiency for the many variables measured, and to ensure that the design and analysis are implemented correctly. We strongly encourage use of probability sampling designs because they guarantee the objectivity of the sample selection process, and they provide an unambiguous identification of the population represented by the sample. Model-dependent spatial analyses ignore the sampling design structure and therefore have no requirement of a probability sample. However, use of a probability sample does not hamper subsequent spatial analyses, and the desirable properties of probability sampling often warrant some minimal increase in effort needed to obtain the data in a manner needed to acquire these properties.

# REFERENCES

**Arbia, G.** The use of GIS in spatial statistical surveys. *International Statistical Review*, 61, 339-359, 1993.

**Bellhouse, D. R.** Some optimal designs for sampling in two dimensions. *Biometrika*, 64, 605-611, 1977.

**Bellhouse, D. R.** Spatial sampling in the presence of a trend. *Journal of Statistical Planning and Inference*, 5, 365-375, 1981.

**Bellhouse, D. R.** Systematic sampling of periodic functions. *Canadian Journal of Statistics*, 13, 17-28, 1985.

**Bellhouse, D. R.** *Systematic sampling. Handbook of Statistics, Vol. 6* (P. R. Krishnaiah and C. R. Rao, Eds.). Elsevier Science Publishers, Amsterdam, 1988.

**Brus, D.J., and De Gruijter, J.J.** Design-based versus model-based estimates of spatial means: Theory and applications in environmental soil science. *Environmetrics*, 4, 123-152, 1993.

**Buckland, S.T., Anderson, D.R., Burnham, K.P., and Laake, J.L.** *Distance Sampling: Estimating Abundance of Biological Populations.* Chapman & Hall, New York, 1993.

**Cochran, W.G.** Relative accuracy of systematic and stratified random samples for a certain class of population. *Annals of Mathematical Statistics*, 17, 164-177, 1946.

**Cochran, W.G.** Sampling Methods (3$^{rd}$ ed.). Wiley, New York, 1977.

**Cordy, C.B.** An extension of the Horvitz-Thompson theorem to point sampling from a continuous universe. *Statistics & Probability Letters*, 18, 353-362, 1993.

**Cordy, C.B., and Thompson, C.M.** An application of the deterministic variogram to design-based variance estimation. *Mathematical Geology*, 27, 173-205, 1995.

**Cressie, N. A. C.** *Statistics for Spatial Data.* Wiley, New York, 1991.

**Das, A.C.** Two-dimensional systematic sampling and the associated stratified and random sampling. *Sankhya*, 10, 95-108, 1950.

**De Gruijter, J.J., and Ter Braak, C.J.F.** Model-free estimation from spatial samples: A reappraisal of classical sampling theory. *Mathematical Geology*, 22, 407-415, 1990.

**Deming, W.E.** *Some Theory of Sampling.* Wiley, New York, 1950.

**De Vries, P.G.** Sampling Theory for Forest Inventory. Springer-Verlag, New York, 1986.

**Dunn, R., and Harrison, A.R.** Two-dimensional systematic sampling of land use. *Applied Statistics*, 42, 585-601, 1993.

**Gilbert, R.O.** *Statistical Methods for Environmental Pollution Monitoring.* Van Nostrand Reinhold, New York, 1987.

**Griffith, D.A.** A spatially adjusted ANOVA model. *Geographical Analysis*, 10, 296-301, 1978.

**Haining, R.P.** *Spatial Data Analysis in the Social and Environmental Sciences.* Cambridge University Press, 1990.

**Jager, H.I., and Overton, W.S.** Explanatory models for ecological response surfaces. In *Environmental Modeling with GIS*, M.R. Goodchild, B.O. Parks, and L.T. Steyaert (ed.). Oxford University Press, pp. 422-431, 1993.

**Jessen, R.J.** *Statistical Survey Techniques.* Wiley, New York, 1978.

**Kish, L.** *Survey Sampling.* Wiley, New York, 1965.

**Kish, L.** *Statistical Design for Research.* Wiley, New York, 1987.

**Linthurst, R.A., Landers, D.H., Eilers, J.M., Brakke, D.F., Overton, W.S., Meier, E.P., and Crowe, R.E.** Characteristics of lakes in the eastern United States. Volume I: Population descriptions and physico-chemical relationships. U.S. Environmental Protection Agency, 401 M Street SW, Washington, DC 20460. (EPA-600/4-86/007a), 1986.

**Mandallaz, D.** (1993). *Geostatistical Methods for Double Sampling Schemes: Application to Combined Forest Inventories. Chair of Forest Inventory and Planning,* Department of Forest and Wood Sciences, ETH-Zentrum, CH-8092, Zürich, 1993.

**Matérn, B.** *Spatial Variation* (2nd ed.). Springer-Verlag, New York, 1986.

**Olea, R.A.** Sampling design optimization for spatial functions. *Mathematical Geology,* 16, 369-392, 1984.

**Overton, W.S., Kanciruk, P., Hook, L.A., Eilers, J.M., Landers, D.H., Brakke, D.F., Blick, D.J., Linthurst, R.A., DeHaan, M.D.** *Characteristics of Lakes in the Eastern United States. Volume II: Lakes Sampled and Descriptive Statistics for Physical and Chemical Variables,* EPA600/486007b, Washington, DC, U. S. Environmental Protection Agency, 1986.

**Overton, W.S., and Stehman, S.V.** Properties of designs for sampling continuous spatial resources from a triangular grid, *Communications in Statistics - Theory and Methods,* 21, 2641-2660, 1993.

**Overton, W.S., and Stehman, S.V.** Variance estimation in the EMAP strategy for sampling discrete ecological resources, *Environmental and Ecological Statistics,* 1, 133-152, 1994.

**Overton, W.S., and Stehman, S.V.** The Horvitz-Thompson theorem as a unifying perspective for probability sampling: with examples from natural resource sampling. *American Statistician* (forthcoming), 1995.

**Overton, W.S., White, D., and Stevens, D.L.** *Design Report for EMAP: Environmental Monitoring and Assessment Program.* Washington, DC, U. S. Environmental Protection Agency (EPA/600/3-91/053), 1991.

**Quenouille, M.H.** Problems in plane sampling. *Annals of Mathematical Statistics,* 20, 355-375, 1949.

**Ripley, B.D.** *Spatial Statistics.* Wiley, New York, 1981.

**Royall, R. M.** On finite population sampling theory under certain linear regression models. *Biometrika,* 57, 377-387, 1970.

**Särndal, C.E., Swensson, B., and Wretman, J.** (1992). *Model Assisted Survey Sampling.* Springer-Verlag, New York, 1992.

**Stehman, S.V., and Overton, W.S.** (1989). Variance Estimation for Fixed-Configuration, Systematic Sampling. *Biometrics Unit Manuscript* Bu-1010-M, Cornell University, Ithaca, NY, 1989.

**Stehman, S.V., and Overton, W.S.** Comparison of variance estimators of the Horvitz-Thompson estimator for randomized variable probability systematic sampling. *Journal of the American Statistical Association,* 89, 30-43, 1994.

**Stuart, A.** *The Basic Ideas of Scientific Sampling* (1984 ed.), Charles Griffin and Company, London, 1962.

**Upton, G.J.G., and Fingleton, B.** *Spatial Data Analysis by Example, Volume I: Point Pattern and Quantitative Data.* Wiley, New York, 1985.

**Upton, G. J. G., and Fingleton, B.** *Spatial Data Analysis by Example, Volume II: Categorical and Directional Data.* Wiley, New York, 1989.

**Webster, R., and Oliver, M.A.** *Statistical Methods in Soil and Land Resource Survey.* Oxford University Press, New York, 1990.

**Wolter, K.** Introduction to Variance Estimation. Springer-Verlag, New York, 1985.

**Yates, F.** *Sampling Methods for Censuses and Surveys* (4th ed.). Charles Griffin and Company, London, 1981.

**Zubrzycki, S.** Remarks on random, stratified and systematic sampling in a plane. *Colloquium Mathematicum,* 6, 251-264, 1958.

# Chapter 4

# SOME GUIDELINES FOR
# SPECIFYING THE GEOGRAPHIC WEIGHTS MATRIX
# CONTAINED IN SPATIAL STATISTICAL MODELS[1]

## Daniel A. Griffith

## I. INTRODUCTION

The choice of a geographic weights matrix specification for spatial statistical modelling is not clear-cut, mostly is done in an ad hoc manner, and seems to be governed primarily by convenience and/or convention. The purpose of this chapter is to provide some explicit guidance on specification of the geographic weights matrix contained in spatial autoregressive models. In so doing, the following three questions will be addressed [after Stetzer, 1982, p. 572]:

- (1) Does the selected specification of a geographic weights matrix make any practical difference in spatial analysis results?
- (2) In what ways does misspecification of a geographic weights matrix influence spatial statistical analysis results?

and

- (3) Are there any rules (-of-thumb) to guide specification of a geographic weights matrix for a given spatial landscape?

Reflections presented here will highlight matters of over- and under-specification of the geographic weights matrix (the posited degree of intensity), as well as the range of a specified geographic field effect.

In outlining answers to these three preceding questions, theorems and conjectures will be cited, and numerical demonstrations will be presented. Proofs of the theorems will be published elsewhere, allowing the discussion here to focus on conceptual issues and intuitive explanations.

---

[1]This chapter was conceptualized while the author was a visiting research professor in the Dipartimento di Statistica, Probabilitá e Statistiche Applicate, Universitá degli Studi di Roma "La Sapienza," jointly funded by the Scientific Research Commission of the University and the Working Group for Cultural Exchanges.

## II. BACKGROUND

Although a perpetual topic of concern among practitioners, few enlightening, helpful treatments of the geographic weights matrix specification problem appear in the literature.  Specification of a geographic weights matrix represents a priori knowledge of the range and intensity of a spatial field effect for a set of areal units constituting a geographic system.  Forms used in this specification frequently include binary or row-standardized contiguity, length of common boundary for juxtaposed areal units, and inter-centroid distance functions; such a variety of specifications dates back at least to Cliff and Ord's [1981] seminal work. Hordijk [1979] supplies a compelling argument in support of using the binary contiguity version in tests for spatial autocorrelation in regression residuals.  Later Stetzer [1982] conducted a simulation experiment that suggests

- (1) using the wrong geographic weights matrix increases standard errors of a mean response estimator, and using the correct geographic weights matrix always renders the minimum standard errors,
- (2) over-specification of the geographic weights matrix inflates the standard errors while under-specification suppresses the standard errors of a mean response estimator,
- (3) matching the effective region over which a spatial field operates appears to be more important than matching the functional form of the weights, and
- (4) small sample sizes magnify differences created by geographic weights matrices, with complications due to misspecification decreasing in severity as sample size increases.

This fourth point is confirmed here only some of the time.  Stetzer also found that misspecification of the weights matrix introduces bias into mean response estimation, a finding that is incorrect, as Theorem 1 subsequently proves.

Florax and Rey [1995] have studied impacts of the misspecification of a geographic weights matrix in terms of statistical power in spatial econometric hypothesis testing.  They conducted a simulation experiment that suggests

- (1) over-specification of the geographic weights matrix causes a loss of power,
- (2) under-specification of the geographic weights matrix causes an increase in power in the presence of positive spatial autocorrelation,

and a decrease in power in the presence of negative spatial autocorrelation,

- (3) the Moran Coefficient seems to retain power better than other spatial autocorrelation test statistics, in the presence of misspecification of the geographic weights matrix, and
- (4) both over- and under-specification of the geographic weights matrix produce an increase in the mean squared error for spatial econometric models.

These findings are fundamentally in concert with those reported by Stetzer, and further uncover some of the risks arising from misspecification, but still offer little guidance to the practitioner concerning the specification task, itself, of the geographic weights matrix.

Upton [1990] has explored an inter-centroid distance specification using a distance decay model of the form $(\text{population}_j)^\alpha (\text{area}_j)^\tau / d_{ij}^{-\tau}$, where both $\alpha$ and $\tau$ are binary exponents, and $d_{ij}$ is the distance separating the centroids of areal units i and j. His recommendation is to use both first and second neighbors (reminiscent of an SAR model specification), with non-zero off-diagonal weights being given by $\text{area}_j/d_{ij}^{-3}$. Of note here is that as the distance exponent $\gamma$ increases, the eigenstructure of its accompanying geographic weights matrix more closely resembles that of its row-standardized binary matrix counterpart.

## III. EVALUATION CRITERIA

Most of the time a spatial scientist is interested in the mean response ($\mu$ for a constant mean, and $X\beta$ for a variable mean), the variance ($\sigma^2$), the spatial autoregressive parameter [$\rho$; or its autocorrelation counterpart for spatial lag (h, k), $\rho_{h,k}$], and/or the frequency distribution of some attribute distributed across geographic space. In order to secure values for these parameter features, the spatial scientist becomes a statistical practitioner, and adopts estimators to compute them. Over the years four cardinal statistical properties of estimators have allowed invaluable ones to be mathematically set off from all others, namely sufficiency, unbiasedness, efficiency, and consistency[2].

---

[2]Today robustness and minimum variance often are appended to this list.

- A sufficiency statistic is one that summarizes all of the relevant information about the parent population that is contained in a sample, seeking to distill attribute information from data.
- An unbiasedness statistic is one for which the arithmetic mean of its sampling distribution equals its population parameter counterpart.
- An efficiency statistic is one having relatively small deviations of its values, obtained from all possible samples for a given sample size n (in the discrete case), about the population parameter. Moreover, as the variance of a sampling distribution decreases, precision of the corresponding unbiased estimator increases.
- A consistency statistic is one that moves closer to its population parameter value with increasing sample size. That is, as the sample size increases, the set of statistic values for all possible samples of size n increasingly concentrates around the population parameter value in the sampling distribution, and hence sampling error decreases toward zero, with probabilities calculated for various features from the set of all possible samples converging upon their theoretical probability counterparts.

For the subsequent discussion, suppose that the true data model is given by

$$\mathbf{Y} = \mathbf{X\beta} + \mathbf{V}^{1/2}\boldsymbol{\epsilon}, \tag{1}$$

while the practitioner posits the following data model:

$$\mathbf{Y} = \mathbf{X\beta} + \mathbf{\Omega}^{1/2}\boldsymbol{\epsilon}, \tag{2}$$

where $\mathbf{V}$ and $\mathbf{\Omega}$ are covariance matrices characterizing the nature, degree and structure of spatial dependency latent in a geo-referenced data set.

## A. MEAN RESPONSE ESTIMATION

In terms of unbiasedness and the mean response parameters of the model defined by equation (1),

THEOREM 1: Misspecification of the geographic weights matrix in a spatial statistical model does not bias the mean response estimator

$$\mathbf{b} = (\mathbf{X^t \Omega X})^{-1} \mathbf{X^t \Omega Y}.$$

This theorem holds even when the determinant $|\mathbf{X^t \Omega X}|$ equals zero, since general linear model theory states that in this problematic case an estimate always can be obtained by employing a generalized inverse. This result, which is exactly the well-known one for the case of OLS (Ordinary Least Squares) estimators computed with geo-referenced data (matrix $\Omega = I$ rather than $V \neq I$), indicates that Stetzer's [1982] simulation experiment may well have been flawed (p. 575: "specifying the wrong weights" *will not cause* "bias in the estimate"). Consequently, regardless of whether the geographic weights matrix is based upon binary contiguity (**C**), row-standardization of this matrix (**W**; $w_{ij} = c_{ij}/\Sigma_{j=1}^n c_{ij}$), either the "rook's" or the "queen's"[3] definition of adjacency, some often standardized declining function of distance [e.g., $d_{ij}^{-\gamma}/\Sigma_{j=1}^n d_{ij}^{-\gamma}$ , or $\exp(-\delta d_{ij})/\Sigma_{j=1}^n \exp(-\delta d_{ij})$], or some other specification, the mean response estimate remains reasonable. And, since $X\beta = \mu 1$ (i.e., replacing a variable mean with a constant mean) is a special case of Theorem 1, this result pertains to $\hat{\mu}$, too (as a corollary).

Does this finding indicate that the selection of a geographic weights matrix specification makes little difference in practice? The answer to this supplemental question lies in implications stemming from the remaining three statistical properties. Griffith [1988] already has shown that the locational information latent in geo-referenced data is not captured by traditional estimators, resulting in their sufficiency being seriously compromised. To recapitulate, regardless of the geographic arrangement of numerical values on a map, conventional statistics (e.g., $\bar{x}$ and $s^2$) yield exactly the same estimates. In other words, when matrix $\Omega$ is misspecified as $I$ (rather than correctly being equated with $V \neq I$), resulting in non-zero off-diagonal elements of matrix $\Omega$ mistakenly being set equal to zero (under-specification), specification of the geographic weights matrix indeed makes a difference.

Next, for simplicity consider the constant mean model. Then, investigating the property of efficiency,

THEOREM 2: Misspecification of the geographic weights matrix in a spatial statistical model tends to suppress statistical efficiency, making the ratio $VAR(\hat{\mu}_\Omega)/VAR(\hat{\mu}_V) \leq 1$ for the constant mean response estimator.

---

[3]Making an analogy with chess moves, the rook's definition of adjacency refers to common boundaries that have non-zero length, while the queen's definition refers to common boundaries that have at least a shared point.

Numerical demonstrations of this theorem can be found elsewhere for the special case of $\Omega = I$ (OLS results, which assume there is no spatial dependency). The following tabulation supplements these results, for a constant mean CAR spatial statistical model using the "rook's" definition of contiguity [after Bartlett, 1975]:

| lag-one spatial correlation $\rho_{1,0}$ | spatial autoregressive parameter $\rho$ | efficiency: $1 - 16\rho^2$ |
|---|---|---|
| 0.0000 | 0.0000 | 1.00000 |
| 0.0506 | 0.0500 | 0.96000 |
| 0.1055 | 0.1000 | 0.84000 |
| 0.1713 | 0.15 | 0.64000 |
| 0.2659 | 0.20 | 0.36000 |
| 0.4898 | 0.245 | 0.03960 |
| 0.7098 | 0.24996 | 0.00032 |
| 1.0000 | 0.25 | 0.00000 |

This tabulation is informative in several ways, namely

- (1) for the level of spatial autocorrelation most commonly encountered in empirical data (positive, in the range 0.15-0.20), the nearby spatial correlation is only roughly 0.2 and the loss of efficiency is roughly equivalent to that associated with a 10% reduction in sample size,
- (2) the relationship between $\rho_{1,0}$ and $\rho$ is complex, and
- (3) moderate-to-strong positive spatial autocorrelation that is ignored virtually destroys efficiency of the OLS estimator.

An additional demonstration that examines the worst case scenario [after Cordey and Griffith, 1993] for the intercept term (a conditional mean response) of a bivariate regression further underscores this third remark:

| $\rho/\rho_{max}$ | 0.1 | 0.5 | 0.9 |
|---|---|---|---|
| CAR | 0.99 | 0.75 | 0.19 |
| SAR | 0.9627 | 0.3769 | 0.0128 |

Moreover, increasing the range of a spatial field (moving from the first-order CAR to the second-order SAR model) exacerbates this situation.

To complete this probe into efficiency, the asymptotic case merits inspection. A tabulation reported in Griffith [1988] suggests that loss of efficiency for a given value of $\rho$ may only be temporary.

THEOREM 3: For the regular square tessellation, if matrices $\Omega$ and $V$ both have $k1$ as one of their eigenvectors, then efficiency is regained asymptotically ($n \to \infty$) in the constant mean CAR or SAR spatial statistical model.

This particular theorem suggests two corollaries, one of which pertains to over-/under-specification of the weights matrix that may be stated as

COROLLARY 3-a: For the regular square tessellation, efficiency is regained as $n \to \infty$ when the "rook's" and the "queen's" adjacency definitions are interchanged, in the constant mean CAR or SAR spatial statistical model.

The following tabulation for a constant mean CAR model indicates that this tendency to regain efficiency begins virtually immediately, with the loss of efficiency tending to be relatively small in the first place:

| n-by-n square tessellation | $\rho/\rho_{max}$ 0.1 | 0.5 | 0.9 |
|---|---|---|---|
| *misspecifying the rook with the queen definition* | | | |
| 3-by-3 | 0.99990 | 0.99640 | 0.97334 |
| 5-by-5 | 0.99992 | 0.99648 | 0.97370 |
| 7-by-7 | 0.99994 | 0.99714 | 0.97607 |
| *misspecifying the queen with the rook definition* | | | |
| 3-by-3 | 0.99990 | 0.99595 | 0.97349 |
| 5-by-5 | 0.99992 | 0.99614 | 0.97377 |
| 7-by-7 | 0.99993 | 0.99688 | 0.97623 |

Over-/under-specification appears to compromise efficiency in a somewhat complicated manner. In the presence of very weak positive spatial autocorrelation, over-/under-specification makes little difference. In the presence of moderate spatial autocorrelation, though, under-specification appears to be slightly more serious. And, in the presence of strong spatial autocorrelation, over-specification appears to be slightly more serious. These complications appear to be magnified by small sample size. These findings are somewhat at odds with those summarized by Stetzer [1982].

The second corollary may be stated as

> COROLLARY 3-b: For the regular linear, triangular, square, or hexagonal tessellations (see Ahuja and Schachter, 1983), efficiency is regained as $n \to \infty$ when matrices $C$ and $W$ are interchanged, in the constant mean SAR spatial statistical model.

In the four cases covered by Corollary 3-b, the limiting situation is $W = (1/n_i)C$, where the number of neighbors $n_i$ is a constant over all areal units i comprising a geographic landscape.

Furthermore, the following conjecture pertains to a limiting situation in the case of irregular surface partitionings:

> CONJECTURE 1: For the maximum connectivity surface partitioning, efficiency *is not* regained as $n \to \infty$ in the constant mean SAR spatial statistical model using matrix $W$.

Inspection of numerous instances of this case reveal that only about half of the eigenvectors of its matrix $W$ are orthogonal to k1. The following tabulation of results from a numerical demonstration exemplifies this conjecture:

| | | $\rho$ | |
|---|---|---|---|
| n | 0.1 | 0.5 | 0.9 |
| 25 | 0.98086 | 0.74173 | 0.54186 |
| 50 | 0.95474 | 0.54307 | 0.32953 |
| 75 | 0.92971 | 0.42751 | 0.23620 |
| 100 | 0.90590 | 0.35236 | 0.18398 |
| 250 | 0.78550 | 0.17129 | 0.07901 |
| 500 | 0.64333 | 0.09212 | 0.04047 |
| 750 | 0.54073 | 0.06275 | 0.02712 |
| 1000 | 0.46538 | 0.04814 | 0.02019 |
| 1250 | 0.40871 | 0.03917 | 0.01632 |

That is, even asymptotically the impact of sample size fails to compensate for the presence of spatial autocorrelation; these complications are not necessarily magnified by small sample size.

Finally, consistency cast in terms of increasing domain asymptotics (the size of a region increases to infinity coupled with stationary spatial processes operating over the infinite geographic landscape) sheds some light on this problem of whether or not selection of a geographic weights matrix specification makes a difference in terms of mean response estimation. Of interest is whether or not the standard error of b goes to zero as n → ∞. Again consider the constant mean CAR spatial statistical model and a regular square tessellation using matrix C. Then the limiting case of the standard error of $\hat{\mu}$ is given by

$$\sigma^2/[n(1 - 4\rho)] \,,$$

which still converges to zero, but at a rate slower than that for independent observations. Of note is that consistency is lost when perfect positive spatial autocorrelation prevails ($\rho = 0.25$). This rate is even slower for the SAR model using matrix W, although again convergence to zero does occur:

$$\sigma^2/[n(1 - \rho)^2] \,.$$

These results can be extended in a fashion parallel to that leading to Corollary 3-a:

THEOREM 4: As n → ∞ for the regular square tessellation, when the "rook's" and the "queen's" adjacency definitions are interchanged, convergence yielding consistency occurs but is slowed in the constant mean spatial statistical models, the standard error becoming, using matrix W and the SAR model,

$$\sigma^2/[n(1 - \rho)^2] \,,$$

and becoming, using matrix C and the CAR model,

$$\sigma^2/[n(1 - 4\rho)] \,,$$

when the "rook" structure is mistakenly replaced with the "queen" structure, and

$$\sigma^2/[n(1 - 8\rho)] \,,$$

when the "queen" structure is mistakenly replaced with the "rook" structure.

This rate slows even further for the worst case bivariate regression situation with the CAR model and a maximum connectivity surface partitioning, using matrix **C**, where $\Omega = \mathbf{I}$:

$$\{1/[(n-1)(n-2)] + \sqrt{1/[2(n-2)]}\}^2/(1 - \rho\lambda_1)$$
$$+ \{1/[(n-1)(n-2)] - \sqrt{1/[2(n-2)]}\}^2/(1 - \rho\lambda_n),$$

where $\lambda_1$ and $\lambda_n$ are the extreme eigenvalues of the affiliated binary geographic weights matrix, and may be approximated by [Griffith and Sone, 1995]

$$\lambda_1 = \{1 + 2COS[\pi/(n-1)]\}/2$$
$$+ \sqrt{2n - \{13 + 4COS[\pi/(n-1)]}$$
$$- 2COS[\pi/(n-1)]\}/4},$$

and

$$\lambda_n = \{1 + 2COS[\pi/(n-1)]\} - \lambda_1.$$

Because the feasible parameter space shrinks to zero when matrix **C** is used in this maximum connectivity case, the following tabulation of results from a numerical demonstration of the squared standard error (i.e., the variance) of $\hat{\mu}$ in the presence of positive spatial autocorrelation, for the constant mean SAR model using matrix **W**, is informative:

| n | ρ 0.1 | 0.5 | 0.9 |
|---|---|---|---|
| 25 | 0.05035 | 0.21571 | 7.38192 |
| 50 | 0.02586 | 0.14731 | 6.06933 |
| 75 | 0.01771 | 0.12475 | 5.64498 |
| 100 | 0.01363 | 0.11352 | 5.43525 |
| 250 | 0.00629 | 0.09341 | 5.06234 |
| 500 | 0.00384 | 0.08684 | 4.94247 |
| 750 | 0.00304 | 0.08499 | 4.91637 |
| 1000 | 0.00265 | 0.08308 | 4.95394[4] |
| 1250 | 0.00242 | 0.08169 | 4.90165 |

---

[4]This anomalous value is not an error.

These values can be compared to those based on independent observations, given by 1/n. Nevertheless, regardless of the extreme degree of geographic connectivity here, convergence is still implied. A principal reason for this slowing down of convergence is that, in the presence of positive spatial autocorrelation, each new observation does not contribute a complete increment of additional information (after the first unit is selected, at least some information contained in each new unit is redundant).

Consequently, the selection of a geographic weights matrix specification makes a difference in terms of mean response estimation. Although parameter estimates remain unbiased, selection of the weights matrix specification affects the sufficiency and efficiency of the mean response estimation, as well as its rate of convergence for consistency.

## B.   VARIANCE ESTIMATION

In terms of unbiasedness and the variance parameter of the model defined by equation (1),

THEOREM 5: Misspecification of the geographic weights matrix in a spatial statistical model yields a biased variance estimator

$$s^2 = (Y - X\beta)'\Omega^{-1}(Y - X\beta)/(n - P - 1) ,$$

where there are $(P + 1)$ variables contained in the matrix $X$.

The following numerical demonstration of this theorem based upon the CAR model, a regular square tessellation, and the posited matrix $\Omega = I$ illustrates that this bias does not disappear as n becomes larger, and that it becomes increasingly problematic as $\rho$ becomes larger:

| n-by-n square tessellation | $\rho/\rho_{max}$ 0.1 | 0.5 | 0.9 |
|---|---|---|---|
| 5-by-5 | 1.00268 | 1.07928 | 1.59132 |
| 10-by-10 | 1.00246 | 1.07141 | 1.44545 |
| 25-by-25 | 1.00245 | 1.07103 | 1.42976 |
| 50-by-50 | 1.00247 | 1.07182 | 1.43684 |
| 75-by-75 | 1.00248 | 1.07220 | 1.44087 |

| | | | |
|---|---|---|---|
| 100-by-100 | 1.00249 | 1.07242 | 1.44323 |
| 1000-by-1000 | 1.00251 | 1.07310 | 1.45087 |

These three trajectories indicate that the variance inflation attributable to the presence of overlooked spatial autocorrelation (under-specification) decreases at first, reaching a minimum, and then begins increasing again, asymptotically reaching a limiting value. Once more, complications appear to be magnified by small sample size. Extending this treatment to other models and spatial field specifications for an infinite surface,

| spatial statistical model | $\rho/\rho_{max}$ 0.1 | 0.5 | 0.9 |
|---|---|---|---|
| *using the rook specification of contiguity* | | | |
| CAR | 1.00251 | 1.07318 | 1.45184 |
| SAR | 1.00757 | 1.24562 | 3.92592 |
| *using the queen specification of contiguity* | | | |
| CAR | 1.00130 | 1.04247 | 1.28747 |
| SAR | 1.00397 | 1.14773 | 2.92489 |

These various results indicate that the variance inflation attributable to the presence of overlooked spatial autocorrelation can be quite considerable, is worse for the SAR than the CAR case, and is worse when a spatial field is less intense (the "rook" versus the "queen" definition of adjacency). In addition, this collection of numerical results suggests that the common spatial autocorrelation situation often may be accompanied by an inflated variance estimate on the order of 10% or more.

Another numerical demonstration, involving substitution of the rook for the queen definition of adjacency for an infinite surface, reinforces this impression:

| spatial statistical model | $\rho/\rho_{max}$ 0.1 | 0.5 | 0.9 |
|---|---|---|---|
| *misspecifying the rook with the queen definition* | | | |
| CAR | 1.00119 | 1.02695 | 1.09490 |
| SAR | 1.00353 | 1.07687 | 1.26756 |
| *misspecifying the queen with the rook definition* | | | |
| CAR | 0.99998 | 0.99787 | 0.98282 |
| SAR | 1.00119 | 1.02408 | 1.05582 |

These results suggest that slight over-specification of the geographic weights matrix is worse than slight under-specification. Stetzer's [1982] conclusion that over-specification leads to inflation while under-specification leads to suppression of variance is reflected here in results for the CAR model, but is not supported by results for the SAR model, which is a second-order model having a greater dependency range. These results are further supplemented by those from a numerical demonstration for the maximum connectivity case, using the SAR model and matrix **W**:

| n | ρ 0.1 | 0.5 | 0.9 |
|------|---------|---------|---------|
| 25   | 1.00572 | 1.22260 | 8.59600 |
| 50   | 1.00543 | 1.20156 | 7.37796 |
| 75   | 1.00534 | 1.19480 | 6.98535 |
| 100  | 1.00529 | 1.19147 | 6.79155 |
| 250  | 1.00521 | 1.18555 | 6.44602 |
| 500  | 1.00519 | 1.18359 | 6.33752 |
| 750  | 1.00519 | 1.18295 | 6.29392 |
| 1000 | 1.00519 | 1.18261 | 6.29048 |
| 1250 | 1.00517 | 1.18243 | 6.27494 |

These results indicate that the orders of magnitude for bias furnished by the regular square tessellation are generally helpful to note, as long as latent spatial autocorrelation is weak-to-moderate in strength.

By establishing bias for $\hat{\sigma}^2$, which seriously affects virtually all facets of spatial statistical analyses, there is at this time no need to investigate the properties of sufficiency, efficiency, or consistency in this context.

## C. SPATIAL AUTOREGRESSIVE PARAMETER ρ ESTIMATION

A final evaluation here is based upon an exploratory simulation experiment, and assesses the impact that misspecification of the geographic weights matrix has on estimation of the spatial autoregressive parameter ρ. This experiment employed a 10-by-10 regular square tessellation, consisted of 25 replications drawn from a normal distribution with a mean of zero and variance of 0.05/99, used as the population attribute surface the third principal eigenvector of matrix **C**, and then corrupted this surface by adding each of the 25 randomly generated independent surfaces, in turn, to it. To help control variation, the 25 error surfaces were orthogonalized, and then modified to have exactly the posited mean and variance. The distribution of these error surfaces reasonably spanned the range of normal distribution possibilities. The third eigenvector of matrix C was used here because it is exactly the same for both the rook and the queen definitions of adjacency, because it is the same for both the AR and CAR models, and because it contains fairly strong spatial autocorrelation (MC = 0.93473 for the rook specification of matrix C; MC = 0.90583 for the queen specification).

Estimation results for this experiment are as follows:

|  | rook specification | | queen specification | |
|---|---|---|---|---|
|  | AR | CAR | AR | CAR |
| $\rho$ | 0.21852 | 0.22897 | 0.09427 | 0.10431 |
| $\bar{x}_\rho$ | 0.18137 | 0.21785 | 0.08259 | 0.09582 |
| bias | 0.03715 | 0.01112 | 0.01168 | 0.00849 |
| $s_\rho$ | 0.01036 | 0.00323 | 0.00405 | 0.00267 |
| Shapiro-Wilk correlation | 0.983 | 0.991 | 0.984 | 0.989 |

With regard to the average estimated values of $\rho$, which have upper limits of 0.26055 and 0.13297 for the rook and the queen adjacency specification, respectively, misspecification results in roughly a 10% error with the AR model, and a 15% error with the CAR model. The true population parameter values of $\rho$, themselves, exhibit almost exactly the opposite error rates. In all four cases, on average, parameter estimates for the corrupted geographic distributions underestimate their true parameter value counterparts. This result is expected, since the minimization problem is of the form

$$(\mathbf{E}_3 - \mathbf{X\beta})^T\mathbf{V}^{-1}(\mathbf{E}_3 - \mathbf{X\beta})/\mathfrak{J}(\rho)$$
$$+ 2(\mathbf{E}_3 - \mathbf{X\beta})^T\mathbf{V}^{-1}(\epsilon - \mu_\epsilon 1)/\mathfrak{J}(\rho)$$
$$+ (\epsilon - \mu_\epsilon 1)^T\mathbf{V}^{-1}(\epsilon - \mu_\epsilon 1)/\mathfrak{J}(\rho) ,$$

where $\mathfrak{J}(\rho)$ is the Jacobian approximation, and $\epsilon$ is the vector of normally distributed independent random error terms. Asymptotically this middle term should go to zero, while this third term should become a function of $-1^T\mathbf{C}1\sigma_\epsilon^2/n$, which would suppress the estimate of $\rho$. As expected, across the 25 simulation replications the regression of $(\mathbf{E}_3 + \epsilon_r)$ on $\mathbf{X}$, $r = 1, 2, ...,$ 25, resulted in roughly a 5% drop in explained variance.

The maximum eigenvalues for the rook and queen adjacency specifications of matrix C are, respectively, 3.83797 and 7.52048. The corresponding relative CAR parameter estimates, then, are

0.22897/(1/3.83797) ≈ 0.87878, and 0.10431/(1/7.52048) ≈ 0.78446. Therefore, one implication here is that over-specifying the adjacency structure will suppress the estimate of the spatial autocorrelation parameter, whereas under-specifying it will inflate the estimate of the spatial autocorrelation parameter. This conclusion is suggested by both the individual estimates based upon vector $E_3$, and by the averaged simulation results. In addition, various ratios of the variances suggest that under-specification has more serious impacts on the estimation of $\rho$ than does over-specification. Apparently a similar finding pertains to the order of the model: increasing a model order has more serious impacts on the estimation of $\rho$ than does decreasing the order. Finally, the relationship between $\hat{\rho}_{AR}$ and $\hat{\rho}_{CAR}$ is closer for the rook than for the queen specification:

rook: $\hat{\rho}_{AR} = -0.47281 + 3.00285 \hat{\rho}_{CAR} + \xi$ , $R^2 = 87.3\%$

queen: $\hat{\rho}_{AR} = -0.04654 + 1.34762 \hat{\rho}_{CAR} + \xi$ , $R^2 = 78.9\%$

The following conjecture can be gleaned from these preceding simulation experiment results:

CONJECTURE 2: The MLE estimate $\hat{\rho}$ is biased by a misspecification of the geographic weights matrix.

An ancillary drawback affiliated with this finding is that the biased $\hat{\rho}$ that is obtained will then propagate complications through a spatial statistical analysis.

Consequently, the selection of a geographic weights matrix specification makes a difference in terms of spatial autocorrelation parameter estimation, too.

## IV. RULES-OF-THUMB IMPLICATIONS

In conclusion, the selected specification of a geographic weights matrix does make a practical difference in spatial analysis results. This misspecification distorts spatial statistical analysis results by affecting the statistical quality of MLE estimators..

Fortunately, from the partial analysis presented in this chapter, several rules-of-thumb can be devised for specifying a geographic weights matrix. Hopefully these rules-of-thumb will prove helpful in guiding specification of geographic weights matrices for a myriad of spatial landscapes. Some of this advice is corroborated by verdicts rendered by Stetzer, Upton, and Florax and Rey. Much of it is not without problematic consequences, though (over-specification will tend to cause a loss in statistical power, and inflate variance estimates, for example). These rules may be stated as follows:

- RULE-OF-THUMB 1: It is better to posit some reasonable geographic weights matrix specification than to assume all entries are zero (the independent observations situation of conventional statistics), the extreme case of under-specification.

- RULE-OF-THUMB 2: It is best to use a surface partitioning that falls somewhere between a regular square and a regular hexagonal tessellation.

- RULE-OF-THUMB 3: Relatively large numbers of areal units [e.g., $(1 - \rho/\rho_{max})n > 30$] should be employed in a spatial statistical analysis; given the commonly encountered circumstance of $\rho/\rho_{max} \approx 0.5$, then n should be at least 60.

- RULE-OF-THUMB 4: Low-order spatial statistical models should be given preference over higher-order ones.

- RULE-OF-THUMB 5: In general, it is better to employ a somewhat under-specified than a somewhat over-specified geographic weights matrix, as long as $\Omega \neq \mathbf{I}$.

## V. LIST OF THEOREMS, COROLLARIES AND CONJECTURES

THEOREM 1: Misspecification of the geographic weights matrix in a spatial statistical model does not bias the mean response estimator
$$b = (X^t \Omega X)^{-1} X^t \Omega Y .$$

THEOREM 2: Misspecification of the geographic weights matrix in a spatial statistical model tends to suppress statistical efficiency, making the ratio $VAR(\hat{\mu}_\Omega)/VAR(\hat{\mu}_V) \leq 1$ for the constant mean response estimator.

THEOREM 3: For the regular square tessellation, if matrices $\Omega$ and $\mathbf{V}$ both have k1 as one of their eigenvectors, then efficiency is regained asymptotically ($n \to \infty$) in the constant mean CAR or SAR spatial statistical model.

COROLLARY 3-a: For the regular square tessellation, efficiency is regained as n → ∞ when the "rook's" and the "queen's" adjacency definitions are interchanged, in the constant mean CAR or SAR spatial statistical model.

COROLLARY 3-b: For the regular linear, triangular, square, or hexagonal tessellations [see Ahuja and Schachter, 1983], efficiency is regained as n → ∞ when matrices C and W are interchanged, in the constant mean SAR spatial statistical model.

CONJECTURE 1: For the maximum connectivity surface partitioning, efficiency *is not* regained as n → ∞ in the constant mean SAR spatial statistical model using matrix W.

THEOREM 4: As n → ∞ for the regular square tessellation, when the "rook's" and the "queen's" adjacency definitions are interchanged, convergence yielding consistency occurs but is slowed in the constant mean spatial statistical models, the standard error becoming, using matrix W and the SAR model, $\sigma^2/[n(1 - \rho)^2]$, and becoming, using matrix C and the CAR model, $\sigma^2/[n(1 - 4\rho)]$, when the "rook" structure is mistakenly replaced with the "queen" structure, and $\sigma^2/[n(1 - 8\rho)]$, when the "queen" structure is mistakenly replaced with the "rook" structure.

THEOREM 5: Misspecification of the geographic weights matrix in a spatial statistical model yields a biased variance estimator
$$s^2 = (Y - X\beta)^t \Omega^{-1}(Y - X\beta)/(n - P - 1),$$
where there are $(P + 1)$ variables contained in the matrix X.

CONJECTURE 2: The MLE estimate $\hat{\rho}$ is biased by a misspecification of the geographic weights matrix.

# REFERENCES

Ahuja, N., and Schachter, B. *Pattern Models.* Wiley, New York, 1983.
Bartlett, M. *The Statistical Analysis of Spatial Pattern.* Chapman and Hall, London, 1975.

Cliff, A., and Ord, J. *Spatial Processes*. London: Pion, London, 1981.

Cordey, C., and Griffith, D. Efficiency of least squares estimators in the presence of spatial autocorrelation. *Communications in Statistics--Simulation*, 22: 1161-1179, 1993.

Florax, R., and Rey, S. The impacts of misspecified spatial interaction in linear regression models, in *New Directions in Spatial Econometrics*, edited by L. Anselin and R. Florax, Springer-Verlag, Berlin, pp. 111-135, 1995.

Griffith, D. *Advanced Spatial Statistics*. Kluwer, Dordrecht, 1988.

Griffith, D., and Sone, A. Trade-offs associated with normalizing constant computational simplifications for estimating spatial statistical models. *Journal of Statistical Computation and Simulation*, 51: 165-183, 1995.

Hordijk, L. Problems in estimating econometric relations in space. *Papers of the Regional Science Association*, 42: 99-115, 1979.

Stetzer, F. Specifying weights in spatial forecasting models: the results of some experiments. *Environment and Planning A*, 14: 571-584, 1982.

Upton, G. Information from regional data, in *Spatial Statistics: Past, Present, and Future*, edited by D. Griffith, Institute of Mathematical Geography, Ann Arbor, pp. 315-359, 1990.

# Chapter 5

## AGGREGATION EFFECTS IN GEO-REFERENCED DATA

David Wong

## I. SPATIAL DEPENDENCY IN SPATIAL DATA ANALYSIS

One of the basic themes of this handbook is that because data used by geographers and spatial scientists usually fail to be composed of independent observations distributed across space, the analysis of spatial data correctly requires techniques beyond classical or traditional statistical ones. Spatial dependency, which is a characteristic of spatial data well-recognized by researchers working in the area of spatial autocorrelation [Cliff and Ord, 1973; Ripley, 1981; Griffith, 1988; and Anselin, 1988], is only one of the major characteristics of spatial data. In this chapter we will focus on another characteristic of spatial data that is rooted in the need to aggregate geo-referenced data in order to make them manageable and to uncover pattern. This aggregation involves various spatial scales and different surface partitioning schemes. This is the so-called modifiable areal unit problem (MAUP).

The term 'modifiable areal unit problem' was first coined by Openshaw and Taylor [1979] when they rediscovered a particular type of spatial dependency latent in geo-referenced data. One aspect of the MAUP was first raised by Gehlke and Biehl [1934] when they discovered the rising trend of correlation coefficients between rental payment and juvenile delinquency when areal units became larger. Later Robinson [1950] placed this methodological issue in the context of the ecological fallacy. The MAUP consists of two major sub-problems: the scale effect and the zoning or aggregation effect.

An area can be partitioned into smaller subareas in a hierarchical manner. First, the area can be divided into relatively large subareas of similar size. Then, large subareas can be further subdivided into smaller subareas of approximately the same size. Thus, layers of partitioning schemes building upon each other can be used to divide the area into subareas of different sizes or resolutions. The scale effect refers to the inconsistency of analytical or statistical results derived from data recorded at, or representing, different levels of partitioning for the same area. Figure 1 is a diagrammatic explanation of the scale effect.

**Aggregation/Zoning Effect**

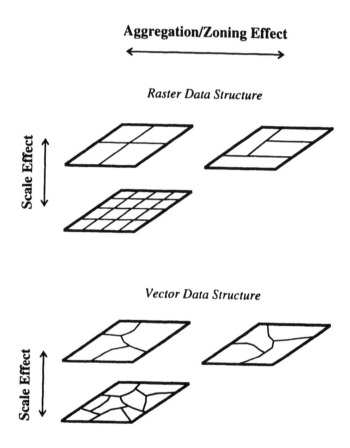

**Figure 1.** Scale effect.

Spatial partitioning can occur along another dimension. Given the number of areal units needed to be formed by subdividing a region, there are many ways, yielding various spatial configurations, to partition this region into the given number of units. From the aggregation perspective, we can aggregate smaller units into fewer larger units in many ways, too. (The upper limit for this number is given by a Stirling number of the second kind.) The zonal or aggregation effect refers to variability of analytical or statistical results derived from data for the same region, but aggregated or partitioned in different ways, with the number of areal units in different partitioning schemes being the same. Figure 1 graphically portrays the two dimensions of spatial partitioning in the MAUP. Both raster and vector data formats are included.

In this chapter we will discuss the source of MAUP, the impacts of the problem, solutions that have been proposed, and conditions under which the MAUP should be acknowledged. The 1990 Census Data for the State of Connecticut, USA, will be used for illustrative purposes here. Socio-economic variables have been extracted from the Summary Tape Files (STFs) 1A and 3A of the U.S. Census database (Bureau of the Census, 1992a and 1992b). Nine variables have been selected from the original battery of census variables. They are population (POP), median house value (MEDHOUVAL), median family income (MFI), percent of white population (PWHITE), percent of population in the age cohort 16-65 (PECONACT), percent of population having received higher education (PHI_ED), percent of white-collar workers (PWH_COL), percent of owner-occupied housing units (POWN), and percent of single (both attached and detached) family housing units (PSING_H). These variables have been aggregated for four levels of census geography: the county level, the town level, census tract level, and census block group level.

## II. SOURCE OF THE MAUP:
### SPATIAL DEPENDENCY AND AVERAGING

Tobler [1979] claims that "everything is related to everything else, but near things are more related than distant things." This 'First Law' of Geography captures a spatial regularity known as the distance decay relationship. It implies that objects close to each other may be very similar in their attributes, pinpointing the source of many stubborn problems that spatial scientists have to deal with in analyzing spatial data. Because objects are spatially related, each observation is no longer independent. Thus spatial autocorrelation has to be taken into account in analyzing spatial data.

Because this 'law' is general, more detailed complexities latent in a spatial distribution of phenomena are not depicted accurately by it. As is the case with the analysis of any phenomenon from a non-spatial perspective, a random error component has to be included to account for stochastic noise. Beyond this stochastic element, the similarity of spatial objects is not homogeneous across geographic space, and may even vary by direction. In addition, different attributes for spatial observations (areal units) tend to have different degrees of similarity across space. That is, different variables/attributes have different degrees of spatial autocorrelation. These characteristics, which are found in the spatial distribution of many phenomena, make MAUP another stubborn problem in spatial science.

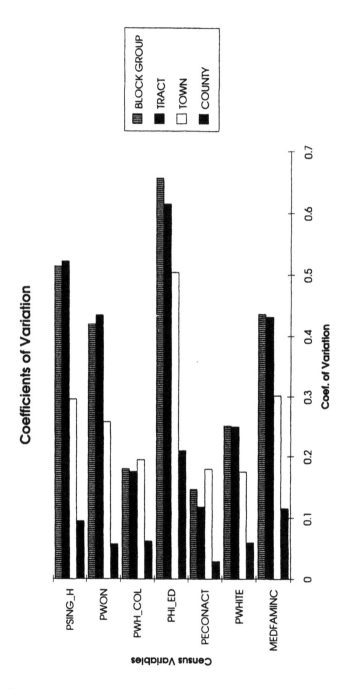

**Figure 2.**

When smaller areal units adjacent to each other are agglomerated to form larger units, the attribute values of all units, regardless of how similar or dissimilar they are, will be represented by a summary measure such as an arithmetic average (perhaps weighted). Thus, both the uniqueness of each areal unit as well as the dissimilarity among units are eliminated, and the variance for the whole area is suppressed when aggregation proceeds. Along the diagonals of the four matrices reported in Table 1 are variances of the selected variables from the Connecticut data. With a few exceptions, the variances of variables at the county levels, the most aggregated level, usually are the lowest among all four levels. Variances increase as the data are more disaggregated, and variances are highest at the block group level. The variances of these variables are transformed to the same scale by deriving the coefficients of variation. These coefficients are shown in Figure 2, and an increasing trend in the variance measure is apparent when areal units become more disaggregated.

This smoothing process is the source of the scale effect. In realistic situations, analysts have to deal with multivariate analysis. The smoothing process applies to all variables/attributes attached to the spatial observations, but the degree of smoothing varies from one variable to another. The magnitude of smoothing taking place depends not only upon the similarity or spatial autocorrelation contained in individual variables, but also upon the cross-correlation of variables in adjacent areal units, or conceptually the multivariate spatial autocorrelation. [Arbia, 1989; Wartenberg, 1985; Griffith, 1988] The impacts of aggregation, which subsequently will be discussed in more detail, are rooted in altering the original variance-covariance matrix when areal units are merged together. Table 1 provides the variance-covariance matrices of all four geographical levels in the Connecticut data. No two matrices are similar to any extent. Though some covariances tend to increase as the level of analysis moves to the more disaggregated levels, this is not an obvious pattern. Comparing these matrices for different geographical levels can easily identify the fluctuations in the covariance measure. Some covariances even shift between positive and negative values. Indeed, the impacts of spatial aggregation upon the variance-covariance matrix do not exhibit a particular pattern, and the results are difficult to predict.

**Table 1.**

| Variance-Covariance Matrices of all Variables at Four Geographical Levels | | | | | | | |
|---|---|---|---|---|---|---|---|
| **COUNTY** | | | | | | | |
| | MEDFAMINC | PWHITE | CONACT | PHI_ED | PWH_COL | POWN | PSING_H |
| MEDFAMINC | 3.17E+07 | -98.613 | 42.6341 | 271.729 | 218.811 | 83.0891 | 88.6296 |
| PWHITE | | 0.002993 | 0.00021 | -0.00118 | -0.00159 | 0.001568 | 0.00275 |
| PECONACT | | | 0.00036 | 0.00034 | 0.00042 | 0.00032 | 0.00057 |
| PHI_ED | | | | 0.002398 | 0.002032 | 0.00046 | 0.00043 |
| PWH_COL | | | | | 0.002136 | 0.00089 | 0.00031 |
| POWN | | | | | | 0.001484 | 0.002254 |
| PSING_H | | | | | | | 0.003821 |
| **TOWN** | | | | | | | |
| MEDFAMINC | 2.40E+08 | 1331.64 | 861.845 | 1705 | 1708.21 | 1980.43 | 2268.08 |
| PWHITE | | 0.02689 | 0.013246 | 0.007338 | 0.015855 | 0.0206 | 0.022284 |
| PECONACT | | | 0.012552 | 0.004033 | 0.011083 | 0.011909 | 0.012092 |
| PHI_ED | | | | 0.017497 | 0.013639 | 0.012132 | 0.015387 |
| PWH_COL | | | | | 0.02037 | 0.016092 | 0.0179 |
| POWN | | | | | | 0.036115 | 0.039406 |
| PSING_H | | | | | | | 0.048616 |
| **CENSUS TRACT** | | | | | | | |
| MEDFAMINC | 4.50E+08 | 2411.57 | -52.331 | 2657.27 | 1838.07 | 4124.82 | 4755.07 |
| PWHITE | | 0.045765 | 0.001332 | 0.014125 | 0.006365 | 0.038922 | 0.044359 |
| PECONACT | | | 0.005733 | 0.00069 | -0.0011 | -0.00295 | -0.0026 |
| PHI_ED | | | | 0.022409 | 0.01362 | 0.022322 | 0.026387 |
| PWH_COL | | | | | 0.017059 | 0.01663 | 0.018989 |
| POWN | | | | | | 0.075214 | 0.081959 |
| PSING_H | | | | | | | 0.10018 |
| **BLOCK GROUP** | | | | | | | |
| MEDFAMINC | 4.90E+08 | 2225.71 | 378.081 | 2793.32 | 1801.47 | 3774.52 | 4544.47 |
| PWHITE | | 0.047941 | 0.003681 | 0.013768 | 0.0099 | 0.034769 | 0.040399 |
| PECONACT | | | 0.008585 | 0.00307 | 0.003375 | 0.001615 | 0.00163 |
| PHI_ED | | | | 0.028014 | 0.01595 | 0.019164 | 0.024193 |
| PWH_COL | | | | | 0.018165 | 0.014195 | 0.017277 |
| POWN | | | | | | 0.077538 | 0.084402 |
| PSING_H | | | | | | | 0.108085 |

Because (dis)similarity among areal units is not uniform over a large region and in all directions, merging smaller areal units using different spatial partitioning schemes is the same as smoothing different combinations of spatial neighbors. Because different combinations have different degrees of similarity, the smoothed results also will differ from each other. Thus, different zoning schemes can render different analytical results.

Many studies have sought to determine when MAUP becomes important, and have verified that the first law of geography (i.e. closer objects are more similar than distant objects) is the source of the problem. When aggregation of areal units is performed in a non-contiguous or spatially random fashion, MAUP does not exist (see for example, Gehlke and Biehl, 1934; Blalock, 1964; Green, 1993). Only when adjacent units are merged, altering the detected level of spatial autocorrelation, does the MAUP effect become marked.

## III. GENERAL IMPACTS OF THE MAUP ON SPATIAL DATA

Most earlier works related to the MAUP focus on impacts of the problem. Presumably, the issue receiving the most extensive scrutiny concerns the impact on correlation coefficients. [Gehlke and Biehl, 1934; Openshaw and Taylor, 1979] In general, correlation coefficients increase as small areal units adjacent to each other are aggregated successively to form larger areal units. This empirical pattern displayed by the correlation coefficient has been explained intuitively, and in a non-mathematical fashion by Fotheringham and Wong [1991]. Let X and Y be the two variables in the analysis. The correlation coefficient between variables X and Y is

$$r_{XY} = \frac{COV(X,Y)}{S_X S_Y}$$

which is a ratio of the covariance of X and Y to the product of the standard deviations of X and Y. When areal units are aggregated or smoothed, the variances (denominator) will become smaller. If the covariance is relatively stable, then the coefficient will rise. This finding has a profound implication for spatial science. It implies that the correlation coefficient, which is a standard statistic commonly used to indicate the degree of similarity of two or more variables, is (spatial) scale dependent. By aggregating smaller areas into larger areas, a researcher may obtain whatever R-squared level is deemed appropriate. In Table 2, R-squared values of the linear regression model adopted to analyze the Connecticut Census data of all four geographical levels are reported. The R-squared increases as data become more aggregated.

**Table 2.**

Multiple Linear Regression Model

| | | INTERCEPT(0) | PWHITE(1) | PECONACT(2) | PHI_ED(3) | PWH_COL(4) | POWN(5) | PSING_H(6) | N= | R-squared= |
|---|---|---|---|---|---|---|---|---|---|---|
| County | Beta Estimates | 8555.28 | -25768.30 | 14367.05 | 95877.71 | -9887.50 | 70946.80 | -13873.70 | 8 | 1 |
| | Prob >\|t\| | 0.0598 | 0.0377 | 0.0455 | 0.0079 | 0.0654 | 0.0096 | 0.0375 | | |
| Town | Beta Estimates | -979.68 | -596.26 | 16071.68 | 71463.06 | 11421.91 | 27755.00 | -6391.17 | 265 | 0.816 |
| | Prob >\|t\| | 0.7081 | 0.8897 | 0.0123 | 0.0001 | 0.0872 | 0.0001 | 0.2684 | | |
| Census Tract | Beta Estimates | 795.19 | 4249.31 | -4367.35 | 80098.82 | 17340.45 | 17401.66 | 6849.75 | 833 | 0.7921 |
| | Prob >\|t\| | 0.8535 | 0.0585 | 0.3569 | 0.0001 | 0.0001 | 0.0001 | 0.0381 | | |
| Block Group | Beta Estimates | -498.99 | 2311.68 | 8177.29 | 70921.49 | 11250.52 | 16193.92 | 10739.85 | 2847 | 0.693 |
| | Prob >\|t\| | 0.8086 | 0.0792 | 0.0017 | 0.0001 | 0.0001 | 0.0001 | 0.0001 | | |

Studies addressing impacts of the MAUP can be divided into two general area: bivariate and multivariate. Clark and Avery [1976] provide a detailed analysis as to how changes in scale affect a simple regression model. Basically, the parameter estimate of a bivariate regression tends to vary across different levels of aggregation. However, there is no obvious way to predict the trend in the parameter changes because the parameter is a function of the correlation coefficient and the ratio of the variance of the two variables. Though the coefficient is likely to increase with scale, the ratio of the two variances may be higher or lower.

Fotheringham and Wong [1991] have investigated extensively the impact of scale effects and zoning effects on multivariate statistical analysis. Multivariate regression and logistic regression models are used in their study. They demonstrate that beyond the impact on a correlation coefficient, the parameter estimates from regression models derived from different scales are widely dispersed and can be significantly different from each other. Furthermore, given the level of scale or resolution, a wide range of results is generated from different spatial partitioning systems. That is, if an analyst selects a spatial partitioning system that has been given a priori, such as the census tracts' spatial configuration, it is only one of many possible configurations and scales a researcher may use; results from the analysis are just one out of many possible sets of results.

Because many geo-referenced data contain significant spatial autocorrelation, the use of conventional multiple linear and logistic regression models may not be appropriate. Moreover, the logistic regression model has the side-effect of constraining the variation of parameter estimates. A researcher interpreting the results from the multiple linear regression and logistic regression models should proceed with caution. However, since these two types of model are commonly used in statistical analysis, using them to depict the impact of the MAUP can provide results that can be applied to many statistical studies.

In contrast, Amrhein [1994] argues that the MAUP effects may not be as unpredictable in spatial analysis as portrayed by Fotheringham and Wong [1991]. Using simulated data to control for both geographical and statistical characteristics of a data distribution, Amrhein generates 10,000 observations as the population, and aggregates the observations into three spatial levels to analyze the scale effect. At each level, different zonal patterns are derived to explore the zoning effect. His findings indicate that the mean and weighted (by population) statistics are not sensitive to the aggregation process (scale and aggregation effects). Most of Amrhein's results support the idea that variance changes according to the degree of aggregation, and thus variances at different levels can be inferred by variance at a given level and the number of areal units. His study also indicates that data with higher variability (variance) are more likely to yield pronounced zoning effects than are more uniformly distributed data. Another set of results confirms previous findings that correlation coefficients increase with increasing aggregation, and parameter estimates for regression are sensitive to aggregation, so much so in fact that they may not even produce any reliable information.

Amrhein's study seems to paint a very optimistic picture that the MAUP, or at least scale effect, is predictable, given particular circumstances. However, only two types of distribution are used in simulating the data: the uniform and the normal. In statistical application, whether it is spatial or traditional, we do not always encounter data distributed according to these two distributions. To what extent the guidelines developed by Amrhein can be applied to data not distributed strictly according to the two distributions is not known. In practice, the multivariate situation is more common than is the bivariate in regression analysis. Thus, some of the findings provided by Amrhein [1994] may have limited utility.

Table 2 reports the results of the multiple linear regression model used to analyze the Connecticut Census data at different levels of aggregation. The regression model is

$$\text{MEDFAMINC} = \alpha + \beta_1 \text{PWHITE} + \beta_2 \text{PECONACT} + \beta_3 \text{PHI\_ED} + \beta_4 \text{PWH\_COL} + \beta_5 \text{POWN} + \beta_6 \text{PSING\_H} + \epsilon.$$

The parameter estimates are also shown graphically in Figure 3. The numbers along the horizontal axis correspond to the intercept and variables reported in Table 2. Parameter estimates fluctuate severely across the four levels of aggregation. Variable POWN has the greatest range in the estimated values. It ranges from 16193.92 at the block group level to 70946.80 at the county level. Though the parameter estimate of POWN follows an increasing trend as the level of aggregation increases, this is not a general pattern followed by other variables. In fact, the fluctuations of all other variables do not follow any significant pattern.

The signs of the estimates also deserve some attention. Except for variables PHI_ED and POWN, all variables have estimates with both positive and negative signs. In other words, depending upon which level of data aggregation the analyst has adopted, relationships between the independent variables and the dependent variable in this linear regression can change in nature. These results are very alarming, especially when this type of statistical analysis is frequently used to formulate public policy. For instance, the expected relationship of variables at the township level may not be valid at the block group level.

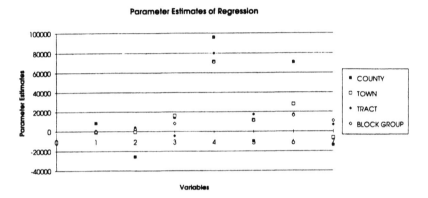

**Figure 3.**

Another interesting observation is that all parameter estimates of the same variable at different levels of aggregation can be significant but have opposite signs. For example, using 0.5 as the confidence level, parameter estimates of variable PSING_H at the block group level (10739.85) and at the county level (-13873.70) are both significant, but they have opposite signs. In general, though, parameter estimates at the block group level are more likely to be significant; this tendency does not constitute a trend. Whether or not an estimate is significant at one level of aggregation has no bearing on its significance at another level. For instance, variable PWHITE is only significant at the county level, but variables PHI_ED and POWN are significant at all levels.

In fact, the impacts of the MAUP go beyond ordinary bivariate and multivariate statistical analysis. They also affect virtually every mathematical model in the spatial sciences using aggregated data. They even transcend the boundaries of human and physical phenomena. [Bian and Walsh, 1993] Their effects can be detected in the calculation of simple indices [Wong, 1994], economic models such as the input-output model [Blair and Miller, 1983], more spatial-oriented models such as the location-allocation models [Current and Schilling, 1990], and spatial interaction models [Putman and Chung, 1989].

## IV. APPROACHES TO 'SOLVING' THE MAUP

A great deal of effort has been spent in assessing the nature, scope, and magnitude of the scale and zoning effects. Only recently have research efforts been directed toward searching for 'solutions' to the MAUP. According to Fotheringham [1991], several methods can be employed to tackle errors generated from changing scales and modifying zonal configurations. They include: "(i) the derivation of 'optimal' zoning systems; (ii) the identification of basic entities; (iii) sensitivity analysis; (iv) abandonment of traditional statistical analysis; and (v) shifting the emphasis of spatial analysis towards relationships that focus on rates of change" (p. 222). In the following discussion, these five categories will be collapsed into three potential approaches for dealing with the MAUP.

### A. THE DATA MANIPULATION APPROACH

One of the basic themes in the MAUP is that given a study area, the zonal configuration adopted to represent spatial data in that region is only one of many possible ways to partition the region into areal units. Most of the time, the reason why one particular zonal system is chosen over the others is not clear, or is simply chosen for administrative convenience. Several researchers suspect that if the chosen zonal system can be justified one way or the other, the MAUP vanishes. That is, the chosen system is the best among all possible spatial partitioning schemes.

Moellering and Tobler [1972] approach this issue from a statistical perspective. They postulate that data at the most disaggregated level is a function of, or is, the outcome of the effects from various aggregation levels. Thus, the variance at the most disaggregated level can be decomposed into effects derived from various aggregation levels, and the level having the most variance is the level at which the spatial process is occurring. Assume that $X_{ijkl}$ is a variable in the lth census block group, the kth census tract, the jth township, and the ith county in Connecticut, Moellering and Tobler have proposed a model stating that,

$$X_{ijkl} = \text{mean}_x + e_{ijkl} + e_{ijk} + e_{ij} + e_i, \qquad (1)$$

where $\text{mean}_x$ is the mean of $X_{ijkl}$. That is

$$\text{mean}_X = \frac{\sum_i \sum_j \sum_k \sum_l X_{ijkl}}{n_{ijkl}}$$

where $n_{ijkl}$ is the total number of block groups in the whole state. Then $e_{ijkl}$ is the effect of the ijkl-th block group, $e_{ijk}$ is the effect of the ijk-th census tract, $e_{ij}$ is the effect of the ij-th township, and $e_i$ is the effect of the i-th county. They argue that

$$SS_{total} = SS_{BG} + SS_{TRT} + SS_{TWN} + SS_{CTY}$$

where $SS_{total}$ is variation around the grand mean, $\text{mean}_x$, $SS_{BG}$ is the sum of squares at the block group level, $SS_{TRT}$ is the sum of squares at the census tract level, $SS_{TWN}$ is the sum of squares at the township level, and $SS_{CTY}$ is the sum of squares at the county level; clearly these effects are nested. In other words, the total sum of squares can be partitioned into sums of squares derived from different levels of the spatial hierarchy. Given the appropriate degrees of freedom for the corresponding sums of squares and the mean squares, the variances derived from different levels of aggregation can be tested for statistical significance. Moellering and Tobler claim that the level at which the variance becomes the most significant is the level where spatial processes are 'in action.' This also is the level at which analysis should be conducted. In order to identify this spatial level, they introduce the scale-variance ratio, which is the sum of squares weighted by the appropriate degrees of freedom, for each level of aggregation.

To illustrate the geographical variance method, the population variable (POP) of the 1990 Connecticut census data is used. For the four levels of aggregation (county, town, tract, and block group), the sums of squares and the scale-variance ratios of corresponding aggregation levels are derived. They are reported in Table 3. According to Moellering and Tobler [1972], the level with the greatest scale-variance ratio is the aggregation level at which the spatial process is the most significant and at which the analysis should be performed. In the Connecticut data, the county level, which has the greatest level of variation, should be the level chosen for analysis.

The geographical variances technique proposed by Moellering and Tobler [1972] is not a complete solution to the MAUP. Not only is the technique unable to accommodate multivariate situations, but in order to identify the aggregation level with the most action, it also requires an a priori definition of the spatial system hierarchy. Thus it fails to deal with the zoning or aggregation effect. However, the major problem with this analysis of geographical variance approach is the theoretical foundation of Equation (1). Equation (1) implies that the data value at the most disaggregated level is a linear combination of the average at the disaggregated level and the effects from various aggregation levels. This conceptualization of the relationship of data among levels of aggregation is not based upon the process of aggregating spatial data. The variation of data at the most aggregated level should have no impact on the disaggregated data, while the opposite relation is more likely to be true. In addition, the claim that the level at which most variation occurs is the appropriate level for spatial analysis is not well justified. The definition of 'best' or 'optimal' can be very subjective, and a zonal system can be optimal for one research endeavor while being suboptimal for others. In addition, spatial autocorrelation is not accounted for in Equation (1).

**Table 3.**

| Spatial Analysis of Variance | | | |
|---|---|---|---|
| | *Sum of Squares* | *DF* | *Scale-Variance Ratio* |
| County | 3.21E+09 | 7 | 4.59E+08 |
| Town | 6.69E+10 | 161 | 4.16E+08 |
| Census Tract | 6.04E+09 | 887 | 6.81E+06 |
| Block Group | 7.31E+08 | 2401 | 3.04E+05 |

Among the methods proposed to handle both the scale and aggregation effects, the optimal zoning approach proposed by Openshaw [1977a, 1977b] is the most controversial. In brief, he specifies how basic areal units should be merged to form larger spatial entities, arguing that this procedure should be dependent upon the performance of the model employed to calibrate the spatial data. The ideal or optimal zonal configuration should provide the best modeling results. That is, basic areal units should be aggregated to maximize or minimize whatever criterion (or criteria) is (or are) used to evaluate the performance of the model. Thus, the aggregation of areal units facilitates spatial modeling instead of being a system existing a priori.

In his demonstration, Openshaw [1977a] uses spatial interaction models to analyze journey-to-work trips among areal units defined by Local Authority boundaries in England. The goodness of fit statistic of the spatial interaction models is reflected in a residual standard deviation statistic. Local Authority units are aggregated according to different principles. Spatial interaction models are used to analyze the aggregated data, and performance of the models is assessed by the goodness of fit statistic. The optimal zonal configuration is the one that yields the best goodness of fit statistic. The whole process of defining an optimal zoning system can be treated as an optimization process in which the objective function is the goodness of fit statistic for the model, with the basic areal units being aggregated according to some principles and constraints (such as the number of aggregated areal units and a contiguity constraint), such that the goodness of fit statistic for the model is maximized or minimized.

## B. A TECHINQUE-ORIENTED APPROACH

Although the scale and zoning effects have been known to geographers and spatial scientists for more than four decades, and simple solutions have been proposed intermittently, most simple methods have not proven to be very useful. One of the earliest methods was proposed by Robinson [1956] who argued that weighting areal units by the population size or number of observations can eliminate the scale effect in aggregated data. However, this simple spatial weighting method fails to correct for the errors propagated by aggregation.

Almost all studies trying to assess the impacts of the MAUP reach the conclusion that scale and zoning effects are real and significant. However, there are several exceptions. Based upon simulation experiments, Green [1994] claims that spatial aggregation does not produce systematic changes in parameter estimates of a specific type of spatial regression model. He proposes that

$$y_i = a + b(l) \times (l)_i + b(r) \times (r)_i + \varepsilon_i$$

where $x(l)_i$ is the independent variable in areal unit i measured at the local scale, and $x(r)_i$ is the independent variable in areal unit i at the regional scale. The two parameter estimates, $b(l)$ and $b(r)$, are for the local and regional independent variables, respectively. Operationally, $x(r)_i$ is the average value of the independent variable $x(l)$ in all adjacent areal units of i excluding the value in unit i. Then the combined parameter from the parameter estimates of the local and regional variables (that is $b(l)+b(r)$) will be free of the scale effect.

But it is likely that Green's simulation exercise is not extensive enough to derive a reliable conclusion. He does not specify which types of variable that the spatial regression model can be applied to, and how those variables should be derived for (or aggregated to) different scales. Using the median house value (MEDHOUVAL) as the dependent variable, population (POP) as the local independent variable, and deriving the regional independent variable of POP from the POP variable at the next level of aggregation in the Connecticut census data, a spatial regression model similar to the one proposed by Green is formulated for all levels of aggregation. The results are reported in Table 4. Green claims that the parameter derived from combining the local and regional parameter estimates is not subject to the scale effect. However, this expected result is not obtained when analyzing the Connecticut data. Significant scale effects are apparent in this spatial regression model. For comparison purposes, results from a standard bivariate regression model without the regional variable also are reported in Table 4.

Another exception to finding no significant zoning or scale effect in multi-scale analysis is the study of Canadian migration reported by Amrhein and Flowerdew [1991]. This pair of researchers aggregated Canadian census divisions into larger areal units using several aggregation algorithms, and employed a Poisson regression model to describe migration flows. A Poisson regression model assumes that migration flows between origins and destinations have a Poisson distribution. The mean of a migration flow becomes an exponential function of the origin population, destination population, and spatial separation measure between the origin and destination. Thus the flow between i and j is given by

$$\hat{y}_{ij} = \exp(b_0 + b_1 \ln p_i + b_2 \ln p_j + b_3 \ln d_{ij}) .$$

This model is very similar to an unconstrained gravity model. Surprisingly, typical effects of the MAUP are not detected in the study. No significant trends or patterns similar to those previously displayed by the statistics reported for the Poisson regression model are uncovered. The authors speculate that the unusual results may be due to the fact that Canadian migration data are not subject to the MAUP effect, the Poisson regression model is not sensitive to changes in scale, or these two reasons acting in concert. Further investigations are necessary to confirm or refute findings of this study.

**Table 4.**

| Spatial Regression Model (Green) | | INTERCEPT | POP(*l*) | POP(*r*) | POP(*l*)+POP(*r*) |
|---|---|---|---|---|---|
| Town | Beta Estimates | 159818 | -0.56 | 0.09 | -0.47 |
| | Prob >|t| | 0.0001 | 0.0086 | 0.0001 | |
| Census Tract | Beta Estimates | 130244 | 16.54016 | -0.345953 | 16.19 |
| | Prob >|t| | 0.0001 | 0.0001 | 0.0001 | |
| Block Group | Beta Estimates | 138897 | 18.393084 | 6.572483 | 24.97 |
| | Prob >|t| | 0.0001 | 0.0001 | 0.0001 | |
| **Bivariate Regression Model** | | INTERCEPT | POP(*l*) | | |
| Town | Beta Estimates | 193004 | -0.09 | | |
| | Prob >|t| | 0.0001 | 0.6612 | | |
| Census Tract | Beta Estimates | 116945 | 15.897562 | | |
| | Prob >|t| | 0.0001 | 0.0001 | | |
| Block Group | Beta Estimates | 158620 | 18.721162 | | |
| | Prob >|t| | 0.0001 | 0.0001 | | |

Based upon their results, Amrhein and Flowerdew [1991] argue that the choice of appropriate models or techniques in analyzing aggregated spatial data is a crucial aspect in obtaining accurate analytical results. They also claim that previous studies encountering the MAUP effect might have done so by choosing the wrong statistical technique for analysis. This argument is the basic tenor of Tobler's [1991] article. He strongly suggests that the MAUP emerges because many studies use inappropriate techniques to analyze spatial data. Many statistical tools, such as correlation coefficients and linear regression, are not appropriate for spatial data analysis. He then proposes to use frame independent spatial analytical techniques, which derive analytical results that do not depend upon the locational coordinate systems employed to describe the data. In his discussion, Tobler treats the zoning problem and scale problem separately. He suggests that the zoning problem, which is a more complicated issue, can be tackled by employing various transformations or interpolation techniques, including the pycnophylactic interpolation, which simultaneously smooths a spatial distribution and maintains the volume of the data. Using migration data for illustrative purposes, Tobler demonstrates that certain types of model will not encounter the MAUP effect. Let $M_{ij}$ be the migration flow between i and j. Suppose migration flows between i and j, and between i and k are given as

$$M_{ij} = (R_i + E_j)l_{ij},$$

and

$$M_{ik} = (R_i + E_k)l_{ik},$$

where $R_i$ and $E_j$ are the push from origin i and the pull from destination j, respectively, and $l_{ij}$ is the length of boundary between i and j. Then he claims that if areal units j and k are merged, the combined migration flows from i to the aggregated unit j+k will be

$$M_{ij} + M_{ik} = M_{i,j+k}$$
$$= R_i(l_{ij} + l_{ik}) + E_j l_{ij} + E_k l_{ik}.$$

However, most of the commonly used statistical tools are not scale-invariant. Even though some techniques seem to possess the scale-invariant property, they need to be scrutinized before they should be used without reservations.

## C. AN ERROR MODELING APPROACH
Another camp of researchers realizes that when analysis moves from one spatial scale to another, relationships among variables and among spatial entities also change. Instead of searching for techniques immune to such scale effects, their approach explicitly documents variations derived from changing scale, and incorporates these changes into modeling and analysis.

Though Arbia [1989] implicitly referred to the role of spatial autocorrelation in his discussion of the impact of alternative spatial configurations on spatial data, Green [1993] acknowledges explicitly that it is the aggregation of similar areal units to form larger spatial entities that is the mechanism that creates the MAUP. As noted earlier, he proposes a spatial regression model in which a component for individual observation and the average of attribute values from surrounding areal units of the observation are incorporated. The latter component captures the spatial autocorrelation effect or the regional effect. Thus, the dependent variable in the spatial regression model becomes a function of individual observations, and the average values of neighboring observations. Green argues that regression parameter estimates at the aggregated level must take into account the observations at the most disaggregated level and the regional effect derived from merging smallest areal units to form larger spatial entities. However, Green [1993] fails to provide a convincing example that his proposed spatial regression model can satisfactorily capture the 'error' of the scale effect.

As was pointed out at the beginning of this chapter, a major impact of the MAUP rests on changes in the variance-covariance matrix or the correlation coefficient matrix when areal units are aggregated. 'Error' accumulates when individual observations are aggregated into regions, and the original variance-covariance matrix structure, upon which many statistical methods are based, gradually disappears during the aggregation process. Using the aggregated data to infer the characteristics or behavior at the individual level unavoidably will lead to the ecological fallacy. Steel, Holt and Tranmer [1994] attempt to model the error created by the aggregation process so that individual information can be estimated from the regional data. Wrigley [1994] provides a summary of their idea. The crux of the whole error modeling exercise lies in the concept of decomposing the conditional expectation of the variance-covariance measure at the regional level into several components: a variance-covariance matrix at the individual level, a bias component accounting for the aggregation effect on a group of variables called grouping variables, and the residuals from within-group correlation. The grouping variables are regarded as characteristics of individual observations upon which individuals are grouped together to form regions. Steel, Holt and Tranmer [1994] believe that individuals sharing similar characteristics are likely to be located close to each other rather than in a scattered manner. Thus, spatial aggregation will likely group similar individuals into the same region.

If the variance-covariance decomposition model is an accurate description of the relationship of individual level information and regional level characteristics, then an estimated individual-level variance-covariance matrix can be derived from the aggregated variance-covariance matrix. However, whether or not this method is feasible depends upon two additional conditions: (1) grouping variables are identifiable at the regional level, and (2) an unbiased estimated of the variance-covariance matrix at the individual level can be derived from a sample of individual observations.

A superficial review of this method may lead one to be skeptical about it, since it seems to apply only classical statistical concepts while failing to deal with the spatial aspect of the MAUP, except in the process of deriving regional level data. Steel, Holt, and Tranmer have captured at least one very important spatial aspect of the MAUP. When the regional variance-covariance matrix is decomposed, one of the terms is the residuals of within-group correlation. The within-group correlation is analogous to a 'local' level multivariate spatial autocorrelation measure in which areal units or individual observations are defined by areal boundaries. [Hubert, Golledge and Costanzo, 1981; Wartenberg, 1985] That is, part of the variance-covariance matrix at the aggregated level is the multivariate spatial autocorrelation of variables within each aggregated unit.

The gist of the MAUP arises when the same geographical region is studied using data gathered at different scales or gathered from different spatial partitioning schemes, and inconsistent results are produced. The error modeling methods suggested by Green [1993], and Steel, Holt, and Tranmer [1994] attempt to link data at different levels  to derive consistent results. This approach implies that analysts have to utilize data from various scales, and the scale component is built into the model. Though this is a theoretically sound method, how feasible this approach can be in application remains questionable for now:  it is a sophisticated approach with immense   computational demands. Considerable time and energy remain to be devoted to investigating the empirical aspects of this method.

## V.  GUIDELINES FOR ANALYZING DATA FROM DIFFERENT SCALES

### A.  USING DATA FROM THE FINEST SCALE
Geographers and spatial scientists are quite often trapped in a dilemma--drawing inferences about individual observations and representing generalized spatial data through cartographic means. Frequently, the goal of an investigation is to find out pertinent information about individual observations or subjects. Thus, using data from the individual or at the most disaggregated level is most appropriate. But analysts may find data from individual observations or data at a very high level of resolution to be too massive; these data become difficult to manage or to represent  by cartographic means. In order to identify patterns or trends, sometimes a certain degree of generalization has to take place in order to summarize the information in a more manageable format or to map the data. Especially when cartographic means are chosen to represent spatial information, the

number of observations has to be reduced to a level that can be comprehended visually or comprehended by human perception. In doing such data reduction, pertinent detail may be lost. Indeed, the results from analyses conducted at the aggregated level are very likely to be inconsistent with the results from the individual level.

Which level of resolution or what degree of details should be chosen, to a large extent, depends upon the purpose or goal of a study. Unless a very legitimate classification scheme is adopted to allocate individual observations to groups, the groups themselves may be of little interest. If the purpose of the investigation is to draw inferences about characteristics of each group, individuals are the subject of statistical inference. Thus, when higher resolution data are available, they should be used. Spatially aggregated data are usually 'smoothed.' 'Error' due to aggregation is part of the data so that individual information cannot be inferred directly from the aggregated level.

An implication of this recommendation is that statistical concerns regarding the scale issue (or the more general issue of MAUP) and cartographic principles in data representation sometimes may not be resolved. Because powerful computing power and massive digital storage devices are available and inexpensive, the cost of losing information due to generalization or aggregation may be too steep.

## B. REPORTING 'ERROR' FROM AGGREGATION

In general, we expect that when scale-sensitive or scale-dependent statistical or analytical techniques are used to study spatial data gathered and reported at different scales, the results derived from data at these different scales will not be consistent. A range of results can be derived. Instead of arbitrarily choosing data from a particular scale or resolution and studying that data set exclusively, it is preferable to replicate the same type of analysis using data from different scales. In other words, analysts should report the sensitivity of their results to changes in scale. To some extent, reporting the scale-sensitivity of results also indicates how reliable the results can be. This is a rather honest strategy in scientific inquiry because it takes into account the intrinsic inaccuracy of aggregated spatial data.

Similar to the above strategy, it is desirable to report 'error' in data due to the aggregation process. This type of information should be included in the spatial metadata--data about the data. Usually quantitative geographers and statisticians are concerned with several types of error in statistical analysis: calculation error, measurement error, specification error, sampling error, and stochastic error [Griffith and Amrhein, 1991]. Spatial aggregation is one source of some of these errors in data. The resulting error is primarily measurement because measures such as averages and proportions are used to represent the data after a merging of spatial units. When aggregated spatial data are used in an analysis, the quality of data or the amount of error due to aggregation should be reported and incorporated into an analysis. In brief, the error due to aggregation can be depicted as the differences between the aggregated data and the data gathered or reported at the most disaggregated level. This is a notion that has not received much attention but should be regarded as a strategy to deal with the scale variation nature of spatial data.

## C.  USING TECHNIQUES INSENSITIVE TO SCALE CHANGES

If the techniques or tools chosen for analysis are not sensitive to scale changes, we can expect consistent analytical and statistical results from using data of different scales. In this chapter, several scale-insensitive models are mentioned. It is very likely that there are other scale-independent analytical tools. Whenever these types of tools or models are available, they should be chosen over the scale-dependent techniques. Many studies have indicated that most commonly used quantitative methods and statistics are (spatially) scale dependent [Fotheringham and Wong, 1991]. However, we still have to explore which of the existing tools are insensitive to scale changes. And equally important is developing new techniques insensitive to scale changes. If no new techniques insensitive to scale are developed, we will be left with very few choices and a rather  narrow scope of analysis within which to study spatial data. In contrast, traditional statistics and spatial statistics have well-developed properties and theories. Abandoning this wealth of information and techniques is too dear a cost.

## VI.  CONCLUSIONS

In this chapter we have discussed the nature of the MAUP, with specific emphasis on the scale effect. We have reviewed several approaches and methods suggested by researchers to deal with the problem. However, no method or model we discussed yields a completely corrected   result. Thus we provide several guidelines for analyzing spatial data when multiple scales are involved.

The MAUP has been regarded as the most stubborn problem in geography and spatial science since Gehlke and Biehl [1934] identified part of the MAUP almost seven decades ago. Many subsequent attempts have failed to produce an acceptable solution. Several studies have indicated that the complexity of MAUP increases as the statistical and mathematical models adopted to analyze the data become more complicated. No simple solutions have  produced acceptable results. A recent revival in this area of research has triggered some glimpses of hope that some solutions may emerge in the near future. However, it may be that these solutions are unlikely to be simple methods and perhaps cannot be applied in a straight-forward manner. Thus, whether or not these potential solutions will be used widely is questionable. Several discussions, including that in this chapter, have confirmed the apparent role of spatial autocorrelation in producing the MAUP. We expect that the solution of the problem is likely to depend upon how we can model the multivariate spatial autocorrelation effect in multi-scale situations. But before a solution for the MAUP is generally available, we should proceed with caution when we deal with spatial data available at various scales.

# REFERENCES

**Amrhein, C. G.** Searching for the elusive aggregation effect: evidence from statistical simulation. *Environment and Planning A* (forthcoming), 1994.

**Amrhein, C. G. and Flowerdew, R.** The effect of data aggregation on a Poisson regression model of Canadian migration. *Environment and Planning A* 24, 1381-1391, 1992.

**Anselin, L.** *Spatial Econometrics: Methods and Models.* Kluwer Academic Publishers, Dordrecht, 1988.

**Arbia, G.** *Spatial Data Configuration in Statistical Analysis of Regional Economic and Related Problems.* Kluwer Academic Publishers, Dordrecht, 1989.

**Bian, L. and Walsh, S.** Scale dependencies of vegetation and topography in a mountainous environment of Montana. *The Professional Geographer* 45, 1-11, 1993.

**Blair, P. and Miller, R. E.** Spatial aggregation in multiregional input-output models. *Environment and Planning A* 15, 187-206, 1983.

**Blalock, H. M.** *Causal Inferences in Nonexperiental Research.* University of North Carolina Press, Chapel Hill, 1964.

**Bureau of the Census.** *Census of Population and Housing, 1990: Summary Tape File 1 on CD-ROM (Connecticut).* The Bureau, Washington, 1992a.

**Bureau of the Census.** *Census of Population and Housing, 1990: Summary Tape File 3 on CD-ROM (Connecticut).* The Bureau, Washington, 1992b.

**Cliff, A.D. and Ord, J. K.** *Spatial Autocorrelation.* Pion, London, 1973.

**Current, J. R. and Schilling, D. A.** Analysis of errors due to demand data aggregation in the set covering and maximal covering location problems. *Geographical Analysis* 22, 116-26, 1990.

**Fotheringham, A. S. and Wong, D. W. S.** The Modifiable Areal Unit Problem in multivariate statistical analysis. *Environment and Planning A* 23, 1025-44, 1991.

**Fotheringham, A. S.** Scale-independent spatial analysis. In *Accuracy of Spatial Databases,* edited by M. F. Goodchild and S. Gopal, pp.221-28, and Francis, London, 1991.

**Gehlke, C. E. and Biehl, K.** Certain effects of grouping upon the size of the correlation coefficient in Census tract material. *Journal of the American Statistical Association Supplement* 29, 169-70, 1934.

**Green, M.** Ecological fallacies and the Modifiable Areal Unit Problem. North West Regional Research Laboratory, Lancaster University, *Research Report No. 27,* 1993.

**Griffith, D. A.** *Advanced Spatial Statistics.* Kluwer Academic Publishers, Boston, 1988..

**Griffith, D. A. and Amrhein, C. G.** *Statistical Analysis for Geographers.* Prentice-Hall, Englewood Cliffs, 1991..

**Hubert, L. J.; Golledge, R. G.; and, Costanzo, C. M.** Generalized procedures for evaluating spatial autocorrelation. *Geographical Analysis* 13, 224-33, 1981.

**Moellering, H. and Tobler, W.** Geographical variances. *Geographical Analysis* 4, 34-50, 1972.

**Openshaw, S.** Optimal zoning systems for spatial interaction models. *Environment and Planning A* 9, 169-84, 1977a.

**Openshaw, S.** A geographical solution to scale and aggregation problems in region-building, partitioning and spatial modelling. *Transactions of Institute of British Geographers* 2, 459-72, 1977b.

**Openshaw, S.** *Concepts and Techniques in Modern Geography, Number 38. The Modifiable Areal Unit Problem.* Geo Books, Norwich, 1984.

**Openshaw, S. and Taylor, P. J.** A million or so correlation coefficients: three experiments on the modifiable areal unit problem. Pages 127-144 in N. Wrigley (ed.) *Statistical Applications in the Spatial Sciences,* Pion Limited, London, 1979.

**Putman, S. H. and Chung, S-H.** Effects of spatial systems design on spatial interaction models. 1: the spatial definition problem. *Environment and Planning A* 21, 27-46, 1989.

**Ripley, B. D.** *Spatial Statistics.* Wiley, New York, 1981.

**Robinson, A. H.** The necessity of weighing values in correlation analysis of areal data. *Annals, Association of American Geographers* 46: 233-236, 1956.

**Robinson, W. S.** Ecological correlations and the behavior of individuals. *American Sociological Review* 15: 351-357, 1950.

**Steel, D.G.; Holt, D.; and, Tranmer, M.** Modelling and adjusting aggregation effects. Paper presented at the Annual Conference of the US Bureau of the Census, 1993.

**Tobler, W.** Cellular geography. In *Philosophy in Geography*, edited by S. Gale and G. Olsson, pp. 379-86, Reidel, Dordrecht, 1979.

**Tobler, W.** Frame independent spatial analysis. In *Accuracy of Spatial Databases,* edited by M. F. Goodchild and S. Gopal, pp. 115-22. Taylor and Francis, London, 1991.

**Wartenberg, D.** Multivariate spatial correlation: a method for exploratory geographical analysis *Geographical Analysis* 17, 263-83, 1985.

**Wong, D.W.S.** Spatial dependency of segregation index. Paper presented at 41st North American Meeting of Regional Science Association International, Niagara Falls, Ontario, Canada, Nov.17-20, 1994.

**Wrigley, N.** Revisiting the Modifiable Areal Unit Problem and the ecological fallacy. In *Festschrift for Peter Haggett*, edited by Cliff, A.D., Gould, P.R., Hoare, A.G. and Thrift, N.J., Blackwell, Oxford, 1994.

Chapter 6

IMPLEMENTING SPATIAL STATISTICS ON PARALLEL
COMPUTERS

Bin Li

## I. INTRODUCTION

Although spatial statisticians have made great progress in the
theoretical and technical domains, the adaptation of spatial statistics to
geographical analysis has been slow. A major obstacle is the high
computational complexity of spatial statistical methods, which is
amplified by the need for timely analyses of large spatial datasets.

Rapid advance in High Performance Computing (HPC) technology
in recent years is likely to overcome the computational bottlenecks in
spatial statistics. With dramatic improvement in processing power,
HPC also may help develop new computational techniques based on
spatial statistical theories. A key issue in realizing the potentials of
HPC, however, is the portability of spatial statistical methods to parallel
processing, a computational model dramatically different from
conventional computing.

This chapter explores the potentials of parallel processing in spatial
statistical analysis. It describes in detail the parallel procedures to
implement major operations in spatial autocorrelation analysis and
spatial autoregressive modeling. The chapter focuses on the data
parallel computing model and its potential for greater speedup and
portability across hardware platforms. A preliminary analysis of
performance is also included to demonstrate the range of speedup we are
able to achieve with parallel computers.

## II. A BRIEF INTRODUCTION TO PARALLEL PROCESSING

Parallel processing is a computational model that allows concurrent
processing of instructions and data so that processing time is minimized.
There are three computer architectures that could employ parallelism.
Michael Flyn [1966] classified these architectures as Single Instruction
Multiple Data (SIMD), Multiple Instruction Multiple Data (MIMD), and
Multiple Instruction Single Data (MISD). Among them, SIMD and
MIMD computers are suitable for a wide range of problems, while

0-8493-0132-7/95/$0.00+$.50
*107*

MISD, also referred to as systolic arrays, is used for pipelined executions of specific algorithms. This section gives a brief introduction to SIMD and MIMD architectures and the corresponding software models.

Figure 1 illustrates a SIMD architecture. SIMD computers use a single controller to command multiple processor units. These processor units, or processor elements (PE), are small but can handle basic arithmetic and logical operations. Each PE has access to either local or shared memories. All PEs are connected in a particular topological network so that they can communicate with each other.

SIMD computers solve problems by decomposing the data domain into small units, each assigned to a single processor. The controller broadcasts instructions to all active PEs, which concurrently execute each stream of instructions on corresponding data elements.

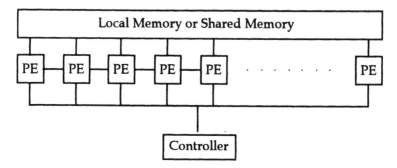

**Figure 1.** Schematic diagram of a SIMD computer

Another major parallel architecture is MIMD. A MIMD computer allows each processor unit to execute different sets of instructions on different sets of data elements. It does so by adding a control unit to each PE. Similar to the SIMD computers, all PEs are connected to each other in a particular topological network and each has access to either local or shared memories. Different from the SIMD model, however, PEs in a MIMD computer are more capable with much more local memory and processing functions. For example, in a Connection Machine Model 2 (CM-2), a SIMD computer, each PE has 8 Kilobytes of local memory and can carry out arithmetic and logical operations. In a CM-5, a MIMD computer, on the other hand, each PE is a RISC processor with 32 Megabytes of local memory and could optionally be attached to four vector units. [Thinking Machine Corporation, 1991a]

To solve a problem on a MIMD system, one must explicitly decompose both computational tasks and the data domain to guarantee equivalent workloads on all processor units. Furthermore, the entire process also has to be coordinated to reflect logical linkages among sub-processes. Scheduling and interprocessor communications are realized by passing messages from one processor to another. For this reason, software implementation of MIMD is often called a message passing model.

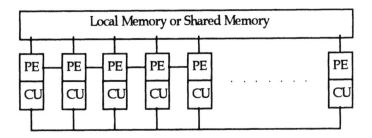

**Figure 2**. Schematic diagram of a MIMD computer

## III. SOFTWARE MODELS FOR PARALLEL PROCESSING

The preceding section describes the characteristics of parallel architectures. Generally, as users of a parallel system, the hardware architecture is seldom our primary concern. We care more about how the parallel computer is presented to us and how we could communicate with the machine. Such an interface is provided by system and programming software based on specific software models. This section describes two primary parallel software models: task parallelism and data parallelism.

The task parallel model represents parallelism in the temporal domain of a process. This model assumes there are separate and independent processes or functions that can be solved simultaneously. In the task parallel model, the entire process is decomposed into a subset of instructions that are then performed concurrently. Data parallelism, on the other hand, represents parallelism in the spatial domain of a process. In this model, parallelism is achieved by executing each instruction on all data elements concurrently. From a programmer's perspective, the data parallel model treats the entire data set or its subsets as a single object. Each piece of instruction is executed on all elements of the data objects simultaneously. For example, to calculate the elemental sum of matrices A and B, in sequential processing, one would use a double DO loop:

*do i=1,n*
  *do j=1,n*
     $D(i,j) = A(i,j) + B(i,j)$
*end do*

which can be simply expressed as a matrix operation in a data parallel language:

$$D = A + B$$

Matrices A, B, and D, instead of elements $A(i,j)$, $B(i,j)$, and $D(i,j)$, are treated as objects in the computation.

Compared with task parallelism, data parallel processing has two advantages. First, data parallelism has more potential for high performance gain because it offers a higher degree of concurrence. Theoretically, performance gain from parallelism is in proportion to the number of items processed concurrently. For control parallelism, this equates to the number of programming steps that can be pipelined, or the number of subtasks a program performs. This number is normally small, and it is likely fixed. Consequently, the performance gain is limited. In data parallelism, however, the number of items processed concurrently is equal to the number of data elements required in the computation. The degree of parallelism increases with the size of a problem, as does the performance gain. This feature is particularly relevant to spatial analyses that involve large amounts of data.

Second, from the perspective of software development, data parallel code is easier to write and to debug than is task parallel code. The latter usually has to use multiple threads of controls to realize parallelism. Data parallel code is also much more readable. As shown in the foregoing examples, a data parallel language expresses and operates on data objects in more natural ways. Finally, because they have a high level of abstraction, data parallel programs can achieve portability across different computer platforms. Data parallel algorithms can be implemented in different languages with few modifications, and a data parallel language can be implemented on different architectures. For example, FORTRAN90, a data parallel language, can run on sequential workstations as well as on several types of parallel systems, including MIMD computers.

Not every problem could map to the data parallel model, however. Only synchronous and loosely synchronous problems are suitable for data parallel processing. [Fox, 1992] Fortunately, most spatial problems seem to be synchronous or loosely synchronous. The following sections describe how one may implement spatial statistical methods with the data parallel model.

Limited space for this chapter does not allow a detailed introduction to the data parallel model. Rather, we encourage the reader to find more information from several references. [Adams, 1992; Blelloch, 1990; Hillis, 1985, 1986, 1987; Maspar Computer Corporation, 1991a, 1991b; Quinn, 1990; Smith, 1993; Thinking Machine Corporation, 1991b, 1991c]

## IV. PARALLEL IMPLEMENTATIONS

From a computational perspective, there are two basic categories of techniques in spatial statistics. The first group involves measuring and testing for spatial autocorrelation, which is used for detecting spatial patterns or to diagnose spatial dependence in regression models. The second group includes methods to model spatial processes and to incorporate spatial structure into classical regression models. This section outlines the computational characteristics of these statistics and the procedures for implementing them on parallel computers.

### A.  ANALYSIS OF SPATIAL AUTOCORRELATION

Three coefficients are used to measure and test for spatial autocorrelation. They are the join-count statistics, Moran's I, and Geary's c. These coefficients indicate the degree of spatial dependency in an observed phenomenon. Definitions of these coefficients and their corresponding distribution statistics are given by Cliff and Ord [1981], and for convenience are included in the appendices to this chapter.

### 1.  Computational characteristics

Despite their lengthy mathematical expressions, the spatial autocorrelation coefficients and their test statistics are simple to compute. Major quantities (expressed in terms of the general matrix $W$, which could then be the symmetric binary matrix $C$) include the following entities as elements of spatial autocorrelation coefficients:

1.  $W$

2.  $S_0 = \sum_{i=1}^{n} \sum_{j=1}^{n} w_{ij}$

    also denoted S0

3. $S_1 = \dfrac{1}{2} \sum_{i=1}^{n} \sum_{j=1}^{n} (w_{ij} + w_{ji})^2$

also denoted S1

4. $S_2 = \sum_{i=1}^{n} \left( \sum_{j=1}^{n} w_{ij} + \sum_{j=1}^{n} w_{ji} \right)^2$

also denoted S2

5. $\sum_{i=1}^{n} \sum_{j=1}^{n} w_{ij} (x_i - \bar{x})(x_j - \bar{x})$

6. $\sum_{i=1}^{n} (x_i - \bar{x})^2$

7. $\sum_{i=1}^{n} \sum_{j=1}^{n} w_{ij} (x_i - x_j)^2$

8. $\sum_{i=1}^{n} \sum_{j=1}^{n} w_{ij} x_i x_j$

Except for the generation of a spatial weights or connectivity matrix, which will be discussed later, all operations are accumulative arithmetic with matrices or vectors. In sequential processing, these quantities arenormally computed with double DO loops. For example, the following FORTRAN77 code calculates $S_1$,

```
    sl = 0.0
    do 10 i = 1,n
        sum1 = 0.0
        do 20 j = 1, n
        sum1 = sum1 + (w(i,j)+w(j,i))**2
20      continue
```

$$s1 = s1 + sum1$$

*10    continue*

This kind of computation has step complexity $O(n^2)$.  That is, the number of execution steps will grow in a quadratic manner as the sample size n increases.  The graph in Figure 3 depicts this relationship. Corresponding data are included in Table 1.

Since computations in each step are simple, calculating the spatial autocorrelation statistics does not impose major computational problems in conventional spatial data analysis.  For instance, an IBM RS6000 calculated S1 for an area with 4096 units in thirty seconds.  For areas with less than 1024 units, the processing times are within one second. This level of performance clearly is sufficient for detecting spatial autocorrelation alone.  However, when spatial autocorrelation is used within a complex framework of data analysis, such as exploratory data analysis (EDA) and real time simulation, further speedup is still preferred.

**Table 1.** Sequential execution time of $S_1$ for a range of problem size

| # Area units (n) | Time (seconds) |
|---|---|
| 32 | 0.01 |
| 64 | 0.01 |
| 128 | 0.02 |
| 256 | 0.09 |
| 512 | 0.38 |
| 1024 | 1.52 |
| 2048 | 5.93 |
| 4096 | 30.5 |

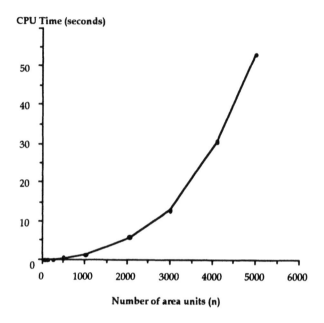

**Figure 3.** Sequential execution time of $S_I$ for **W** matrices in a range of sizes. The size of the **W** matrix is *n*-by-*n*.

## 2. Data parallel implementation

Accumulative matrix operations can be accomplished by the prefix-scan algorithm, implemented as the REDUCTION operators in data parallel software models. As one of the reduction operators, array summation is performed by an intrinsic function SUM in FORTRAN90:

*SUM(Array, Dim,Mask)*

where **Array** is the name of the numeric array, **Dim** is an integer constant indicating the rank of the array, and **Mask** is a logical constant to control where the operation should occur. With this function, S0 and $\sum_{i=1}^{n}(x_i - \bar{x})^2$ are readily expressed as *SUM (W)* and *SUM((X-Xmean)\*\*2)*, respectively. S2 is also straightforward to compute. Described in FORTRAN77, the procedure to calculate S2 is a double DO loop:

*S2 = 0*
*DO i = 1, n*
   *rowsum = 0.0*
   *colsum = 0.0*
   *Do j = 1, n*
      *rowsum = rowsum + W(i,j)*
      *colsum = colsum + W(j,i)*
   *End Do*
      *S2 = S2 + (rowsum + colsum)\*\*2*
*End Do*

The procedure can be mapped directly to a five-step parallel operation: row-wise reduction, column-wise reduction, element-wise addition, element-wise multiplication, and vector reduction, which is expressed with the single statement in FORTRAN90

$$S2 = SUM\ ((SUM(W,\ 1) + SUM(W,\ 2))\ast\ast2).$$

The calculation of S1 is only slightly different. The procedure involves adding each element in each row with the corresponding element in each column; hence it is difficult to compute without intermediate operations. The strategy here is to compute the n inner sums and store them in a vector on which reduction is then performed:

*FORALL (i=1:n) temp(i) = SUM ((W(:, i) + W(i, :))\*\*2)*

*S1 = SUM (temp)*

FORALL is a parallel construct that facilitates array assignment, selection, and various other operations (Thinking Machine Corporation, 1991b). The above FORALL statement carries out the same task as the following DO loop,

   *Do i=1,n*
      *tmp = 0.0*
      *Do j=1,n*
         *temp(i) = tmp + (W(i,j)+W(j,i))\*\*2*
   *End Do*

except that the order of execution in the FORALL statement is undefined, which makes it possible for concurrent processing.

Calculations for the rest of the three terms involve the sum of some cross-products. This seemingly sequential process can be made parallel by transforming the attribute vector X to two 2-dimensional matrices, Xi and Xj, i.e.,

| X | $X_i$ | | | $X_j$ | | |
|---|---|---|---|---|---|---|
| 1 | 1 | 2 | 3 | 1 | 1 | 1 |
| 2 | 1 | 2 | 3 | 2 | 2 | 2 |
| 3 | 1 | 2 | 3 | 3 | 3 | 3 |

A data parallel operation called SPREAD performs the above transformation. It has the same syntax as SUM, i.e., Xi = SPREAD(X, 2, n), Xj = SPREAD(X, 1, n). With these transformations, parallel expressions of all three spatial autocorrelation coefficients are straightforward, e.g., in FORTRAN90,

```
integer n
real W(n, n), X(n), Xi(n, n), Xj(n, n)
real S0, xmean, BB, BW, Geary, Moran

S0 = sum(W)
xmean = sum(X)/n

Xi = spread(X, 2, n)
Xj = spread(X, 1, n)

BB = sum(W*Xi*Xj)/2
BW = sum(W((Xi - Xj)**2))/2

Geary = ((n-1)/(2*S0))*(sum(w*((xi-xj)**2)))/(sum((x-xmean)**2))
Moran = (n/S0)*sum(w*(xi-xmean)*(xj-xmean))/(sum((x-
xmean)**2))
```

## 3. Connectivity matrix

Matrix **W** is a row standardized form of the connectivity matrix. Each element in **W** is calculated by dividing the corresponding entry with the respective row sum in the original weights matrix. Clearly, each row of **W** sums to unity.

Given locations of the sampled data, it normally takes 4 steps to generate **W**:

- 1) partitioning the surface into Voronoi Diagrams or Thiessen Polygons;
- 2) constructing the polygon topology;
- 3) generating the connectivity matrix;
- 4) standardizing the connectivity matrix.

Steps one and two are computationally complex, but fortunately many GIS packages provide functions for these tasks. For example, ARC/INFO and GRASS generate the Thiessen Polygons with commands THIESEN and *s.voronoi*, respectively. [ESRI, 1991; McCauley, 1993] Parallel algorithms for constructing the Thiessen Polygons for a point coverage are also available but too complex to include here. Interested readers are referred to Fang and Piegl [1993], ElGindy [1990], and Puppo *et al.* [1994] for detailed discussions.

Since polygon topology is essential in GIS operations and readily available as part of the database, we focus on how to extract the connectivity matrix and perform the standardization using a data parallel approach. Li [1993] found that the left-right topology, normally stored in a GIS, can accommodate data parallel construction of the connectivity matrix for a polygon coverage. Given vector LEFT and RIGHT, where LEFT(i) and RIGHT(i) are neighboring polygons, vectors LEFT and RIGHT then can serve as the parallel indices for the two dimensional connectivity matrix. The following FORALL construct expresses the relations between the connectivity matrix and vectors LEFT and RIGHT:

*FORALL (i=1:N) C(LEFT(i), RIGHT(i)) = 1*
where C is an N-by-N matrix, with 1 indicating connectivity, and 0 otherwise. Due to the symmetrical nature of C, a complete form can be constructed by combining its transpose, i.e.,

*C= TRANSPOSE(C) + C*
where TRANSPOSE is an intrinsic function in FORTRAN90. Finally, W can be constructed with three steps,
- 1) row-wise reduction on C to calculate the row-sum;
- 2) spreading entries in the row-sum vector to their corresponding entries in C;
- 3) dividing each entry in C by its corresponding row-sum.
  expressed in FORTRAN90:

*row-sum = SUM (C( 2, n))*

*W = W / SPREAD(row-sum, 2, n)*


The data parallel algorithms described above are highly efficient. With parallel prefix scan, the complexity of matrix summation is reduced from $O(n^2)$ to $O(log(n))$. Further, with parallel elemental operations, calculations of cross products become a one-step operation, regardless of the size of the matrix. In both cases, maximal parallelism is achieved.

So far, we have emphasized exploiting data parallelism. Of note is that task parallelism also exists in computations of spatial autocorrelation coefficients. For example, Rokos and Armstrong [1993] developed an MIMD algorithm for Moran's I. As a general strategy, however, we should explore data parallelism first since it has the potential to achieve the greatest speedup.

## B.  ESTIMATING SPATIAL AUTOREGRESSIVE MODELS
## 1.  Definitions

Measuring and testing for spatial autocorrelations could reveal spatial pattern in the data but are not capable of modeling a spatial process.      Modeling    is    facilitated    with    spatial    autoregressive specifications.  There are four major models, namely, the simultaneous autoregressive model (SAR), the conditional autoregressive model (CAR), the moving average model (MA), and the autoregressive response model (AR).  Since these models share similar computational procedures, the following section describes only the SAR model.

Given an *n-by-1* vector of observations *y*, and the *n-by-n* row standardized connectivity matrix *W*, the variable mean SAR model has the form:

$$y = x\beta + \rho W(y - x\beta) + \xi,$$

or

$$y = x\beta + (I - \rho W)^{-1}\xi,$$

where *r* is the spatial autoregressive coefficient, **x** is the *n-by-k* matrix of observations on the explanatory variables, **xb** is the variable mean, *b* is a *k-by-1* vector of regression coefficients, and $x \sim MVN(0, s^2I)$ is an *n-by-1* vector of the error terms.  Under this condition, a classical regression model, $y = X\beta + \xi$, and OLS are no longer appropriate for the inferential analysis of the variable mean, **xb**.  One strategy is to use maximum likelihood procedures.  The log-likelihood based upon the normal probability density function for the SAR model is:

$$\ln(L) = \ln|I - \rho W| - \frac{n\ln(2\pi)}{2} - \frac{n\ln(\sigma^2)}{2} - $$
$$\frac{1}{2\sigma^2}(y - \rho Wy - X\beta + \rho WX\beta)^T (y - \rho Wy - X\beta + \rho WX\beta),$$

with estimators

$$\hat{\beta} = \left[(X - \hat{\rho}WX)^T (X - \hat{\rho}WX)\right]^{-1}(X - \hat{\rho}WX)^T (y - \hat{\rho}Wy),$$

$$\hat{\sigma}^2 = \frac{1}{n}(\mathbf{y} - \hat{\rho}\mathbf{W}\mathbf{y} - \mathbf{X}\hat{\beta} + \hat{\rho}\mathbf{W}\mathbf{X}\hat{\beta})^T (\mathbf{y} - \hat{\rho}\mathbf{W}\mathbf{y} - \mathbf{X}\hat{\beta} + \hat{\rho}\mathbf{W}\mathbf{X}\hat{\beta}),$$

which are functions of the $\hat{\rho}$ that minimizes the concentrated log-likelihood function:

$$L_C = -\frac{n}{2}\ln\left(\frac{\mathbf{e}^T\mathbf{e}}{n}\right) + \ln|\mathbf{I} - \hat{\rho}\mathbf{W}|,$$

where

$$\mathbf{e}^T\mathbf{e} = (\mathbf{y} - \hat{\rho}\mathbf{W}\mathbf{y})^T (\mathbf{y} - \hat{\rho}\mathbf{W}\mathbf{y}) - (\mathbf{y} - \hat{\rho}\mathbf{W}\mathbf{y})^T (\mathbf{X} - \hat{\rho}\mathbf{W}\mathbf{X}) - \left[(\mathbf{X} - \hat{\rho}\mathbf{W}\mathbf{X})^T (\mathbf{X} - \hat{\rho}\mathbf{W}\mathbf{X})\right]^{-1} (\mathbf{X} - \hat{\rho}\mathbf{W}\mathbf{X})^T (\mathbf{y} - \hat{\rho}\mathbf{W}\mathbf{y}),$$

and the estimate $\hat{\rho}$ optimizes:

$$MIN : \left[\prod_{i=1}^{n}(1 - \hat{\rho}\lambda_i)\right]^{-\frac{2}{n}} (\mathbf{y} - \mathbf{x}\beta)(\mathbf{I} - \hat{\rho}\mathbf{W})^T (\mathbf{I} - \hat{\rho}\mathbf{W})(\mathbf{y} - \mathbf{x}\beta) .$$

[Anselin and Hudak, 1992; Griffith, 1988a]

## 2. Computational characteristics

The computational procedures for estimating the spatial autoregressive models are well-understood. Cliff and Ord [1981] provided the original methods. Griffith [1988b, 1989, 1990a], and Anselin and Hudak [1992] have implemented the models with either FORTRAN77 or standard statistical packages, as well as a vector language. In the following two sections we describe the major quantities required in estimating an SAR model and how to implement them for data parallel processing.

Following the description presented in the previous section, estimating a constant mean SAR model (**xb** reduces to **m1**) with the maximal likelihood approach involves three terms, m, $s^2$, and r. The key is to find the value for r that maximizes the probability density function. After a logarithmic transformation, the log-likelihood function takes the following form:

$$-\frac{2\sum\limits_{i=1}^{n}\ln(1-\hat\rho\lambda_i)}{n}+\ln\left(\begin{array}{c}\sum\limits_{i=1}^{n}(y_i)^2-2\hat\mu\sum\limits_{i=1}^{n}y_i+n\hat\mu^2\\[2mm]-2\hat\rho\left(\sum\limits_{i=1}^{n}y_i\sum\limits_{j=1}^{n}w_{ij}y_j-\hat\mu\sum\limits_{i=1}^{n}y_i-\hat\mu\sum\limits_{i=1}^{n}\sum\limits_{j=1}^{n}w_{ij}y_j+n\hat\mu^2\right)\\[2mm]+\hat\rho^2\left(\sum\limits_{i=1}^{n}\sum\limits_{j=1}^{n}(w_{ij}y_j)^2-2\hat\mu\sum\limits_{i=1}^{n}\sum\limits_{j=1}^{n}w_{ij}y_j+n\hat\mu^2\right)\end{array}\right)$$

where

$$\hat\mu=\frac{\sum\limits_{i=1}^{n}y_i-\hat\rho\sum\limits_{i=1}^{n}\sum\limits_{j=1}^{n}w_{ij}y_j}{n(1-\hat\rho)}$$

Finding the value of r that maximizes the log-likelihood function requires three major groups of operations:

- (1) Calculating the eigenvalues of matrix **W**, i.e., **l**;
- (2) Calculating the spatially lagged variables and other quantities in the log-likelihood function:

$$\sum_{i=1}^{n}\ln(1-\rho\lambda_i)$$

$$\sum_{i=1}^{n}y_i$$

$$\sum_{i=1}^{n}y_i^2$$

$$\sum_{i=1}^{n}\sum_{j=1}^{n}w_{ij}y_j$$

$$\sum_{i=1}^{n}y_i\left(\sum_{j=1}^{n}w_{ij}y_j\right)$$

$$\sum_{i=1}^{n}\left(\sum_{j=1}^{n}w_{ij}y_j\right)^2$$

- (3) Nonlinear optimization (derivative optimization or direct search).

Sone and Griffith developed a FORTRAN77 program for the above procedures. Their program replaces matrix W with the symmetrical C and utilizes a direct search approach to nonlinear optimization. The appendices include a more readable version of the code. The following section describes strategies to translate the sequential procedure into a data parallel algorithm.

## 3. The eigenvalues of matrix W

Calculating the eigenvalues of matrix **W** is a numerically intensive operation. There are three steps involved, transforming **W** into a symmetrically equivalent form, reducing the symmetrical version of **W** to the symmetric tridiagonal form, and then calculating the eigenvalues (normally with the so called QL algorithm and other simplified methods). Sequential algorithms for these tasks are well-developed. The symmetrical transformation is based on **W\*** = **DCD**, where **D** is a diagonal matrix with elements composed of the inverse of the square root of the row sums of matrix **C** [Ord, 1975], i.e.,

$$D = diag \left[ ( \_C_{ij} )^{-1/2} \right].$$

**W\*** then is obtained from two matrix multiplications. Further reduction of **W\*** to a symmetric tridiagonal form is generally accomplished by the Householder Transformation described in Press, *et al.* [1988]. In sequential processing, matrix multiplication and the Householder Transformation are both $O(n^3)$, constituting the major computational bottlenecks in using spatial autoregressive models. Figure 4 depicts the CPU time spent on calculating the eigenvalues of a symmetric matrix in a range of sizes. The cubic form of the curve is clear. Table 2 highlights the time profile for the estimation of the SAR model. It shows that reducing a symmetric matrix to a symmetric tridiagonal form took more than 95% of total CPU time, and this proportion simply increases with the size of the matrix. For instance, for a 2500-by-2500 matrix, calculating the eigenvalues occupies 99% of total CPU time spent on estimating the SAR model.

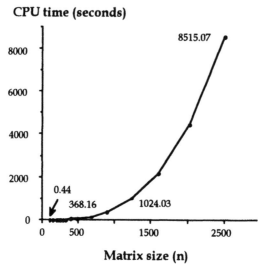

Matrix size (n)

**Figure 4.** Sequential calculation of eigenvalues for a range of problem sizes

**Table 2.** Proportions of execution time required for estimating an SAR model. Note that **tred2** is the subroutine for transforming a symmetric matrix to a symmetric tridiagonal matrix, and **tql1** is the subroutine that calculates the eigenvalues with the QL algorithm. GPROF generated the program profile.

$n = 400$

| Percent | Name |
|---|---|
| 95.9 | tred2 |
| 1.3 | tql1 |
| 3.8 | other routines |

$n = 2500$

| Percent | Name |
|---|---|
| 99.2 | tred2 |
| 0.2 | tql1 |
| 0.6 | other routines |

    To overcome this bottleneck, a logical approach is to find analytical solutions for the eigenvalues. Cliff and Ord [1981] show that for regular lattices eigenvalues of a binary connectivity matrix can be reduced to a simple summation of trigonometric functions. Observing the converging behavior of eigenvalues, Griffith [1990b] further simplified the calculations with approximation methods and extended his findings to irregular surface partitioning [Griffith and Sone, 1993]. In contrast to lattice partitioning, however, irregular partitioning of space is more difficult to generalize. Griffith has found approximations for selected geographic landscapes but has yet to uncover general approximations.

Matrix multiplication and eigenvalue calculation are major operations in linear algebra and statistics and have attracted much effort from researchers in parallel computing to speed up the operations. They also are the primary functions to be included in the mathematics libraries on parallel computers. For example, the Connection Machine Scientific Software Library (CMSSL) developed by Thinking Machine Corporation for its parallel computers includes comprehensive functions for matrix manipulation and eigenvalue analysis [1993b]. As we show later in this chapter, parallel calculation of eigenvalues achieves dramatic speedup over sequential processing, allowing timely solutions for spatial statistical analysis.

## 4.  Spatially lagged variables

These are matrix summations, with step complexity $O(n^2)$. They are perfectly suited for data parallel processing, as shown in the previous section on computing spatial autocorrelation statistics. With a strategy developed in the previous section, we first generate an n-by-n matrix $q$ by spreading vector y along dimension 1, i.e., in FORTRAN90,

$$q = spread(y, 1, n)$$

| y | | q | | | |
|---|---|---|---|---|---|
| 1 | | 1 | 2 | 3 | 4 |
| 2 | | 1 | 2 | 3 | 4 |
| 3 | | 1 | 2 | 3 | 4 |
| 4 | | 1 | 2 | 3 | 4 |

With matrix $q$, we obtain a new *n*-by-*1* vector *tempwy*:

$$(tempwy)_i = \sum_{j=1}^{n} w_{ij} q_{ij} .$$

With *tempwy*, all the terms are easily expressed:

$$\sum_{i=1}^{n} \ln(1 - \rho \lambda_i) \qquad \Rightarrow \quad \text{sum(dlog(1.0d0 - rho*eigenn)}$$

$$\sum_{i=1}^{n} y_i \qquad \Rightarrow \quad \text{sum(y)}$$

$$\sum_{i=1}^{n} y_i^2 \qquad \Rightarrow \quad \text{sum(y**2)}$$

$$\sum_{i=1}^{n} \sum_{j=1}^{n} w_{ij} y_j \qquad \Rightarrow \quad \text{sum(tempwy)}$$

$$\sum_{i=1}^{n} y_i \left( \sum_{j=1}^{n} w_{ij} y_j \right) \qquad \Rightarrow \quad \text{sum(y*tempwy)}$$

$$\sum_{i=1}^{n} \left( \sum_{j=1}^{n} w_{ij} y_j \right)^2 \qquad \Rightarrow \quad \text{sum(tempwy**2)}$$

## 5.  Nonlinear optimization

There are several approaches to nonlinear optimization. [Anselin and Hudak, 1992; Griffith, 1989; Griffith, *et al.*, 1990c]  We adopt the well-known Golden Section search algorithm here.  This direct search approach involves simple but iterative scalar operations. [Press, *et al.*, 1988]  It is possible to parallelize the procedure with the task parallel approach.  However, since this procedure consumes a small fraction of the execution time, it can be left as a sequential process without introducing a communication penalty that probably exceeds the speedup in CPU time.

## 6.  Data parallel implementation

The FORTRAN 77 code for the SAR model constitutes a function **fwn** and the following subroutines (Figure 5 and Appendix):

- **getdata**—open file and read data;

- **wfirst**—construct W matrix

- **tred2**—Householder transformation

- **tql1**—QL algorithms for eigenvalues

- **wsecond**—construct W matrix

- **staterms**—calculate terms for the log-likelihood function

• **sumeigenn**—calculate the sum of and the sum of squared eigenvalues

• **golden**—golden section search

With the strategy discussed above, porting the sequential codes to the data parallel FORTRAN90 is straightforward. Subroutine **tred2** and **tql1** are replaced with a data parallel subroutine **sym_tred_gen_eigensystem**, a combination of **sym_tred** and **tridiagonal_eigensystem**, included in CMSSL. Figure 7 compares the FORTRAN90 implementation of key subroutines with their sequential counterparts. Except for new operators for array processing, other parts of the program remain the same. The FORTRAN90 code is clearly more readable than is the sequential code on account of the elimination of the double DO loops.

The *sar.f* simulates a linear geographic landscape. The spatial structure is reflected in the **W** matrix where only the sub-diagonal elements have non-zero values. The **W** matrix is constructed with the parallel selection operator *where* (Figure 7).

## V. PERFORMANCE

This section presents comparisons of processing times on a sequential and a data parallel platform, to show the potentials of parallel processing in spatial statistical analysis. An IBM RS6000 Model 550 has been used to generate the sequential performance data. It has 64 MB of RAM and a 41 Mhz Clock, with a peak performance of 23.0 MFLOPS and 54.3 SPECmarks. The AIX operating system provides a number of system commands for performance measurement. [IBM, 1992]

Performance data for data parallel processing have been produced with a Connection Machine Model 5. The CM5 has 32 processors, each has 32 MB of local memory and four vector units. It supports a number of data parallel languages including FORTRAN90, C*, and *LISP, as well as the comprehensive mathematics library CMSSL. The CM5 operating system provides CM_timers, a set of utilities for performance measurement.

```
        program SAR

        implicit double precision (a-h,o-z)

        real*8 y(400),e(400),z(400,400 ),w(400,400),eigenn(400)
        real*8 mu
        common /first/  sumy,sumy2,sumwy,sumywy,sumywwy,
       +        /second/ sumeign, sumesqn
       +        /third/ m
       +        /fourth/ w
       +        /fifth/ y
       +        /sixth / eigenn
       +        /seventh/ n
       +        /eighth/ niter

        open(unit=2,file='sar.out', status='new')
        print *, 'input an icount value = sqrt(n)'
        read *, icount
        n = icount**2
        m = dble(n)
        call getdata()
        call wfirst()
        call tred2(n, n, w, eigenn, e, z)
        call tql1(n, eigenn, e, ierr)
        call wsecond()
        call stateoms()
        call sumeigenn()
c
c Estimation of SAR model: (ax < bx < cx)
c        bx: Starting value
c        cx: Upper bound of rho
c        xtol: convergence criterion
c        xfgld: Minimized LLF
c        hrho: Rho that Minimizes LLF
c        itmax: No. of iterations
c
        xtol=1.0d-15
        xax= 1.0d0/eigenn(1)
        xcx= 1.0d0/eigenn(n)
        xbx= 0.9d0
        niter = 0
        call golden(xax,xbx,xcx,xfgld,xtol,hrho)
        mu = (sumy - hrho*sumwy)/(m-m*hrho)
        stop
        end
```

**Figure 5.** The main program in FORTRAN77 to estimate an SAR model with a golden section search. Based on the program developed by Sone.

```
        program SAR

        include '/usr/include/cm/CMF_def.h'
        include '/usr/include/cm/cmssl_cmf.h'

        implicit double precision (a-h,o-z)

        real*8 y(400),d(1,400),q(400,400),w(400,400),eigenn(400), mu
        common /first/ sumy,sumy2,sumwy,sumywy,sumywwy,
      +        /second/ sumeign,sumesqn
      +        /third/ rn /fourth/ w /fifth/ y
      +        /sixth/ eigenn /seventh/ n
      +        /eighth/ niter /nineth/ q

        open(unit=2,file='sar.out', status='new')
        print *, 'input an icount value = sqrt(n)'
        read *, icount

        n = icount**2
        rn = dble(n)
        call getdata()
        call wfirst()
        nblock = 2
        tol = 1.0d-15
        eigen_flag = 0
        group = 1.0d-4
        call sym_tred_eigensystem(d, q, w, n, 1, 2, nblock,
      +                           eigen_flag, tol, group, ier)
         eigenn = d(1,:)

        call wsecond()
        call staterms()
        call sumeigenn()
c
c Estimation of SAR model: (ax < bx < cx)
c
        xtol=1.0d-15
        xax= 1.0d0/eigenn(1)
        xcx= 1.0d0/eigenn(n)
        xbx= 0.9d0
        niter = 0
        call golden(xax,xbx,xcx,xfgld,xtol,hrho)
        mu = (sumy - hrho*sumwy)/(rn-rn*hrho)
        stop
        end
```

**Figure 6.** The main program in CM FORTRAN to estimate an SAR model with a golden section search. Note that array **z** and **e** in Figure 5 are replaced by **d** and **q**.

**Fortran77**

```
      subroutine staterms()
c
c  Calculate terms for the
c  log likelihood function
c
      implicit double precision (a-h, o-z)
      real*8 y(400), w(400,400)
      common /first/sumy,sumy2,sumwy,sumywy,
     +         sumywwy
     +         /fourth/ w /fifth/y /seventh/n

      sumy    = 0.0d0
      sumy2   = 0.0d0
      sumwy   = 0.0d0
      sumywy  = 0.0d0
      sumywwy = 0.0d0
      do 35 i=1,n
         sumy   = sumy  + y(i)
         sumy2  = sumy2 + y(i)**2
         tempwy = 0.0d0
         do 33 j=1,n
            tempwy = tempwy + w(i,j)*y(j)
33       continue
         sumwy   = sumwy + tempwy
         sumywy  = sumywy + y(i)*tempwy
         sumywwy = sumywwy + tempwy**2
35    continue

      return
      end

      subroutine wfirst()
c
c  Stochastic matrix w for eigenvalues:
c  linear case
c
      implicit double precision (a-h, o-z)
      real*8 w(400,400)
      common /fourth/ w /seventh/ n

      do 7020 i=1,n
      do 7016 j=1,n
      w(i,j) = 0.0d0
7016  continue
      do 7017 j=1,n
      if(j.eq.i+1.and.j.le.n) then
          w(i,j) = 0.5d0
      end if
      if(j.eq.i-1.and.j.gt.0) then
          w(i,j) = 0.5d0
      end if
7017  continue
7020  continue
      w(1,2) = 1.0d0/dsqrt(2.0d0)
      w(2,1) = 1.0d0/dsqrt(2.0d0)
      w(n,n-1) = 1.0d0/dsqrt(2.0d0)
      w(n-1,n) = 1.0d0/dsqrt(2.0d0)

      return
      end
```

**CM Fortran**

```
      subroutine staterms()
c
c  Calculate terms for the
c  log likelihood function
c
      implicit double precision (a-h, o-z)
      real*8 y(400), w(400,400), q(400,400)
      real*8 tempwy(400)
      common /first/sumy,sumy2,sumwy, sumywy
     +         sumywwy
     +         /fourth/ w /fifth/y /seventh/n
     +         /nineth/ q

      sumy    = 0.0d0
      sumy2   = 0.0d0
      sumwy   = 0.0d0
      sumywy  = 0.0d0
      sumywwy = 0.0d0
      tempwy = 0.0d0
      q = spread(y,1,n)
      sumy = sum (y)
      sumy2 = sum(y*y)

      tempwy = sum(w*q,2)
      sumwy = sum(tempwy)
      sumywy = sum(y*tempwy)
      sumywwy = sum(tempwy*tempwy)

      return
      end

      subroutine wfirst()
c
c  Stochastic matrix w for eigenvalues:
c  linear case
c
      implicit double precision (a-h, o-z)
      real*8 w(400,400)
      integer row(400,400), col(400,400)
      common /fourth/ w /seventh/ n

      forall (i=1:n, j=1:n) row(i,j) = i
      forall (i=1:n, j=1:n) col(i,j) = j

      w = 0.0d0
      where((col.eq.row+1.and.col.le.n) .or.
     +      (col.eq.row-1 .and. col.gt.0))
          w = 0.5d0
      end where

      w(1,2) = 1.0d0/dsqrt(2.0d0)
      w(2,1) = 1.0d0/dsqrt(2.0d0)
      w(n,n-1) = 1.0d0/dsqrt(2.0d0)
      w(n-1,n) = 1.0d0/dsqrt(2.0d0)

      return
      end
```

**Figure 7.** A comparison of FORTRAN77 and CM FORTRAN implementations.

## A. ANALYSIS OF SPATIAL AUTOCORRELATION

The Moran's I is selected for performance comparison. The experiment used a simple control structure to simulate the geographic landscape. The number of areal units increases in the $2^n$ sequence, with n = 4, 5, ..., 13, forming a set of **W** matrices with sizes ranging from 32-by-32 to 4096-by-4096. The random number generator on the CM5 produced attribute data for the experiment. Table 3 summarizes the results.

Table 3 and the associated graphic plots (Figure 8) reveal information useful for evaluating the parallel approach. First, the parallel computer clearly delivers extremely high speed. The CM5 calculated the Moran's I with $n = 4096$ in about one sixtieth of a second, 270 times faster than the RS6000. Second, there are no significant performance gains when problem size is small. For $n$ less than 64, the performance ratio RS6000/CM5 is less than 10. Third, the performance ratio RS6000/CM5 increases with the matrix size, indicating that speedup is greater with a larger problem size. As Figure 8 depicts, the relationship is logarithmic rather than linear. The ratio increases quickly from $n = 128$ to $n = 512$ but slows down beyond that point. This behavior might be related to the grain size, the ratio between total memory requirement and the physical memory of the parallel computer. [Fox, 1992]

**Table 3.** Performance comparisons with Moran's I on the RS6000 and the CM5.

| # Area units (n) | RS6000 (seconds) | CM5 (seconds) | RS6000/CM5 |
|---|---|---|---|
| 32 | 0.01 | 0.001 | 10.00 |
| 64 | 0.01 | 0.001 | 10.00 |
| 128 | 0.02 | 0.001 | 20.00 |
| 256 | 0.09 | 0.001 | 90.00 |
| 512 | 0.38 | 0.002 | 190.00 |
| 1024 | 1.52 | 0.007 | 217.14 |
| 2048 | 5.93 | 0.027 | 219.63 |
| 4096 | 30.5 | 0.114 | 267.54 |

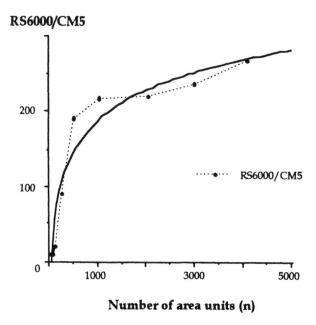

**Figure 8.** Performance comparisons with Moran's I on the RS6000 and the CM5. The curve that fits the performance data is a logarithmic function

## B.  ESTIMATING SPATIAL AUTOREGRESSIVE MODELS

We measured the processing time for sar.f and sar.fcm on the RS6000 and the CM5 respectively. Table 4 compares the total execution time for the SAR model on the RS6000 and the CM5. Two sample sizes are selected, $n = 400$ and $n = 2500$. The CM5 is about ten times and 178 times faster than the RS6000.

**Table 4.** Performance comparisons between sequential and parallel estimation of an SAR model

| Matrix size | RS6000 | CM5 | Ratio |
|---|---|---|---|
| 400 | 38.19 | 3.6 | 10.6 |
| 2500 | 8566.50 | 48 | 178.47 |

Further comparisons of performance for the spatial autoregressive model focus on the calculation of eigenvalues for a range of problem sizes, for a number of reasons. First, as Table 4 shows, calculating the eigenvalues takes up more than 95 percent of the total execution time for estimating the SAR model. Second, performance improvements for the calculations of those terms in the log-likelihood functions are easy to predict. The reduction operation is $O(log(n))$. Data parallel processing should achieve dramatic speedup with these operations, particularly with large problem sizes. But for eigenvalues, the step complexity of the data parallel algorithms adapted in CMSSL is not yet available from the software developer.

Eigenvalues are computed for matrices with sizes ranging from $n =$ 100, 200, ..., 2500, to provide further insight into the performance by varying problem size. Figure 9 and Table 5 summarize the performance data. Several tendencies concerning performance are conspicuous here. First, the relations between matrix size and speedup are linear. The larger the matrix size, the higher the speedup ratio. Second, the processing speed achieved with data parallel processing is sufficiently high for interactive analysis. It took the CM5 only 46 seconds to find all the eigenvalues for a 2500-by-2500 matrix, 186 times faster than the RS6000. However, there is no advantage to using parallel processing for small matrices. For example, for matrices smaller than 100-by-100, the RS6000 calculates the eigenvalues faster than the CM5.

Performance measure has much to do with whether or not a compiler takes full advantage of the hardware. A comparison between the above performance data with those reported by Griffith and Sone [1993] further reveals such potentials. Using the ESSL, the Engineering and Scientific Subroutine Library developed by IBM for the mainframe and workstations, Griffith and Sone were able to achieve 80% of the peak performance of the RS6000 Model 330 and the IBM 3090 VF. The ESSL is coded with macro-languages and optimized for specific hardware produced by IBM. In contrast, the FORTRAN program on the RS6000 Model 550 reached less than 2 MFLOPS. On the CM5, a near TelaFLOP parallel computer, the eigenvalue routines performed at only 230 MFLOPS, a fraction of the peak performance.

## VI. SUMMARY

This chapter demonstrates that the data parallel model can efficiently implement all the operations in spatial statistics defined in two primary areas: detecting spatial patterns with spatial autocorrelation coefficients and modeling spatial processes with spatial autoregressive specifications. In fact, data parallel languages are more natural than sequential languages in expressing operations in spatial statistics. Data parallel operations treat spatial variables as entities and not as a sequence of array elements. With array sectioning, the FORALL and the WHERE constructs set the context for parallel operations, allowing nested DO and IF loops mostly to be eliminated.

RS6000/CM5

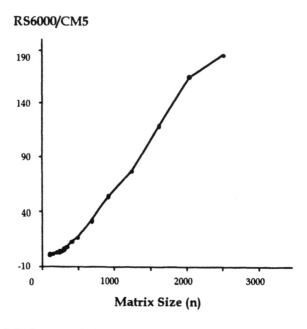

**Matrix Size (n)**

**Figure 9.** Performance ratios between sequential and parallel computations of eigenvalues

**Table 5.** Performance comparisons between sequential and parallel computations of eigenvalues

| Matrix size $(n^{0.5})$ | RS6000 | CM5 | RS6000/CM5 |
|---|---|---|---|
| | | | |
| 10 | 0.44 | 0.53 | 0.82 |
| 12 | 1.38 | 0.77 | 1.78 |
| 14 | 3.15 | 1.23 | 2.57 |
| 15 | 4.70 | 1.38 | 3.42 |
| 16 | 6.89 | 1.53 | 4.51 |
| 17 | 9.64 | 1.71 | 5.63 |
| 18 | 13.76 | 1.86 | 7.42 |
| 20 | 27.62 | 2.19 | 12.59 |
| 22 | 55.71 | 3.51 | 15.86 |
| 26 | 153.10 | 4.81 | 31.82 |
| 30 | 368.16 | 6.82 | 53.97 |
| 35 | 1024.03 | 13.3 | 76.95 |
| 40 | 2166.12 | 18.02 | 120.19 |
| 45 | 4459.35 | 27.00 | 165.16 |
| 50 | 8515.07 | 45.78 | 185.99 |

Porting existing sequential codes to a data parallel language is straightforward. Software overhead is minimal. All sequential expressions have parallel counterparts that are much shorter and clearer. For more complicated algorithms, mathematical subroutines are available. Software vendors for parallel computers gradually are providing subroutines and functions for standard operations in matrix analysis and linear algebra. For example, Thinking Machine Corporation has CMSSL for its Connection Machines, and Digital Equipment Corporation offers the Massively Parallel Math Library (MPML) for its SIMD computer DECmpp. Since professional software engineers have often coded the best-known parallel algorithms, general users, including geographers who appreciate spatial statistics, can use these libraries to solve their computational problems with guaranteed high performance. For software vendors in spatial data analysis, these libraries should make it technically easy to develop packages for spatial statistics on parallel computers.

Data parallel processing has the potential to enhance geographers' capability to analyze spatial data. First, researchers now are able to apply spatial statistical techniques to large size problems. The short response time achieved by data parallel processing will enable researchers to explore alternative sampling designs and model specifications. The rapid processing speed also opens up possibilities for integrating simulation and visualization with spatial statistics.

Second, data parallel software models better facilitate integration of spatial statistics and GIS than does the current approach (Goodchild, 1990). Presently, to conduct spatial statistical analysis, a user first must export topological and attribute data from a GIS package to an external file with formats specific to the statistical package. Then the user must run a sequence of macros in the statistics program to perform the analysis. Finally, the results must be exported to a file, reformatted with another user-developed program so that they can be read back to the GIS program. This tedious process will be greatly simplified with a data parallel language. The reason is simple: a data parallel language, such as CM FORTRAN, not only has the syntax simplicity of macro languages in standard statistical packages, but also the full-capability of low level programming languages such as C and FORTRAN77. The latter will allow a direct extension of a GIS package to include spatial statistical functions without the complication of a software interface and user intervention.

# REFERENCES

**Adams, J., *et al.** Fortran90 Handbook, Complete ANSI/ISO Reference*, McGraw-Hill Book Company, New York, 1992.

**Anselin, L.** *Spatial Econometrics: Methods and Models*, Kluwer Academic Publisher, Dordrecht, The Netherlands, 1988.

**Anselin, L., and Hudak S.** Spatial econometrics in practice: a review of software options, *Regional Science and Urban Economics*, Vol. 22, 509-536, 1992.

**Blelloch, G.** *Vector Models for Data-Parallel Computing*, MIT Press, Cambridge, MA, 1991.

**Cliff, A. D., and Ord, J. K.** *Spatial Processes, Models & Applications*, Pion, London, 1981.

**ElGindy, H.** Optimal parallel algorithms for updating planar triangulations, *Proceedings of the Fourth International Symposiums on Spatial Data Handling*, Zurich, Switzerland, 200-208, 1990.

**ESRI.** *ARC Command References*, Redlands, CA, 1991.

**Fang, T. P., and Piegl, L.** Delaunay Triangulation Using a Uniform Grid, *IEEE Computer Graphics and Applications*, Vol. 13, 36-47, 1993.

**Flynn, M. J.** Very high-speed computing systems, *Proceedings of the IEEE*, Vol. 54, 1901-1909, 1966.

**Fox, G.** *Parallel Computers and Complex Systems*, SCCS-370, Syracuse University, Syracuse, NY, 1992.

**Griffith, D.** *Spatial Autocorrelation, A Primer*, Resource Publications in Geography, American Association of American Geographers, Washington, DC, 1987.

**Griffith, D.** *Advanced Spatial Statistics, Advanced Studies in Theoretical and Applied Econometrics*, Kluwer Academic Publishers, Dordrecht, The Netherlands, 1988a.

**Griffith, D.** Estimating spatial autoregressive model parameters with commercial statistical packages, *Geographical Analysis*, Vol. 20, 176-186, 1988b.

**Griffith, D.** *Spatial Regression Analysis on the PC: Spatial Statistics Using Minitab*, Discussion Paper No. 1. Institute of Mathematical Geography, Ann Arbor, 1989.

**Griffith, D.** Supercomputer and spatial statistics: a reconnaissance, *Professional Geographer*, Vol. 42, 481-492, 1990a.

**Griffith, D.** A numerical simplification for estimating parameters of spatial autoregressive models, in *Spatial Statistics, Past, Present, and Future*, edited by D. Griffith, Institute of Mathematical Geographers, Ann Arbor, MI, 183-197 1990b.

**Griffith, D., *et al.** Developing minitab software for spatial statistical analysis: a tool for education and research," *Operational Geographer*, Vol. 8, No. 3., 28-34, 1990c.

**Griffith, D., and C. Amrhein, C.** *Statistical Analysis for Geographers*, Prentice-Hall, Englewood Cliffs, NJ, 1991.

**Griffith, D.** Which spatial statistics techniques should be converted to GIS functions?" in *Geographic Information Systems, Spatial Modelling, and Policy Evaluation*, edited by M. Fisher and P. N. Nijkamp, Springer-Verlag, Berlin, 103-114, 1992a.

**Griffith, D.** Simplifying the normalizing factor in spatial autoregressions for irregular lattices, *Papers in Regional Science*, Vol. 71, 71-86, 1992b.

**Griffith, D., and Sone, A.** *Some Trade-offs Associated with Computational Simplifications for Estimating Spatial Statistical/Econometric Models: Preliminary Results*, Discussion Paper No. 103. Department of Geography, Syracuse University, Syracuse, NY, 1993.

**Haining, R.** *Spatial Data Analysis in the Social and Environmental Sciences*, Cambridge University Press, Cambridge, 1990.

**Haining, R., and Wise, S. M.** *GIS and Spatial Data Analysis: Report on the Sheffield Workshop*, Regional Research Laboratory Initiative Discussion Paper Number 11, University of Sheffield, UK, 1991.

**Hillis, D.** *The Connection Machine*, MIT Press, Cambridge, MA, 1985.

**Hillis, D., and Steele, G. L. Jr.** Data parallel algorithms, *Communications of ACM*, Vol. 29, pp. 1170-1183, 1986.

**Hillis, D.** The connection machine, *Scientific American*, Vol. 256, No. 6, 108-115, 1987.

**IBM.** *Engineering and Scientific Suboutine Library Guide and Reference*, 1992.

**Li, B.** Opportunities and challenges of parallel processing in spatial data analysis: initial experiments with data parallel map analysis, *GIS/LIS '92*, 445-458, 1992a.

**Li, B.** Prospects of parallel processing in geographic data analysis, in *Development and Potentials of GIS: Inside and Outside China*, Hui Lin (eds), Science Press, Beijing, China, 76-89, 1992b.

**Li, B.** Suitability of topological data structures for data parallel operations in computer cartography, *Auto-Carto 11*, 434-443, 1993a.

**Li, B.** Developing network-oriented GIS software for parallel computing, *GIS/LIS '93*, 403-413, 1993b.

**Maspar Computer Corporation.** *Maspar Parallel Application Language (MPL) User Guide*, Sunnyvale, CA, 1991a.

**Maspar Computer Corporation.** *Maspar FORTRAN User Guide*, Sunnyvale, CA, 1991b.

**McCauley, J., and Engel, B.** Spatial statistics and interpolation procedures for GRASS, paper presented to *The 8th Annual GRASS Users' Conference and Exhibition*, March 14-16, 1993, Reston, Virginia, 1993.

**Ord, J. K.** Estimation methods for models of spatial interaction, *Journal of the American Statistical Association*, Vol. 70, 120-126, 1975.

**Press, W.,** *et al.* *Numerical Recipes in C, the Art of Scientific Computing*, Cambridge University Press, Cambridge, 1988.

**Puppo, E., L. Davis, D. DeMenthon, and Y. A. Teng,** Parallel Terrain Triangulation, *Proceedings of the 5th International Symposium on Spatial Data Handling*, Charleston, SC, 632-641, 1992.

**Quinn, M. J.** Data-parallel programming on multicomputers, *IEEE Software*, Sept. 1990, 69-76, 1990.

**Rose, J.R. and Steele, G. L. Jr.** C*: An Extended C Language for Data Parallel Programming, *Technical Report PL 87-5*, Thinking Machines Corp., Cambridge, MA, 1986.

**Sandhu, J. S. and Marble, D.** An investigation into the utility of the Cray X-MP supercomputer for handling spatial data, *Proceedings, Third International Symposium on Spatial Data Handling*, Sydney, Australia, 253-267, 1988.

**Smith, J. R.** *The Design and Analysis of Parallel Algorithms*, Oxford University Press, New York and Oxford, 1993.

**Thinking Machines Corporation.** *The Connection Machine CM-5 Technical Summary*, Cambridge, MA, 1991a.

**Thinking Machines Corporation.** *Programming in FORTRAN*, Cambridge, MA, 1991b.

**Thinking Machines Corporation.** *Programming in C\**, Cambridge, MA, 1991c.

**Thinking Machines Corporation.** *CM FORTRAN Reference Manual,* Cambridge, MA, 1992.

**Thinking Machines Corporation.** *CM FORTRAN Utility Library Reference Manual*, Cambridge, MA, 1993a.

**Thinking Machines Corporation.** *CMSSL for CM Fortran: CM-5 Edition*, Cambridge, MA, 1993b.

**Trew, A., and Wilson, G.** *Past, Present, Parallel: A Survey of Available Parallel Computer Systems*, Springer-Verlag, London, 1992.

**Upton, G., and Fingleton, B.** *Spatial Data Analysis by Example*, John Wiley & Sons, New York, 1989.

## APPENDIX I:  TEST STATISTICS FOR SPATIAL AUTOCORRELATION COEFFICIENTS

To simplify the expressions, the following notions are used:

$$S_0 = \sum_{i=1}^{n} \sum_{j=1}^{n} w_{ij},$$

$$S_1 = \frac{1}{2} \sum_{i=1}^{n} \sum_{j=1}^{n} (w_{ij} + w_{ji})^2,$$

$$S_2 = \sum_{i=1}^{n} (\sum_{j=1}^{n} w_{ij} + \sum_{j=1}^{n} w_{ji})^2,$$

$W$        general (weighted) spatial connectivity matrix;
$n_r$       the number of observed values with attribute $r$;
$n$        total number of observed values;
$X_i$       the ith observed value;

$$k = \frac{m_4}{m_2^2} \text{ and } m_r = \frac{1}{n} \sum_{i=1}^{n} (x_i - \bar{x})^r.$$

### 1. Join-count statistics

$$S_{rr} = \frac{1}{2} \sum_{i=1}^{n} \sum_{j=1}^{n} w_{ij} x_i x_j,$$

$$D_{rs} = \frac{1}{2} \sum_{i=1}^{n} \sum_{j=1}^{n} w_{ij} (x_i + x_j)^2,$$

where
    $S_{rr}$ is the number of joins between contiguous places colored in the same way (with color $r$);
    $D_{rr}$ is the number of joins between contiguous places colored in a different way (with colors $r$ and $s$).

$$E(S_{rr}) = \frac{S_0 n_r^{(2)}}{2n},$$

$$\text{var}(S_{rr}) = \frac{1}{4}\left[\frac{S_1 n_r^{(2)}}{n^{(2)}} + \frac{(S_2 - 2S_1)n_r^{(3)}}{n^{(3)}} + \frac{(S_0^2 + S_1 - S_2)n_r^{(4)}}{n^{(4)}} - 4\{E(S_{rr})\}^2\right],$$

$$E(D_{rs}) = \frac{S_0 n_r n_s}{n^{(2)}},$$

$$\text{var}(D_{rs}) = \frac{1}{4}\left[\frac{2S_1 n_r n_s}{n^{(2)}} + \frac{(S_2 - 2S_1)n_r n_s (n_r + n_s - 2)}{n}\right.$$

$$\left. + \frac{4(S_0^2 + S_1 - S_2)n_r^{(2)} n_s^{(2)}}{n} - 4\{E(D_{rs})\}^2\right].$$

## 2. Moran's I

$$I = \frac{n}{S_0} \times \frac{\sum\limits_{i=1}^{n}\sum\limits_{j=1}^{n} w_{ij}(x_i - \bar{x})(x_j - \bar{x})}{\sum\limits_{i=1}^{n}(x_i - \bar{x})^2},$$

$$E(I) = -\frac{1}{(n-1)},$$

$$\text{var}(I) = \frac{n}{(n-1)^3 S_0^2}\{[n^2 - 3n + 3)S_1 - nS_2 + 3S_0^2]$$

$$- k[(n^2 - n)S_1 - 2nS_2 + 6S_0^2]\} - \frac{1}{(n-1)^2}.$$

## 3. Geary's c

$$c = \frac{n-1}{2S_0} \times \frac{\sum\limits_{i=1}^{n}\sum\limits_{j=1}^{n} w_{ij}(x_i - x_j)^2}{\sum\limits_{i=1}^{n}(x_i - \bar{x})^2},$$

$$E(c) = 1,$$

$$\mathrm{var}(c) = \{(n-1)S_1[n-3n+3-(n-1)k]$$

$$-\frac{1}{4}(n-1)S_2[n+3n-6-(n^2-n+2)k]$$

$$+S_0^2[n^2-3-(n-1)^2k]\}\frac{1}{n(n-2)^2S_0^2}.$$

## APPENDIX II:  SOURCE CODE

FORTRAN77 and CM FORTRAN code for estimating the SAR model is included here.  *sar.f* is a structured version of *dlw3gx.f* written by Griffith and Sone (1993).  *tql1* and *tred 2* from EISPACK replace the ESSL subroutines for eigenvalue calculation.  *sar.fcm* is a data parallel implementation of *sar.f.*  It uses CMSSL routines to calculate the eigenvalues.  The two programs share two sequential subroutines: *golden* and *getdata,* which are appended in the end.

### sar.f

```
c
c
c     FORTRAN77 program to stimate the SAR model using eigenvalue decomposition
c     and Golden Section Search.
c
c     Original code was written by Akio Sone.
c
c
      implicit double precision (a-h,o-z)

      real*8 y(400),e(400),z(400,400),w(400,400),eigenn(400)
      real*8 mu
      common /first/ sumy,sumy2,sumwy,sumywy,sumywwy,
     +        /second/ sumeign,sumesqn /third/ rn
     +        /fourth/ w /fifth/ y /sixth/ eigenn
     +        /seventh/ n /eighth/ niter

      open(unit=2,file='dlw3gx-m.out', status='new')

      print *, 'input an icount value = sqrt(n)'
      read *, icount

      n = icount**2
      rn = dble(n)
c
c Read in data
c
      call getdata()
c
c Generate W matrix for eigenvalue calculation
c
      call wfirst()
c
c Numerical calculation of eigen values with EISPACK library
```

```
c
      call tred2(n, n, w, eigenn, e, z)
      call tql1(n, eigenn, e, ierr)
c
c Generate the stochastic matrix W: linear case
c
      call wsecond()
c
c Calculation of basic statistics: y,yy,wy,ywy,ywwy
c
      call staterms()
c
c Calculation of the sum and squared sum of eigen values
c
      call sumeigenn()
c
c Estimation of SAR model: (ax < bx < cx)
c        bx: Starting value
c        cx: Upper bound of rho
c        xtol: convergence criterion
c        xfgld: Minimized LLF
c        hrho: Rho that Minimizes LLF
c        itmax: No. of iterations
c
      xtol=1.0d-15
      xax= 1.0d0/eigenn(1)
      xcx= 1.0d0/eigenn(n)
      xbx= 0.9d0
      niter = 0
c
c Print out some parameters
c
      write(2,270)xtol,xcx,xax,xbx
270   format('tolerance =      ',d23.16/'upper bound =    ',d23.16/
     c'lower bound =    ',d23.16/'starting value = ',d23.16/)
      write(2,275)
275   format(21x,'rho',20x,'sumj',16x,'log l.f.')
c
c Golden section search
c
      call golden(xax,xbx,xcx,xfgld,xtol,hrho)
c
c Calculate mu
c
      mu = (sumy - hrho*sumwy)/(rn-rn*hrho)
c
c Print out results
c
      write(2,280)niter,hrho,xfgld,mu
280   format(/'iterations = ',i5/'rho    hat = ',d23.16/
     c'min    llf = ', d23.16/'mu     hat = ',d23.16//)
c
      stop
      end


      double precision function fwn(rho)
c
c     Log lilkelihood function for SAR model
c
      implicit double precision (a-h,o-z)
      real*8 mu, eigenn(400)
      common /first/ sumy,sumy2,sumwy,sumywy,sumywwy
     +       /third/ rn
```

```
      +        /sixth/ eigenn
      +        /seventh/ n
      +        /eighth/ niter

       mu = (sumy - rho*sumwy)/(rn-rn*rho)

       sumj=0.0d0
       do 10 i=1,n
 10    sumj = sumj + dlog(1.0d0 - rho*eigenn(i))

       fwn = -2.0d0*sumj/rn+dlog(sumy2 -2.0d0*mu*sumy+rn*mu**2
      c  - 2.0d0*rho*(sumwy - mu*sumy - mu*sumwy+rn*mu**2)
      c  + (rho**2)*(sumywwy-2.0d0*mu*sumwy+rn*mu**2))
      c
       niter = niter + 1
      c
       write(2,15) rho,sumj,fwn
 15    format(3d24.16)

       return
       end


       subroutine wfirst()
      c
      c
      c  Generate W matrix for eigenvalue calculation.
      c
      c
       implicit double precision (a-h, o-z)
       real*8 w(400,400)
       common /fourth/ w /seventh/ n

       do 7020 i=1,n
       do 7016 j=1,n
       w(i,j) = 0.0d0
 7016  continue
       do 7017 j=1,n
       if(j.eq.i+1.and.j.le.n) then
          w(i,j) = 0.5d0
       end if
       if(j.eq.i-1.and.j.gt.0) then
          w(i,j) = 0.5d0
       end if
 7017  continue
 7020  continue
       w(1,2) = 1.0d0/dsqrt(2.0d0)
       w(2,1) = 1.0d0/dsqrt(2.0d0)
       w(n,n-1) = 1.0d0/dsqrt(2.0d0)
       w(n-1,n) = 1.0d0/dsqrt(2.0d0)

       return
       end


       subroutine wsecond()
      c
      c
      c  Generate stochastic matrix W: linear case
      c
      c
       implicit double precision (a-h, o-z)
       real*8 w(400,400)
       common /fourth/ w /seventh/ n
```

```
      do 1020 i=1,n
      do 1016 j=1,n
      w(i,j) = 0.0d0
 1016 continue
      do 1017 j=1,n
      if(j.eq.i+1.and.j.le.n) then
         w(i,j) = 0.5d0
      end if
      if(j.eq.i-1.and.j.gt.0) then
         w(i,j) = 0.5d0
      end if
 1017 continue
 1020 continue
      w(1,2) = 1.0d0
      w(2,1) = 0.5d0
      w(n,n-1) = 1.0d0
      w(n-1,n) = 0.5d0

      return
      end



      subroutine sumeigenn()
c
c
c  Calculate sum(eignn) and sum(eignn**2)
c
c
      implicit double precision (a-h, o-z)
      real*8 eigenn(400)
      common /sixth/ eigenn /second/ sumeign, sumesqn /seventh/ n

      sumeign = 0.0d0
      sumesqn = 0.0d0
      do 10 i=1,n
          sumeign = sumeign + eigenn(i)
         sumesqn = sumesqn + eigenn(i)**2
 10      continue

      write(2,260)
 260  format (11x,'sum eignn',9x,'sumsq eignn',
     c11x,'max eigen',11x,'min eigen')
      write(2,262) sumeign,sumesqn,eigenn(n),eigenn(1)
 262  format (4d20.12/)

      return
      end



      subroutine staterms()
c
c
c  Calculate terms for the log likelihood function
c
c
      implicit double precision (a-h, o-z)
      real*8 y(400), w(400,400)
      common /first/sumy,sumy2,sumwy,sumywy,sumywwy /fifth/y /seventh/n
     +        /fourth/ w

      sumy   = 0.0d0
      sumy2  = 0.0d0
      sumwy  = 0.0d0
```

```
         sumywy  = 0.0d0
         sumywwy = 0.0d0

         do 35 i=1,n
           sumy   = sumy   + y(i)
           sumy2  = sumy2  + y(i)**2
           tempwy = 0.0d0
             do 33 j=1,n
               tempwy = tempwy + w(i,j)*y(j)
33         continue
           sumwy   = sumwy + tempwy
           sumywy  = sumywy + y(i)*tempwy
           sumywwy = sumywwy + tempwy**2
35       continue
c
c Print out some results
c
         write(2,250) n,sumy,sumy2,sumwy,sumywy,sumywwy
250      format('n=',i5/
        c'sum y=    ',d23.16/'sum y2=   ',d23.16/
        c'sum wy=   ',d23.16/'sum ywy=  ',d23.16/
        c'sum ywwy= ',d23.16/)

         return
         end
```

## sar.fcm

```
c
c
c       CM FORTRAN program for estimating the SAR Model using
c       eigenvalue decomposition and Golden section search.
c
c       Ported from "dlw3gx.f", a FORTRAN77 program written by Akio Sone.
c
c
        implicit double precision (a-h,o-z)

        include '/usr/include/cm/CMF_defs.h'
        include '/usr/include/cm/cmssl-cmf.h'

        parameter(mats=400)

        real*8 y(mats), q(mats,mats),d(1,mats),w(mats,mats)
        real*8 eigenn(mats)
        real*8 mu
        common /first/ sumy,sumy2,sumwy,sumywy,sumywwy
       +       /second/ sumeign,sumesqn /third/ rn
       +       /fourth/ w /fifth/ y /sixth/ eigenn
       +       /seventh/ n /eighth/ niter /nineth/ q
CMF$    layout y(:news), w(:news,:news)
CMF$    layout q(:news,:news),eigenn(:news),d(:serial,:news)
c
        print *, 'input an icount value = sqrt(n)'
        read *, icount
c
        open(unit=2,file='dlw3gx.out', status='new')
c
        write(2,20)
```

```
 20     format('program name: dlw3gx.f'/)

        n = icount**2
        rn = dble(n)
c
c  Read data
c
        call getdata()
c
c  Generate W matrix for eigenvalue calculation
c
        call wfirst(n)
c
c  Numerical calculation of eigenvalues with CMSSL routine
c
        nblock = 2
        tol = 1.0d-15
        eigen_flag = 0
        group = 1.0d-4

        call sym_tred_eigensystem(d, q, w, n, 1, 2, nblock,
       +     eigen_flag, tol, group, ier)

        eigenn = d(1,:)
c
c  Calculate the sum and square sume of eigenvalues
c
        call sumeigenn()
c
c  Generate a stochastic matrix W: linear case
c
        call wsecond(n)
c
c  Calculate basic statistics: y, yy, wy, ywy, ywwy
c
        call staterms()
c
c  Estimate the SAR model: (ax < bx < cx)
c
        xtol=1.0d-15
        xax= 1.0d0/eigenn(1)
        xcx= 1.0d0/eigenn(n)
        xbx= 0.9d0
        niter = 0
c
c  Print out some parameters
c
        write(2,270)xtol,xcx,xax,xbx
 270    format('tolerance =        ',d23.16/'upper bound =     ',d23.16/
       +'lower bound =      ',d23.16/'starting value = ',d23.16/)
        write(2,275)
 275    format(21x,'rho',20x,'sumj',16x,'log l.f.')
c
c  Golden section search
c
        call golden(xax,xbx,xcx,xfgld,xtol,hrho)
c
c  Calculate mu
c
        mu = (sumy - hrho*sumwy)/(rn-rn*hrho)
c
c  Print out results
c
        write(2,280)niter,hrho,xfgld,mu
```

```
280   format(/'iterations = ',i5/'rho    hat = ',d23.16/
     +'min    llf = ', d23.16/'mu    hat = ',d23.16//)

      stop
      end


      double precision function fwn(rho)
c
c
c     log lilkelihood function for SAR model
c
c
      parameter (mats = 400)
      implicit double precision (a-h,o-z)
      real*8 mu, eigenn(400)
      common /first/ sumy,sumy2,sumwy,sumywy,sumywwy
     +       /third/ rn
     +       /sixth/ eigenn
     +       /seventh/ n
     +       /eighth/ niter

      mu = (sumy - rho*sumwy)/(rn-rn*rho)

      sumj=0.0d0
        sumj = sum (dlog(1.0d0 - rho*eigenn))

      fwn = -2.0d0*sumj/rn+dlog(sumy2 -2.0d0*mu*sumy+rn*mu**2
     c  - 2.0d0*rho*(sumywy - mu*sumy - mu*sumwy+rn*mu**2)
     c  + (rho**2)*(sumywwy-2.0d0*mu*sumwy+rn*mu**2))
c
      niter = niter + 1
c
      write(2,15) rho,sumj,fwn
  15  format(3d24.16)

      return
      end


      subroutine wfirst(isize)
c
c
c  Generate W matrix for eigenvalue calculation.
c
c
      include '/usr/include/cm/CMF_defs.h'

      integer row(isize, isize), col(isize, isize)
      real*8 w(400,400)
      common /fourth/ w /seventh/ n

      forall (i=1:isize, j=1:isize) row(i,j) = i
      forall (i=1:isize, j=1:isize) col(i,j) = j

      w = 0.0d0
      where((col.eq.row+1.and.col.le.n) .or.
     +    (col.eq.row-1 .and. col.gt.0))
        w = 0.5d0
      end where

      w(1,2) = 1.0d0/dsqrt(2.0d0)
      w(2,1) = 1.0d0/dsqrt(2.0d0)
      w(n,n-1) = 1.0d0/dsqrt(2.0d0)
```

```
      w(n-1,n) = 1.0d0/dsqrt(2.0d0)

      return
      end


      subroutine wsecond(isize)
c
c
c  Generate stochastic matrix W: linear case.
c
c
      include '/usr/include/cm/CMF_defs.h'

      integer row(isize, isize), col(isize, isize)
      real*8 w(400,400)
      common /fourth/ w /seventh/ n

      forall (i=1:isize, j=1:isize) row(i,j) = i
      forall (i=1:isize, j=1:isize) col(i,j) = j

      w=0.0d0
      where((col.eq.row+1.and.col.le.n) .or.
     +    (col.eq.row-1 .and. col.gt. 0))
         w = 0.5d0
      end where

      w(1,2) = 1.0d0
      w(2,1) = 0.5d0
      w(n,n-1) = 1.0d0
      w(n-1,n) = 0.5d0

      return
      end


      subroutine sumeigenn()
c
c
c  Calculate the sum(eigenn) and sum(eigenn**2)
c
c
      implicit double precision (a-h, o-z)
      include '/usr/include/cm/CMF_defs.h'

      real*8 eigenn(400)
      common /sixth/ eigenn /second/ sumeign,sumesqn /seventh/ n

      sumeign = 0.0d0
      sumesqn = 0.0d0
      sumeign = sum (eigenn)
      sumesqn = sum ((eigenn**2))

      write(2,260)
260   format (11x,'sum eignn',9x,'sumsq eignn',
     c11x,'max eigen',11x,'min eigen')
      write(2,262) sumeign,sumesqn,eigenn(n),eigenn(1)
262   format (4d20.12/)

      return
      end


      subroutine staterms()
```

```
c
c
c  Calculate terms for the log likelihood function
c
c
      implicit double precision (a-h, o-z)
      include '/usr/include/cm/CMF_defs.h'

      real*8 y(400),w(400,400),q(400,400)
      real*8 tempwy(400)
      common /first/sumy,sumy2,sumwy,sumywy,sumywwy /fifth/y /seventh/n
     +        /fourth/ w /nineth/ q

      sumy    = 0.0d0
      sumy2   = 0.0d0
      sumwy   = 0.0d0
      sumywy  = 0.0d0
      sumywwy = 0.0d0
      tempwy  = 0.0d0

      q = spread(y,1,n)
      sumy = sum(y)
      sumy2 = sum(y**2)
      tempwy = sum(w*q, 2)
      sumwy = sum(tempwy)
      sumywy = sum(y*tempwy)
      sumywwy = sum(tempwy**2)

      write(2,250) n,sumy,sumy2,sumwy,sumywy,sumywwy
250   format('n=',i5/
     +'sum y= ',d23.16/'sum y2= ',d23.16/
     +'sum wy= ',d23.16/'sum ywy= ',d23.16/
     +'sum ywwy= ',d23.16/)

      return
      end
```

# Shared subroutines

```
      subroutine golden(ax,bx,cx,fgld,tol,xmin)
c
c
c     One-dimensional golden section search
c
c
      implicit double precision (a-h,o-z)
      external fwn
      parameter (r=.61803399d0,c=1.0d0-r)
      x0=ax
      x3=cx
      if(dabs(cx-bx).gt.dabs(bx-ax)) then
        x1=bx
        x2=bx+c*(cx-bx)
      else
        x2=bx
        x1=bx-c*(bx-ax)
      endif
```

```
      f1=fwn(x1)
      f2=fwn(x2)
10    if (dabs(x3-x0).gt.tol*(dabs(x1)+dabs(x2))) then
        if(f2.lt.f1) then
           x0=x1
           x1=x2
           x2=r*x1+c*x3
           f0=f1
           f1=f2
           f2=fwn(x2)
        else
           x3=x2
           x2=x1
           x1=r*x2+c*x0
           f3=f2
           f2=f1
           f1=fwn(x1)
        endif
      goto 10
      endif
      if(f1.lt.f2) then
        fgld=f1
        xmin=x1
      else
        fgld=f2
        xmin=x2
      endif

      return
      end


      subroutine getdata()
c
c
c  Get data from external file
c
c
      implicit double precision (a-h, o-z)
      real*8 y(400)
      common /fifth/ y /seventh/ n

      open(unit=5,file='nrandx.data', status='old')

        do 10 i = 1, n
              read(5,100) y(i)
10        continue
100       format(d20.13)

      return
      end
```

Chapter 7

# SPATIAL STATISTICS AND GIS APPLIED TO INTERNAL MIGRATION IN RWANDA, CENTRAL AFRICA

Daniel G. Brown

## I. INTRODUCTION

Integration of spatial statistical analysis and geographic information systems (GIS) is an important next step in the development of spatial analysis technologies. Alternative modes of implementing such integration include adding spatial statistical analyses as GIS functions [Anselin and Getis, 1992; Griffith, 1992; Can 1993], providing transfer routines between spatial statistical analysis and GIS software [Anselin *et al.*, 1993a], and adding mapping and spatial data management functions to spatial statistical analysis software [Haslett *et al.*, 1991; SAS Institute, 1994].

The strength of GIS lies in its ability to maintain absolute and relative spatial location information about geographic phenomena. This facility has led to the development of a wide array of data processing and analysis tools that are explicitly spatial in character. Spatial queries, overlay of multiple variables for the determination of spatial coincidence, distance and buffer calculations, and map generation are several core GIS functions. The spatial data management, transformation, and analysis tools within GIS packages simplify the preparation and exploration of a spatial data set for statistical analysis. Display functions provide for the visualization of spatial patterns of raw data and derived statistics. Tests of relationships between spatial variables stored in a GIS are currently lacking in most GIS software. Adding them would greatly enhance the spatial analysis toolkit.

Spatial statistics can be employed for description, inference, and modeling/prediction. [Anselin and Getis, 1992] Descriptive measures of spatial pattern include Moran's Coefficient (MC), point pattern analysis routines, semivariograms, and spatial join or weight matrices. Descriptions of spatial pattern in geographic data can be used to suggest appropriate scales of analysis given the base scales of input data [Brown *et al.*, 1993], reveal artifacts resulting from methods of data production [Brown and Bara, 1994], and suggest hypotheses regarding the processes that have created a certain pattern of heterogeneity or spatial dependence. [e.g., Oliver *et al.*, 1989] The descriptive analyses may be ends in themselves, or they may contribute to inference and/or predictive models. The effects of spatial dependence and heterogeneity on statistical inference have been outlined elsewhere. [Anselin and Griffith, 1988; Haining, 1990; Cordy and Griffith, 1993] Spatially dependent data and errors render significance tests unreliable because variances are underestimated in such situations. Spatial lag regression and spatial autoregression are appropriate alternatives to traditional regression analysis for spatial data because they eliminate the dependencies that cause problems, as is demonstrated in the applications chapters of this handbook. Although many categorical statistical methods have not yet been developed to deal with spatial autocorrelation (e.g., for Poisson or logistic regression), Griffith [1978; 1992] describes methods for spatially adjusted analysis of variance (ANOVA). Further development of spatial categorical statistics will enhance the utility of spatial statistics in the GIS environment.

Links between GIS and descriptive spatial statistics are best developed. Arc/Info v6.1 supports the rudiments of semivariogram analysis, but is very limited. [ESRI, 1992] IDRISI v4.1 provides MC calculations for raster maps. [Eastman, 1993] Other researchers have written modules using macro languages or programming subroutines for addition to existing packages. Anselin *et al.* [1993b] and Can [1993] have written routines in Arc Macro Language (AML) and C for computing the MC and Geary's ratio from an Arc/Info coverage. Lowell [1991] has written an ERDAS toolkit-based program for calculating the join count statistic and MC from raster maps grouped into polygons. A module for computing MC in GRASS has been formulated at Purdue University. An interactive exploratory spatial data analysis package, called REGARD, supports spatial descriptive analysis through interactive histograms and semivariograms linked to a "map view". [Haslett *et al.*, 1991]

Inferential spatial statistics have not been as well integrated into GIS. Spatial autoregression and spatial lag regression have been implemented in SPACESTAT, a stand alone spatial statistics package by Anselin [1991], and as MINITAB and SAS programs by Griffith [1988, 1993] and Griffith *et al.* [1990]. Each of these packages requires data transfer between GIS and statistical packages.

The purpose of this chapter is to illustrate how spatial statistics can support GIS analyses, and vice versa. This chapter illustrates the importance of efforts to further integrate the technologies, by presenting a spatial statistical analysis performed with data stored in a GIS database on societal and environmental characteristics of Rwanda, Central Africa. I attempt to model intra-national net migration between 1978 and 1991 on the basis of push and pull factors. The analysis contributes to a larger project described by Berry *et al.* [1994] and

Campbell *et al.* [1993], which explores the gap between data availability and data needs for resource management in Sub-Saharan Africa. This study is based on data collected before disruptive events in Rwanda, beginning in April 1994, dramatically changed the political and social landscapes. These events altered the situation for the people of Rwanda. Nevertheless, it is hoped that the lessons learned from the research will be of use to help in the rebuilding of Rwanda in the future.

The analyses presented in this paper were not conducted entirely with commercial GIS packages because all the functionality is not currently available. Many of the analyses required that data be exported from a GIS package (Arc/Info) to spatial statistical software (SpaceStat) and back to the GIS for display of results. As the spatial analysis toolkit in commercial GIS packages is further developed, such data structure transformations should become unnecessary.

## II. STUDY AREA

Rwanda has an area equivalent to that of the state of Maryland and has had the highest national population density in Africa. Despite its high population density, over 90 percent of Rwanda's population has relied on farming for its livelihood. The distinctly rural character of Rwanda's population suggests that environmental quality, particularly as it relates to agricultural productivity, plays an important role in the lives of the people. The physical environment in Rwanda is very diverse, with high rainfall totals and mountains in the west and drier, undulating savannas in the lowland east.

Although Rwanda is very small, several data sets available at the sub-national level provide very fine resolution information about agricultural activities, population, and the natural environment. The coarsest sub-national administrative level includes 10 prefectures. In the next finer level are 143 communes, which are the smallest administrative units for which data were available (Figure 1). Three national parks have been closed to settlement and are displayed in Figure 1.

Historically, population densities have been highest in Rwanda's western mountains and foothills. Two ethnic groups, the Hutu majority and the Tutsi minority, shared Rwanda for hundreds of years under a tightly organized political system with Tutsi leadership. The savanna in most of the eastern portion of the country was reserved for the cattle of the Tutsi king. The social revolution, 1959-62, saw the overthrow of the royal leadership, the election of a Hutu-led government, and independence from colonial rule. This revolution led to the opening of much of the land to the east for settlement by farmers. The issues of land availability and quality will become critical as refugee resettlement takes place in the aftermath of the current revolution.

**Figure 1.** Map of Rwanda with commune boundaries (lighter lines) and prefecture boundaries (darker lines). Note: North is at the top of this and all other maps in this chapter.

## III. DATABASE DESCRIPTION

Empirical investigations of broad-scale people-environment interactions, especially in developing countries, are often hampered by data paucity. Rwanda is a data-rich country relative to other African nations. However, it is important to note that data availability remains a concern. A primary goal of the Rwanda Society-Environment Project has been to explore the limits of the available data for description, inference, and modeling. In many cases it has not been possible to evaluate the absolute quality and accuracy of the data. This chapter, rather, focuses on the methods available for making the most of available data *assuming the data are correct*. The issues of spatial data quality and accuracy are critical, and they have been and are being addressed in other forums (e.g., NCGIA Initiatives 1 and 12). Error handling needs to be better developed for GIS and spatial statistical methods to take full advantage of their potential.

## A. GIS DATABASE

Table 1 lists the major components of the GIS compiled for the Rwanda project. Each data set was digitized from paper maps and converted to a common coordinate system--Universal Transverse Mercator (UTM)--to facilitate overlay. The database is maintained in Arc/Info v. 6.1 (ESRI, 1992) and IDRISI v. 4.0 [Eastman, 1990] formats. Each of the data sets includes an attribute file containing variables collected about the spatial objects. The most extensive of the attribute sets is linked to the political boundaries. For example, several census variables were collected on the basis of communes.

**Table 1.** GIS data sets.

| Topic | Source | Date | Base Scale |
|---|---|---|---|
| Political Boundaries | Carte Administrative et Routiere (Rep. du Rwanda) | 1985 | 1:250,000 |
| Elevation | Atlas du Rwanda (Prioul and Sirven) | 1981 | 1:2,000,000 |
| Ave. Annual Isohyets | Bulletin Climatologique Annee 1991 MINITRAP | 1929-88 | 1:3,000,000 |
| Rivers/Lakes | Carte Administrative et Routiere (Rep. du Rwanda) | 1985 | 1:250,000 |
| Roads | Same as previous. | 1985 | 1:250,000 |
| National Parks | Same as previous | 1985 | 1:250,000 |
| Soil Capability | Carte Pedologique du Rwanda (MINAGRI) | 1992 | 1:250,000 |
| Agro-climatic Zones | Bulletin Agricole du Rwanda (Delepierre) | 1975 | 1:3,000,000 |

| Cultural Regions | Bulletin Agricole du Rwanda (Nzisabira) | 1989 1:3,000,000 |
| Farming Systems Regions | Rwanda Society- Environment Working Paper (Olson) | 1994 1:250,000 |

## B.  POPULATION AND AGRICULTURAL CENSUS DATA

The population censuses, taken in 1978 and 1991, include multiple census variables collected by commune. [MINIPLAN, 1982; MINIPLAN, 1992] Variables include population counts, and age and sex structures.  Net migration for each commune was calculated by Olson [1994b] as the difference between the population growth and the population's natural increase between the two dates:

$$M = (P_1 - P_0) - (B - D) \qquad (1)$$

where M is the number of migrants, $P_1$ is the 1991 population, $P_0$ is the 1978 population, B is the number of births and D is the number of deaths that occurred during the interval. [Shryock and Seigel, 1976] The numbers of births and, especially, deaths in Rwanda have not been consistently recorded.  However, the Rwandan Direction of the Census calculated crude birth rates (CBR) and crude death rates (CDR) using the life table method for each prefecture, including the capital city Kigali as a separate prefecture, from the 1978 census, and the CBR for each prefecture and the CDR for the country from the 1991 census.  The 1991 CDR for each prefecture was estimated from its 1978 CDR and adjusted by the change in death rate for the entire country between 1978 and 1991.  The crude rate of natural increase (CRNI) was then calculated for each prefecture for 1978 and 1991 by subtracting CDR from CBR.  The natural increase for each prefecture was subtracted from the population change in each commune within the prefecture to arrive at an estimate of net migration for each commune.  The migration rate was calculated by assuming that the migration rate between 1978 and 1991 was constant, but the population growth rate varied with the CRNI (this section is based on Olson 1994b).

The Ministry of Agriculture [INAGRI, 1989] conducted an agricultural census in 1987 from which estimates have been produced of the total commune production in a variety of crops (in kilograms, kilocalories, proteins, and lipids) and the numbers and food values of livestock.  Estimates of rural income were made by Ben Chaabane and Cyiza [1992] on the basis of the agricultural census data plus off-farm income estimates. Olson [1994a] described the patterns of agricultural land use in Rwanda on the basis of these data and found that current farming practices are as much a reflection of historical and cultural influences, political decisions, and economic development as they are of the natural environmental constraints.

## IV.  GIS DATA MANAGEMENT

The GIS functions for overlay, distance and area calculation, and reselection can be used in the preparation and maintenance of a database for spatial statistical analysis. In the case of the Rwanda commune-level data, initial data preparations were made in Arc/Info. One important assumption is that no people have been living in the three national parks. The park polygons were added to the commune coverage using overlay (Figure 1). The resulting area of the communes was used for a more realistic calculation of population density values.

Spatial data from different sources and/or of different types are easily integrated using most GIS packages. In some cases it is too easy to combine data that are incompatible. However, the flexibility to transform spatial data is an important advantage that GIS brings to the statistical analysis of spatial data. One class of transformations that is important is the ability to translate and/or summarize data collected using one scheme of regionalization to another scheme. For example, satellite data collected in pixels might be summarized by digitized soil units, or data collected by census tract may be re-systematized using postal code regions. The former process, where the units of one scheme fit within the other units, is referred to as areal unit aggregation while the latter, where unit boundaries overlap, is areal interpolation. Areal unit aggregation may involve calculation of an area weighted mean value, the calculation of sub-unit proportions of nominal classes, and/or the estimation of within unit variation.

In the case of areal interpolation a common approach to converting from one regionalization scheme to another is to assume that the spatial distribution of the variable of interest is uniform within spatial units. To calculate population of a subset of any spatial unit for which population is known, for example, the unit population value is multiplied by the proportion of the unit area covered by the subset. The subsets can then be re-aggregated to form a new set of spatial units. More elaborate approaches include pycnophylactic interpolation [Tobler, 1979], use of ancillary data [Martin and Bracken, 1991; Flowerdew and Green, 1989], and use of "control zones" [Anselin *et al.*, 1992]. Areal interpolation methods should be built into GIS packages to improve ease of implementation.

For this project, census data, collected for communes, were integrated with environmental data, which were collected for a variety of spatial units (i.e., soil polygons, satellite image pixels, and surfaces of elevation and rainfall variation). Given the focus of the analysis on the census variables, and the secondary nature of the environmental variables, the environmental data were summarized by commune. Area-weighted averages of elevation, rainfall, and soil capability were calculated by overlaying the commune boundaries on each of these coverages. The Arc/Info FREQUENCY routine was used to calculate area-weighted averages by commune.

## V.  THE EFFECTS OF AGGREGATION ON POPULATION DATA

Aggregation of data, though often necessary in order to uncover pattern in detail or integrate data from various sources, has the undesirable effect of masking finer resolution variability in the data. However, where "natural regions"--contiguous groupings of spatial units with low internal variance--occur within a data set, judicious aggregation can, in fact, result in a quite reasonable representation of a variable. In cases where the analyst can choose the aggregation units, it is possible to control the loss of information by minimizing the within-unit variance at the coarser scale. A univariate classification procedure, with the added constraint of contiguity, might be used to identify regions with low internal variance. [Semple and Green, 1984] Griffith [1982] has explored the relationship between aggregate and regional shape and spatial interaction within regions, i.e., cities in his example. Regional shape might be expected to have an effect on internal regional variance as well. However the relationship may be counterintuitive. If the natural regions were always compact in shape, as regions based on a distance-decay model might be, then more compact natural regions would be expected to have lower internal variance, assuming they are properly located. However, it is reasonable to assume that (a) distance-decay does not perfectly explain the spatial interaction between places in every case, and (b) spatial interaction is but one cause of spatial variability. Hence, natural region shapes are not likely to be the most compact shapes. The shapes will depend, instead, on the underlying spatial distributions of variables that affect the distribution of the variable in question and spatial interaction. Further, when data are aggregated into natural regions for a given variable, the degree of spatial autocorrelation should tend to diminish. If the internal homogeneity of regions is minimized and multiple regions exist, then the variation between adjacent units should increase with aggregation, thereby lowering spatial autocorrelation.

In the case of Rwanda, three possible regionalizations on the basis of aggregated communes were examined (Figures 1 and 2). Prefectures are sub-national administrative units for which data are often collected and in turn are used for policy making. Eight farming systems regions (FSR) were identified by Olson [1994a] on the basis of similar agricultural activities and cropping systems as reflected in the crop and animal data reported in the agricultural census. Agroclimatic zones (ACZ) were defined by Delepierre [1975] on the basis of environmental variability, especially elevation, rainfall, and soils. The variable analyzed in Table 2 is total commune population in 1991. Each regionalization was constructed by aggregating communes into between 8 and 11 regions. The internal variation in each of the regions was measured using the average coefficient of variation of commune populations within each region. The between regions variation was calculated as the CV of all regional population values ($CV = SD / \overline{x}$). The spatial autocorrelation was measured using the Moran Coefficient. Regional shapes were characterized for the regionalization system as a whole, using the fractal dimension of the regions. A fractal dimension is calculated as a function of the slope of the relationship between the log of regional area and the log of regional perimeter (FD = 2 / slope), and characterizes the degree to which the boundaries of regions are complex.

The results reported in Table 2 indicate that Prefectures have higher within-regions rather than between-regions variation of commune population. While they are the most compact regions, they also display a slightly higher level of spatial autocorrelation than the communes. Although the difference in autocorrelation is not likely significant, what is important is that the aggregation did not result in a reduction of the level of spatial autocorrelation, implying that the prefectures are not natural regions in terms of 1991 population levels. The two other regionalization schemes display lower internal variance and higher between regions variance in 1991 population. They tend to be less compact, while exhibiting a lower level of spatial autocorrelation.

The results are clearly variable dependent. Other variables may form natural regions in other ways. However, the results have implications for analyses on the basis of aggregated data. Natural patterns of variability will tend to be masked when data are collected and analyzed using units with high internal variances. Furthermore, a compact shape does not guarantee an internally homogenous region. Regions formed on the basis of actual patterns in the landscape or on land use patterns (e.g., farming systems regions or agroclimatic zones) are likely to be more representative than arbitrarily demarcated regions (e.g., most administrative units). Where a choice between regionalization schemes exists, a check of internal homogeneity is likely to be helpful.

**Table 2.** Comparison of central tendencies, Moran Coefficient values, and fractal dimensions for 1991 population aggregated using alternative regionalizations of Communes in Rwanda.

| Regionalization | N | C.V. by Region of Commune Pop | C.V. of Region Pop | MC† | FD |
|---|---|---|---|---|---|
| Communes | 143 | 45.62 | N/A | 0.187 | 1.146 |
| Prefectures | 10 | 34.36* | 27.84 | 0.190 | 1.480 |
| FSR | 8 | 40.14* | 56.36 | -0.004 | 1.764 |
| ACZ | 11 | 35.25* | 62.75 | -0.163 | 1.672 |

Note.  C.V. = coefficient of variation, MC = Moran coefficient, and FD = fractal dimension.

\* - Means and standard deviations calculated by region and then averaged.

† - Calculated using a weight based on the proportion boundary lengths joined and the inverse of distance from geographical center of polygons raised to the power of 3.

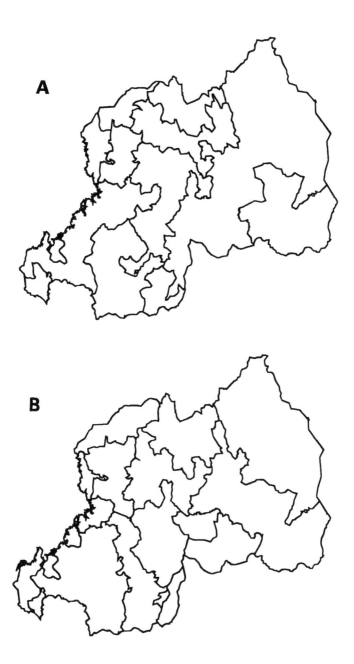

**Figure 2**.  Boundaries for alternative regionalizations of communes.  A.  Farming systems regions defined by Olson [1994a].  B.  Agroecological zones defined by Delepierre [1975].

## VI. TRADITIONAL REGRESSION ANALYSIS

Permanent migration is an important indicator of how people in an area perceive relative opportunities in different places.  In this analysis the net migration in a given commune in Rwanda is assumed to be a function of push and pull factors related to environment, land availability, income, and facilities.  The model describing the net migration rate between the 1978 and 1991 censuses may be specified as follows:

net migration =   f [cumulative population density (people-years/ha),
            land availability (ha),
            income (francs/year),
            development (index),
            mean rainfall (mm/year),
            mean soil capability (index)  ].

The cumulative population density was the area under a graph of population density change between 1948 (the earliest date for which data were available) and 1991.  Time in years was plotted on the x-axis and population density was plotted on the y-axis.  Four data points were available to define the curves for each commune: 1948, 1970, 1978, and 1991.  By accumulating population density over a period of time, the variable provides a better measure of population pressure (the continued habitation and use of land for farming) than does a single time measurement of population density.

Land availability was estimated by Campbell [1994] as the proportion of arable land in the commune (i.e., land area minus an estimate of land area lost to homes, roads, lakes, swamps, and rock outcrops) that was not used for agricultural activities, including cultivation, pasture, and fallow.  The area under cultivation was summed for all crops and, in many cases, multiple crops were grown on the same land, yielding several negative values for available land area.  This situation illustrates a problem that can arise out of integrating data from a variety of sources, collected for a variety of purposes.  Nevertheless, the variable provides a relative measure of the land use intensity in an area.

The income variable was estimated by Ben Chaabane and Cyiza [1992] and represents rural income only, not employment in urban areas.  The development index is from Ben Chaabane et al. [1991], who extracted principal component scores from a battery of primary variables including access to markets, transportation routes, education, and overall infrastructural facilities. Higher values of the development variable suggest more urban functions with access, economic activity, and infrastructure. Communes with towns, especially the capital Kigali, scored highly on the development index. Mean rainfall was summarized by commune through area-weighted averaging from the digitized isohyet map. [MINITRAP, 1992] Mean soil capability was summarized from the soil capability map [CPR, 1993] through the same method, and is interpreted as soil limitation. Higher values of soil capability reflect greater limitations to agricultural production on the basis of soils.

## A. DESCRIPTIVE ANALYSIS OF VARIABLES

The visualization of spatial patterns in a set of data can play an important role in inferential analysis. In most cases the production of "map views" does not require the full analytical capabilities of a GIS. SAS-GRAPH [SAS, 1985] and REGARD [Haslett *et al.*, 1991] demonstrate the utility of a simple mapping capability in the context of statistical analysis. However, given that GIS combine mapping and data management capabilities, their linkage with spatial statistical analyses facilitates integrated analysis and management of spatial data.

A map of the net migration rate between 1978 and 1991 illustrates the impact of historical settlement patterns (Figure 3). The western half of the country, with traditionally higher density settlements, tends to have higher out-migration than the east, which has opened to settlement only in the past 30 years. Several communes around parks have a net positive migration rate estimate. The former Gishwati Forest, in the northwest, provided land for in-migration in several communes (southeast of the Parc des Volcans) when it was opened for settlement in the 1980s. The communes in the east have experienced the highest in-migration rates. They had low population densities to begin with and a continued stream of settlers from the west. [Olson, 1990]

The basic descriptive statistics for the statistical and spatial distributions of each of the variables are given in Table 3. The primary statistics of interest are those relating to normality and spatial autocorrelation, as these properties affect the assumptions of traditional statistical tests. Non-normality compromises the interpretability of significance tests based on the normality assumption. Spatial autocorrelation results in underestimated variance values and, therefore, bias towards rejecting a correct null hypothesis of no significant relationships. Each of the variables had statistical distributions that were significantly different from the normal distribution, indicated by the Shapiro-Wilk statistic and its associated probability value (Table 3). The average annual rainfall variable is the most similar to the normal distribution. Each variable, except development, exhibits significant positive spatial autocorrelation, as measured by the Moran Coefficient under the randomization assumption. These issues may become important. However, in regression analysis, the normality and spatial autocorrelation of errors are two of the paramount concerns regarding the soundness of the significance tests.

In order to avoid the possibility of non-normal errors, the dependent variable (migration rate) was transformed with a natural logarithm function. Because the natural log of negative values is undetermined and the minimum value of the net migration rate variable was -2.1, the constant 3 was first added to the migration rate for each commune. This provides a simple solution to a common problem, but the value of the constant could be estimated to optimize the normality in the resulting distribution (Owen, 1988). Test statistics for the transformed variable are listed as the last entry in Table 3. The transformed variable exhibits a non-normal distribution at p=0.05, but the strength of the Shapiro-Wilk statistic was reduced markedly such that the distribution is not significantly different from normal at p=0.01.

**Table 3.** Descriptive statistics of variables used in regression analysis.

| Variable | Mean | Skew -ness | Shapiro -Wilk | Prob | MC | Prob |
|---|---|---|---|---|---|---|
| migration rate | -0.493 | 2.09 | 0.827 | 0.00 | 0.474 | 0.00 |
| cumulative pop. density | 10692.2 | 4.81 | 0.716 | 0.00 | 0.289 | 0.00 |
| spare land | -11.38 | -0.67 | 0.265 | 0.00 | 0.281 | 0.00 |
| income | 499079.2 | 1.92 | 0.833 | 0.00 | 0.704 | 0.00 |
| development | 1505.5 | 10.15 | 0.269 | 0.00 | 0.027 | 0.13 |
| rainfall | 1202.9 | -0.47 | 0.756 | 0.00 | 0.874 | 0.00 |
| soil capability | 4.31 | 0.65 | 0.199 | 0.00 | 0.512 | 0.00 |
| ln (migration rate + 3) | 0.852 | 0.56 | 0.967 | 0.03 | 0.507 | 0.00 |

## B. MODEL RESULTS

The fit of the ordinary least squares (OLS) regression model was measured using four statistics: $R^2$, adjusted $R^2$, log-likelihood, and Akaike Information Criterion (AIC). Higher values of the first three statistics indicate a better fitting model, while a lower AIC indicates a better fit. The $R^2$ and log-likelihood values both will increase as variables are added to the model, regardless of whether or not the variables improve the explanation in the model. In contrast, the adjusted $R^2$ and AIC take into account the over-estimated level of fit. The log-likelihood and AIC are presented for comparison with the spatial statistical models.

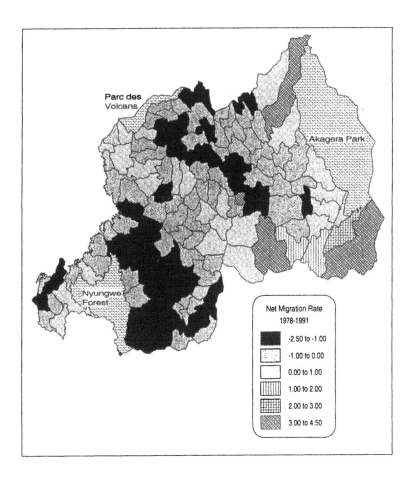

**Figure 3**. Spatial pattern of Net Migration Rate variable (1978 to 1991).

SPACESTAT provides diagnostic tests for four possible assumption violations of traditional OLS regression:    non-normal errors, heteroskedastic (non-constant variance) errors, multicollinear predictors and spatially autocorrelated errors. The residuals in the OLS regression model for the migration rate were not distributed significantly differently from a normal distribution (Kiefer-Salmon test, Table 4).

Use of the untransformed dependent variable, however, resulted in significantly non-normal errors (Kiefer-Salmon=32.92); so, the transformation was justified. Anselin (1991) suggests that variance in errors is often related to some measure of unit size. The heteroskedasticity of errors was tested with respect to 1991 commune population and was not significant (Breusch-Pagan test). The multicollinearity condition number is one measure of the degree to which the predictor variables are redundant. Anselin (1991) suggests that values larger than 20 or 30 may indicate the standard errors are being unduly affected by calculation error. The model may have a problem with multicollinearity (Table 4). One solution is to remove some of the variables that have too much duplicate attribute information.

Subsequent models, therefore, are calculated with rainfall and soil capability removed, each of which did not contribute significantly to the model. The natural environment variables were not significant for several possible reasons: 1) non-environmental factors may be more important than environmental ones; 2) the soil capability variable did not reflect very well the impoverishment of soils in some areas, especially in the Gikongoro and Kibuye prefectures in the southwest; and, 3) rainfall, for the most part, has not yet been a limiting factor on agricultural activities in Rwanda. Removing rainfall and soil capability from the model reduced the condition number to a value of 10.1, indicating a decreased likelihood that multicollinearity is a problem. Another approach to reducing multicollinearity is to compute orthogonal principal components from the matrix of predictor variables. Finally, spatial autocorrelation is clearly present in the model residuals (Moran Coefficient). A generalized weight matrix was used for the calculation of the MC and for all other spatial dependence tests, unless otherwise indicated. As suggested by Upton (1990), weights were calculated as

$$W_{ij} = l_{ij} \cdot d_{ij}^{-3} \qquad (2)$$

where $W_{ij}$ is the calculated weight for the interaction between polygons $i$ and $j$, l is the length of border and d is the distance between geographical centers. The weight matrices always are row standardized before use. Before interpreting the model, the spatial autocorrelation should be accounted for; otherwise the significance tests are suspect.

## VII.  MAPPING RESIDUALS

Residual mapping is an important component of any spatial statistical analysis. Residuals from the OLS regression model of the migration rate using the four significant variables in Table 4 (i.e., cumulative population density, spare land, rural income, and development) have been mapped (Figure 4). Analysis of the residual pattern can provide important localized information about the controls on migration that may not apply nationwide. Additionally, such an analysis may suggest that additional variables are missing from the model specification.

**Table 4.** Results of traditional regression analysis with diagnostics for spatial autocorrelation, non-normality, and heteroskedasticity in residuals and multicollinearity in predictors.

---

Dependent Variable: *ln (migration rate + 3)*

$R^2 = 0.419$    Adj. $R^2 = 0.393$    log-likelihood = -14.99    AIC = 43.98

| Variable | Coefficient | Std Err | t-value | Prob |
|---|---|---|---|---|
| intercept | 1.362 | 0.250 | 5.46 | 0.000 |
| cumulative pop. density | -2.4E-05 | 7.5E-06 | -3.23 | 0.002 |
| spare land | 0.004 | 0.001 | 3.93 | 0.000 |
| income | 2.7E-07 | 8.7E-08 | 3.13 | 0.002 |
| development | 4.9E-05 | 1.2E-05 | 4.03 | 0.000 |
| rainfall | -0.0003 | 0.0001 | -1.78 | 0.077 |
| soil capability | -0.012 | 0.037 | -0.32 | 0.751 |

Kiefer-Salmon (error normality) = 3.20 (p=0.202)
Breusch-Pagan (heteroskedasticity) = 0.001 (p=0.975)
Multicollinearity condition number = 29.87
MC error = 0.42  (p=0.000)
Lagrange Multiplier test (error spatial autocorrelation) = 38.60
    (p=0.000)

---

The spatial autocorrelation in the residuals indicates that communes with higher in-migration than predicted by the model (i.e., underpredicted) tend to occur in groups; for example in the southeast, and northwest. The high positive residuals in two communes in the south central region (called Bugesera) suggest an underestimation of the in-migration by the model (A in Figure 4) This is an area that is not very densely populated because the land is quite swampy and not suitable, as is, for agricultural production. It may be that settlers are coming to the area and draining some of the swampland, which is not included in the spare land variable, and bringing it into production. Several communes in the northwest (in the area of B in Figure 4) have higher in-migration than predicted because they include the former Gishwati Forest, which previously had been closed to settlement but now is open. Similarly, communes bordering on the other parks, south of Parc des Volcans and west of the Nyungwe Forest, experienced higher than predicted in-migration, possibly due to land clearance near or in the parks by settlers.

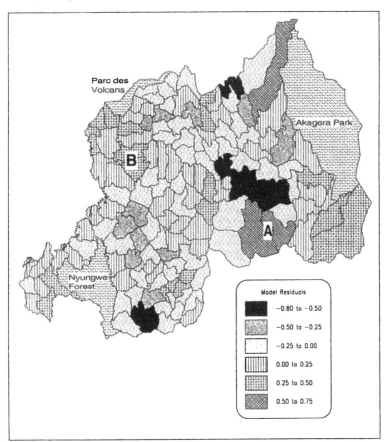

**Figure 4**. Residuals from OLS regression of Net Migration Rate. Predictors were cumulative population density (1948-1991), spare land, income, and development.

## VIII. SPATIAL STATISTICAL MODEL

Because migration is a spatially dependent process, explanation is not complete without some characterization of spatial interaction. The positive spatial autocorrelation in the variables (Table 3) and in the model residuals (Table 4) suggests similar destination and source communes may occur in spatial groups. If a given commune is perceived as a desirable destination and/or has available land, its neighbors also may be perceived as desirable destinations and/or have available land. Inclusion of a spatial lag variable, i.e., a variable representing the neighborhood effect of the net migration rate, may help explain some of the residual variation. By accounting for the spatial effects in the model, we can interpret the significance of the other, non-spatial variables. Otherwise the significance tests are suspect. A model including an estimated spatial lag variable, termed the regressive spatial autoregressive model (AR), was calculated through maximum likelihood estimation in SPACESTAT.

The spatial lag term was highly significant and, more importantly, its addition reduced the spatial autocorrelation in the model residuals to an insignificant level (Table 5). The pattern of residuals resulting from the spatial lag model (Figure 5) appears similar to that from the OLS model (Figure 4), even though the spatial autocorrelation was reduced to a non-significant level. As expected, the overall fit of the model was improved with the addition of the spatial lag variable; the log-likelihood increased and the AIC decreased. The $R^2$ value given in Table 5 is only an estimate and should not be compared with the OLS model value in Table 4. The residual sum of squares is also lower for the spatial lag model (7.74 compared with the OLS model 10.38). The estimates of the model parameters in the spatial lag model are all more precise, i.e., the standard errors of the estimates are lower.

Spatial autoregressive models are sensitive to the specification of the weights matrix. Significant spatial autocorrelation remained in the model residuals when the weights matrix was calculated with the exponent of distance in Equation 2 changed to -1 (Lagrange Multiplier test = 6.24, p=0.01). The use of a row standardized binary connectivity matrix also resulted in significant residual spatial autocorrelation (LM test = 32.54, p=0.00). Therefore, the generalized weights matrix with the $d^{-3}$ term suggested by Upton [1990], which renders a tighter pattern, characterized the spatial interaction in the net migration field better than the other weighting schemes.

Given that the error normality assumption was probably met, heteroskedasticity was not apparent in the residuals, multicollinearity was likely not a problem, and the spatial autocorrelation in residuals was accounted for, the relationships between the dependent and independent variables can be interpreted with some degree of confidence. Approximately one-half of the variation in the net migration rate between 1978 and 1991 is accounted for by the push and pull factors of cumulative population density, spare land, income, and development once spatial dependency is taken into account.

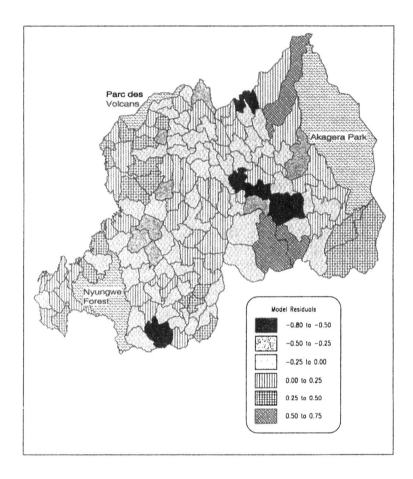

**Figure 5.** Residuals from spatial lag regression of Net Migration Rate. Model is shown in Table 5.

As the cumulative population density in a commune increases, the net migration rate is increasingly likely to be negative (i.e., more out-migration than in-migration). This suggests that population pressure is a real force affecting the migration behavior of the population of Rwanda. The issue of population pressure is directly related to land availability, especially in a society that is as dependent on agriculture as Rwanda's. The spare land variable indicates that communes with more arable land not currently under production are more likely to be destinations for migrants. Additionally, even seemingly non-arable land may be brought under production by migrants, as seen in the marshlands of Bugesera. Each of the income and development variables describes the attractiveness of destination communes for a migrating population in terms of economic opportunities. Communes with higher total rural incomes are more likely to be net destinations than net sources of migrants. Similarly, communes with higher scores on the development variable have higher net migration rates, reflecting the pull of urban employment possibilities.

**Table 5.** Results of spatial lag regression analysis, solved through maximum likelihood, with diagnostics for spatial autocorrelation and heteroskedasticity in residuals.

Dependent Variable: *ln (migration rate + 3)*

psuedo $R^2$ = 0.472    log-likelihood = -2.50    AIC = 16.99

| Variable | Coefficient | Std Err | z-value | Prob |
|---|---|---|---|---|
| intercept | 0.648 | 0.094 | 6.92 | 0.000 |
| lagged migration rate | 0.402 | 0.062 | 6.52 | 0.000 |
| cumulative pop. density | -2.36E-05 | 6.38E-06 | -3.70 | 0.000 |
| spare land | 0.003 | 0.0001 | 3.92 | 0.000 |
| income | 1.71E-07 | 6.93E-08 | 2.46 | 0.014 |
| development | 5.15E-05 | 1.02E-05 | 5.06 | 0.000 |

Breusch-Pagan (heteroskedasticity) = 0.0003 (p=0.986)
Lagrange Multiplier test (spatial autocorrelation) = 2.21  (p=0.137)

## IX. CONCLUSIONS

By way of example, linkage with spatial statistical analysis is shown to be a very helpful next evolutionary step in the expansion of the spatial analysis toolkit available in GIS.   Given that the assumptions of traditional regression analysis are often violated when applied to spatial data, and that correlation of spatial variables is a central tool in the testing of geographical hypotheses, tests for and model specifications embracing spatial correlation should be made readily available in GIS packages.  Furthermore, given that spatial data require explicit spatial referencing and topology management, and that the management and visualization of spatial data are central in a geographic analysis, certain GIS functions should be available within a full function statistical package. Neither of these two developments will likely take precedence over the other, but they are both unfolding. Given the increasing availability of spatial statistical methods, up until recently a fairly arcane methodology, analysts are finally able to take advantage of these methods and can no longer ignore the assumption violations in traditional regression analysis.

In the Rwanda example a fairly strong relationship is demonstrated between the net migration rate and population pressure, land availability, and socioeconomic push and pull factors. Although spatial autocorrelation proves to be a significant component in the resulting spatial lag model, the statistical significance decisions as to which variables are important remain unchanged. Residuals from the OLS net migration rate model suggest that the way in which the actual commune migration rate differs from the predicted rate tends to be spatially autocorrelated. Although not true for this example, in some cases it may be possible to suggest missing variables on the basis of an analysis of the residual patterns. In this case, accounting for the spatial autocorrelation in the data resulted in a model that better extracts information from the variables and has more precise estimates of model coefficients than does the OLS model.

Population pressure, as it affects land availability, among other things, appears to be one of the driving forces in the internal migration in Rwanda. However, it supplies only part of the story. The degree of development, i.e., access to markets and infrastructure, has a positive effect on the desirability of a region for in-migration, as does total income. Whereas population pressure and lack of available land may be pushing out-migrants, the opportunity for land and economic security are probably important pull factors for in-migrants. These push and pull factors seem to override the importance of any environmental variability that may affect migration decisions.

## Acknowledgments

This work was supported in part by the Consortium for International Earth Science Information Network (CIESIN) under contract CSN #415-93 (Dr. David J. Campbell, principal investigator). The author wishes to acknowledge the help of Jennifer M. Olson, David J. Campbell, and Gichana Manyara in various phases of project conceptualization and implementation.

## REFERENCES

**Anselin, L.** *SpaceStat: A program for the analysis of spatial data.* Department of Geography, University of California, Santa Barbara, CA, 1991.

**Anselin, L., Dodson, R. F., and Hudak, S.** Linking GIS and spatial data analysis in practice. *Geographical Systems,* 1, 2-23, 1993a.

**Anselin, L. and Getis, A.** Spatial statistical analysis and geographic information systems. *Annals of Regional Science,* 26, 19-33, 1992.

**Anselin, L. and Griffith, D. A.** Do spatial effects really matter n regression analysis? *Papers of the Regional Science Association,* 65, 11-34, 1988.

**Anselin, L., Hudak, S., and Dodson, R. F.** Spatial Data Analysis and GIS: Interfacing GIS and econometric software. *NCGIA Technical Paper 93-7.* University of California, Santa Barbara, CA, 1993b.

**Ben Chaabane, H., Cyiza, P., and Rushengura, A.** *L'Indice des Priorités Communales: Un Indice Synthétique du Taux de Développement des Communes Rwandaises.* Kigali, Rwanda: MINIPLAN, Rép. Rwandaise, 1991.

**Ben Chaabane, H., and Cyiza, P.** *Méthodologie d'Elaboration de Données sur le Revenu Rural des Communes Rwandaises.* Kigali, Rwanda: MINIPLAN Direction de la Planification, Division des Stratéties de Développement Communal et Régional, 1992.

**Berry, L., Olson, J. M., Campbell, D. J., and Brown, D. G.** The Rwanda Society-Environment Project: a pilot study to link socioeconomic and physical data within an environmental information system. An interim report. *Proceedings of the Second Annual Meeting of the CIESIN User's Group,* Atlanta, GA, 1994.

**Brown, D. G., Bian, L., and Walsh, S. J.** Response of a distributed watershed erosion model to variations in input data aggregation levels. *Computers and Geosciences.* 19(4), 499-509, 1993.

**Brown, D. G. and Bara, T. J.** Recognition and reduction of systematic error in elevation and derivative surfaces from 7-1/2 minute DEMs. *Photogrammetric Engineering and Remote Sensing.* 60(2), 189-194, 1994.

**Campbell, D. J.** Environmental stress in Rwanda No. 1: Preliminary analysis. *Rwanda Society-Environment Project Working Paper,* 4, Department of Geography, Michigan State University, East Lansing, MI, 1994.

**Campbell, D. J., Olson, J. M., and Berry, L.** Population pressure, agricultural productivity and land degradation in Rwanda: an agenda for collaborative training, research and analysis. *Rwanda Society-Environment Project Working Papers,* 1, Department of Geography, Michigan State University, East Lansing, MI, 1993.

**Can, A.** Residential quality assessment: alternative approaches using GIS. *Annals of Regional Science,* 26, 97-110, 1992.

**CPR (Carte Pedologique du Rwanda).** *Carte d'Aptitude des Sols du Rwanda.* Kigali, Rwanda: MINAGRI and Cooperation Technique Belge, 1993.

**Cordy, C. and Griffith, D.** Efficiency of least squares estimators in the presence of spatial autocorrelation. *Communications in Statistics* B, 22, 1161-1179, 1993.

**Delepierre, G.** Les Régions Agricoles du Rwanda. *Bulletin Agricole du Rwanda*, 8(4), 216-225, 1975.

**DSA (Division des Statistiques Agricoles).** *Enquête Nationale Agricole 1989: Production, Superficie, Rendement, Elevage et Leur Evolution 1984-1989.* Kigali, Rwanda: MINAGRI, Rép. Rwandaise, 1991.

**Eastman, J. R.** *IDRISI: A Grid-Based Geographic Analysis System*, version 3.2. Clark University School of Geography, Worcester, MA, 1990..

**ESRI.** *Arc/Info, User's Guide.* Environmental Systems Research Institute, Redlands, CA, 1992..

**Flowerdew, R., and Green, M.** Statistical methods for inference between incompatible zonal systems. In M. F. Goodchild and S. Gopal, Eds., *Accuracy of Spatial Databases.* Taylor and Francis, London, 239-247, 1989.

**Griffith, D. A.** A spatially adjusted ANOVA model. *Geographical Analysis*, 10, 296-301, 1978.

**Griffith, D. A.** Geometry and spatial interaction. *Annals of the Asociation of American Geographers*, 72(3), 332-346, 1982.

**Griffith, D.A.** Estimating spatial autoregresive model parameters with commercial statistical packages. *Geographical Analysis*, 20, 176-186, 1988.

**Griffith, D. A.** A spatial adjusted N-way ANOVA model. *Regional Science and Urban Economics*, 22, 347-369, 1992.

**Griffith, D. A., Lewis, R., Li, B., Vasiliev, I., Knight, S., Yang, X.** 1990. Developing Minitab Software for spatial statistical analysis: a tool for education and research. *Operational Geographer*, 8, 28-33, 1990.

**Haining, R.** *Spatial Data Analysis in the Social and Environmental Sciences.* Cambridge University Press, Cambridge, 1990.

**Haslett, J, Bradley, R., Craig, P., Unwin, A., and Wills, G.** Dynamic graphics for exploring spatial data with application to locating global and local anomalies. *American Statistician*, 45(3), 234-242, 1991.

**Lowell, K.** Utilizing discriminant function analysis with a geographical information system to model ecological succession. *International Journal of Geographical Information Systems*, 5(2), 175-191, 1991.

**Martin, D. and Bracken, I.** Techniques for modelling population-related raster databases. *Environment and Planning A*, 23, 1069-1075, 1991.

**MINAGRI.** *Production Agricole en 1987: Bilan d'Autosufficance Alimentaire par Commune et par Habitant.* Kigali, Rwanda: MINAGRI, Rép. Rwandaise, 1989.

**MINIPLAN.** *Syntése des principaux résultats du recensement général de la population et de l'Habitat 1978.* Kigali, Rwanda: Rép. Rwandaise, 1982.

**MINIPLAN.** *Recensement Général de la Population et de l'Habitat au 15 aout 1991: Résultats Préliminaires Echantillon au 10e.* Kigali, Rwanda: MINIPLAN, Rép. Rwandaise, 1992.

**MINITRAP.** *Bulletin Climatologique Année 1991.* Kigali, Rwanda: MINITRAP, Rép Rwandaise, 1991.

**Nzisabira, J.** Accumulation du Peuplement Rural et Ajustements Structurels du Système d'Utilisation du Sol au Rwanda Depuis 1945. *Bulletin Agricole du Rwanda*, 2, 117-127, 1989.

**Oliver, M., Webster, R., and Gerrard, J.** Geostatistics in physical geography. Part 2: Applications. *Transactions, Institute of British Geographers*, 14, 270-286, 1989.

**Olson, Jennifer M.** *The Impact of Socioeconomic Factors on Migration Patterns in Rwanda.* Unpublished M.A. Thesis. Department of Geography, Michigan State University, 1990.

**Olson, Jennifer M.** Farming systems of Rwanda: echoes of historic divisions reflected in current land use. *Rwanda Society-Environment Project Working Paper*, 2, Department of Geography, Michigan State University, East Lansing, MI, 1994a..

**Olson, Jennifer M.** Demographic responses to resource constraints. *Rwanda Society-*

*Environment Project Working Paper*, 7, Department of Geography, Michigan State University, East Lansing, MI, 1994b.

**Owen, D.** The Starship. *Communications in Statistics*, 17, 315-341, 1988.

**Prioul, C. and Sirven, P.** *Atlas du Rwanda*. Kigali, Rwanda: Imprimérie Moderne Nantasie Coueron, 1981.

**Semple, R. K. and Green, M. B.** Classification in human geography. . In G. L. Gaile and C. J. Willmott, Eds., *Spatial Statistics and Spatial Models*. Dordrecht: Reidel, 133-145.

**Shryock, H. S. and Seigel, J. S.** *The Methods and Materials of Demography*, Academic Press, Inc., New York, 1976.

**Tobler, W.** Smooth pycnophylactic interpolation for geographical regions. *Journal of the American Statistical Association*, 74(367), 519-536, 1979.

**Upton, G. J. G.** Information from regional data. In, D. A. Griffith, Ed., *Spatial Statistics: Past, Present, and Future*, Monograph Series, Institute of Mathematical Geography, Ann Arbor, MI, Monograph #12, 315-360, 1990.

Chapter 8

SPATIAL STATISTICAL MODELING
OF EXPLICIT AND IMPLICIT DETERMINANTS OF
REGIONAL FERTILITY RATES:
A CASE STUDY OF HE-NAN PROVINCE, CHINA

H. Michael Feng

I. INTRODUCTION

The demographic literature establishes that many factors influence
human fertility behavior within a region. Fertility transition theory may
be seen as a grand scheme to address this issue. In its simplest form, this
theory states that a region's fertility rate decreases as its
industrialization and urbanization advance. [Notestein, 1953, 1954;
Davis, 1963; Goldscheider, 1971; Freedmen, 1979] Positive social
changes are considered determinants that lead people to curtail their
fertility. Focusing on behavior at an individual rather than an aggregate
level, some researchers have constructed a theoretical economic
framework to explain the influence of fertility determinants. These
scholars see a couple's reproductive decision as a balance among three
forces: demand for children, supply of children, and regulation costs.
[Becker, 1960, 1965, 1974; Nambookiri, 1972; Willis, 1973; Schultz,
1981; Easterlin, 1978; Bulatao and Lee, 1983] A regional fertility rate
results from accretion of individual acts. More recently there has been a
resurgence of work from the perspectives of political economy and
cultural studies. [Greenhalgh, 1990; Johnsson, 1991]

There is an important difference between the first two and the third
of these perspectives. It has to do with how visible the process is for the
macro determinants, from societal modernization to state population
policies, to be translated into micro fertility decisions of reproductive
couples. Despite criticisms to the contrary, the variables in fertility
transition theory and economic frameworks are intrinsically observable
and eventually measurable. Indeed, they are often collected and
computed in official and unofficial surveys, including population
censuses at various levels. This quantifiability can lead to a more
widespread use of formal statistical models. The determinants
considered in cultural and political-economic perspectives, however, are
often too broad or vague to be quantitatively measured. This lack of

0-8493-0132-7/95/$0.00+$.50
© 1996 by CRC Press, Inc.

systematic measurement can render invalid numerous conventional statistical modeling techniques. [Smith, 1989] Thus, alternative statistical techniques need to be adopted.

This chapter attempts to use spatial statistical techniques to simultaneously model socio-economic factors and effects of state policies. Its begins with a determination of two types of variables: explicit and implicit.

- *Explicit variables* refer to those whose values may be measured and indeed are computed at various levels of surveys and publications.
- *Implicit variables* refer to those whose values may not be collected in such a manner.

This division of variables corresponds to a dichotomy in research methods: that of transition theory and economic frameworks, and that of political-economic and cultural perspectives.

On both fronts, China is an interesting candidate in which to field test methodology. It is a country in which intertwining socio-economic and cultural-political factors have apparently induced the fertility transition, although perhaps not in a demographically "smooth" manner. Indeed, the most important factor appears to be government population policies and the forceful and effective implementation of these policies, especially at the local level. [Feeney, *et al.*, 1989; Greenhalgh, *et al.*, 1994]

During the mid-1980s serious economic, fiscal, and political obstacles emerged in the countryside to block the achievement of the ambitious demographic goals of the "one-child-per-couple" policy introduced in 1979. As implementation faltered, fertility began to rise, climbing from 2.27 in mid-1985 to 2.46 in mid-1987. [Luther *et al.*, 1990] In this two-year span, economic development was accelerating rapidly, suggesting the possibility that increases in fertility might be pronatalist in nature. However, that trend did not go unchecked. The Chinese leadership reduced the fertility rate by 25% in five years, from 2.46 in 1987 to 1.90 births per woman in 1992, by which time the birth rate reached its replacement level. [P. Peng, 1993] During this five-year span, economic development continued to expand, as well. Whether or not economic development is related to fertility, in this case, remains a question of continuing debate. The Chinese fertility transition is an important case in which to test the synthetic relationship betwee fertility rates with various economic and cultural-political determinants.

## II. PRELIMINARY CONSIDERATIONS OF THE
## SPATIAL STATISTICAL APPLICATION

In this chapter, more attention is focused on implicit variables since they are the ones that have not been incorporated in formal models. I contend that rigorous statistical analysis of them may be performed when the implicit variables are associated with spatial patterns. That is, implicit variables may be analyzed by spatial statistical methods when they possess three characteristics:

- [1] The causal relationship between such implicit variables and their response variable is established through theory;
- [2] the objects these implicit variables act on may be collectively measured as relatively homogeneous areal units, and their effects are generally measurable, although it may not be immediately separable from the effects of explicit variables; and,
- [3] the mechanism by which these variables take effect consists of either human groups responding simultaneously to a common source, or interacting with each other.

The effects of such variables may be found either as a spatial trend, or as spatial correlation and autocorrelation.

In the context of Chinese fertility determinants, one implicit variable that meets the above three criteria is governmental population policy. [Wolf, 1986; Yi, 1989; X. Peng, 1989; Pye, 1981] Governmental policies act on couples of child-bearing age. Policy effects are reflected by changes in the fertility rate of an area. The most widely used areal unit, for purposes of partitioning, is the administrative region. Implementation of policies usually emamates from the central government and cascades down its administrative hierarchy. [Feng, 1993; Greenhalgh, *et al.*, 1994] This characteristic implies that effects may be seen as simultaneous responses of areal units to a common source or group of sources: they may be regarded as creating one form of spatial interaction. Thus, one might expect to see a spatial trend in the effects of government policies once they are separated from effects caused by of other variables.

Therefore, the point of departure for this examination of implicit variables is the separation of their effects from those of explicit variables. This goal usually can be achieved with regression procedures if theory establishes that such procedures are valid. More specifically, the response variable may be regressed against the explicit explanatory variables and the residuals should represent the effects of the implicit variables. The critical point is to establish that no specification error occurs:  all possible effects in the response variable that are supposed to be caused by the explicit variables are exhausted.

One needs to note that in this approach, the residuals from regression analysis are not further separable. Hence when the effects latent in residuals are found to be statistically significant, it may become necessary to separate effects of different implicit variables. This is not necessary when undertaking spatial autoregressive modeling, but interpretation is rendered problematic when the issue is not addressed. Such separation requires that substantive theory supports the models.

The problem is greatly simplified when substantive theory identifies only one important implicit variable. This happens to be the case with the contemporary Chinese fertility decline. As is well documented, in recent years the Chinese economy is growing at a very fast pace and people's living standards are rising quickly. At the same time, the Chinese government has been rigorously carrying out a nationwide population control policy . Accompanying all these events, the nation's fertility rates have been dropping. A current debate focuses on whether the fertility decline results from socio-economic development or from the joint effects with population policy. The only factor that matches the socio-economic determinants in importance is the governmental population policy. Thus, this latter policy is considered to be the only implicit variable in this study.

### III. THE DATASET AND THE MODEL SPECIFICATION

The technique will be demonstrated through a case study of He-Nan Province of China (see Map 1 for the location of the province). The dataset is based on the Fourth National Population Census publications. [He-Nan Province Census Bureau, 1992] He-Nan was selected as representative of  the entire country in population density, living standards, economic structure, and cultural characteristics. The census took place at 0:00 a.m., July 1, 1990, for all provinces of China. Indices are enumerated at the county level and may be aggregated into levels of prefecture and province, which represent the three basic areal units of the Chinese administrative hierarchy.

Currently the country has thirty province-level administrative regions (excluding Taiwan, Hong Kong, and Macao), among which are three municipalities (Beijing, Shang Hai, and Tian Jin). A typical province covers about fifteen prefectures. A prefecture usually embraces about ten counties. Some cities (*shi*) may have jurisdiction over several neighboring counties. In that case a *shi* may be considered a prefecture-level administrative region. However, the city proper, which is only composed of its districts (*qu*) is still considered equivalent to a county.

In census publications some statistics are only reported for urban districts without being aggregated to the level of the cities themselves. In that situation an average is calculated for the districts and this mean value is taken as the value for the city proper. In this study of He-Nan Province, such procedures are performed on a total of 17 observations. The eventual dataset consists of 130 county-level areal units.

The raw data employed in this study may be categorized under three general headings, namely,

- (1) demographic structure,
- (2) economic structure, and
- (3) educational attainment.

They are briefly described in Appendix I. In the ensuing spatial analysis a fifth group of variables is added, which mainly pertains to the effects of government population policies.

The specified model is composed of two sub-models. The first is the following general linear regression model, which ignores relative geographic locations:

$$[\textbf{Fertility Rate}] = [\textbf{GroupI}]\beta_1 + [\textbf{GroupII}]\beta_2 + [\textbf{GroupIII}]\beta_3 + \varepsilon,$$

where [**Fertility Rate**] is the vector whose elements are the fertility rates for each county, and [**GroupI**] through [**GroupIII**] are variable matrices where the sets of vectors are the groups of variables described in Appendix I. $\beta_1$, $\beta_2$ and $\beta_3$ are parameter vectors, and $\varepsilon$ is an error vector.

The second, or spatial sub-model, starts with an analysis on this error vector, $\varepsilon$. The values of $\varepsilon$ may be considered as the portion of variation in [**Fertility Rate**] that is *not* associated with any of the explicit variables in the first model. It is the variation associated with implicit variables as well as random noise. At this juncture, the relative location and interaction of the regions is introduced.

The first question addressed in building the spatial sub-model asks whether or not any spatial pattern is latent in the error terms. If the answer is "no," then a case can be made for collectively denying the relevance of effects of implicit variables. But if the answer is "yes," then further investigation is in order. Detection of a spatial pattern relies on two indices, namely the Moran Coefficient (I) and Geary Ratio (c). [Griffith, 1987; Griffith and Amrhein, 1991, pp. 115-43] The first step is to construct a connectivity matrix. Two counties or county level administrative regions are considered connected if they share a common boundary. This boundary may be either a line or a point. In the matrix, each row as well as each column represents an areal unit. If the row-represented areal unit and the column-represented areal unit are connected, then the cell defined by the intersecting row and column is coded 1; otherwise it is coded 0. Their computational formulæ are

$$I = \frac{n}{\sum\limits_{i=1}^{n}\sum\limits_{j=1}^{n} c_{ij}} \times \frac{\sum\limits_{i=1}^{n}\sum\limits_{j=1}^{n} c_{ij}(y_i - \bar{y})(y_j - \bar{y})}{\sum\limits_{i=1}^{n}(y_i - \bar{y})^2},$$

and

$$c = \frac{(n-1)}{2\sum\limits_{i=1}^{n}\sum\limits_{j=1}^{n} c_{ij}} \times \frac{\sum\limits_{i=1}^{n}\sum\limits_{j=1}^{n} c_{ij}(y_i - y_j)^2}{\sum\limits_{i=1}^{n}(y_i - \bar{y})^2}$$

where i and j denote areal units, $y_i$ and $y_j$ are the attribute values for areal units i and j, and $c_{ij} = 1$ if units i and j are adjacent and 0 otherwise.

When a spatial pattern is detected, we need to analyze the particular form of the pattern and to provide an explanation in light of it. There are few quantitative studies on the effects of governmental policies because there are no associated explicit statistics. Here, spatial autocorrelation will be incorporated into a model of spatial variation for the fertility measurement itself, as well as the error term from a classical regression model that has the fertility rate as its response variable.

To begin, suppose the form of the autocorrelation model is specified as

$$\mathbf{B(Y\text{-}\mu)=De} \tag{1}$$

or

$$\mathbf{Y=\mu+B^{-1}De} \tag{2}$$

where $\mathbf{Y}$ is n-by-1 vector that is the spatially correlated variable under study. The mathematical objects $\mathbf{B}$ and $\mathbf{D}$ are nonsingular parameter matrices with $\mathbf{B}=\{b_{ij}\}$ and $b_{ii}=1$ for all i. The quantity $\mathbf{e}$ is an n-by-1 vector of residuals that are spatially independent of each other. It should have mean zero and covariance matrix $\sigma^2 \mathbf{V_e}$. Note that since its elements are independent of each other, $\mathbf{V_e}$ must be a diagonal matrix.

In addition, if we assume a constant variance across them, the diagonal elements of $V_e$ also should have identical values of one. Such an assumption is not necessary in this case but will add great convenience to future computations.

Practically speaking, the model says that after certain operations on $(Y-\mu)$ we should be able to filter out its spatial autocorrelation structure. That is, we should be able to transform it to a new vector $De$ that does not contain any spatial autocorrelation. This transformation is done through matrix $B$. The theoretical information we are seeking is revealed by the format of the transformation.

One format starts with the assumption that $D=I$ and $B=(I-S)$, where $I$ is the identity matrix. The meaning of $S$ will become clearer below. For now, assume that $(I-S)$ is invertible, and that the diagonal elements of $S$ are all zero. Thus, equation (2) may be rewritten as

$$Y=\mu+S(Y-\mu)+e. \tag{3}$$

In individual observation notation equation (3) is equivalent to

$$y_i=\mu_i+\Sigma_j s_{ij}(y_j-\mu_j)+e_i. \tag{4}$$

This last equation makes the interdependency structure a bit more intuitive. The value at areal unit $i$ depends on the mean of its distribution at that areal unit $(\mu_i)$, plus a weighted sum of the errors at the areal units connected to it $(\Sigma_j s_{ij}(y_j-\mu_j))$.

In the preceding discussion, the meaning of matrix $S$ has not been fully explained. In fact it is similar to the connectivity matrix discussed earlier. It determines whether areal units $i$ and $j$ are correlated (the connectivity matrix) and how strong that correlation is ($\rho$). One particular form of $S$ that is widely adopted among spatial statisticians is:

$$S=\rho W. \tag{5}$$

In equation (5), $\rho$ is a constant spatial autocorrelation parameter and $W$ retains its original definition as a row-standardized version of connectivity matrix $C$, with $\{w_{ij}\}=c_{ij}/(\Sigma_j c_{ij})$, and $\{c_{ij}\}=1$, if $j$ is connected to $i$, and $\{c_{ij}\}=0$ otherwise. Substituting equation (5) into equations (3) and (4) leads, respectively, to equations (6) and (7) below:

$$Y=\mu+\rho W(Y-\mu)+e, \tag{6}$$

$$y_i=\mu_i+\rho\Sigma_{j\epsilon N(i)} w_{ij}(y_j-\mu_j)+e_i, \tag{7}$$

where N(i) denotes the set of areal units that are connected to areal unit i, that is, the set for which $w_{ij} \neq 0$. Equations (6) and (7) specify a constant mean autoregressive model, or an autoregressive model with no explanatory variable. A full autoregressive model, or an AR model, arises when explanatory variables are required. It states that the realization of response variable **Y** at areal unit i is a function of its expected realization at i (**Xβ**), plus realizations of **Y** at locations connected to i ($\rho$ **WY**), plus an error term **e**. This relationship may be expressed by the following equation:

$$Y=X\beta+\rho WY+e, \tag{8}$$

where **X** is the matrix of explanatory variables, $X_{(n-by-p)} = (1_{(n-by-1)}, X_{1(n-by-1)}, X_{2(n-by-1)}, ..., X_{p(n-by-1)})$, and β is the parameter vector, $\beta = (b_0, b_1, b_2, ..., b_p)^T$, p being the number of explanatory variables. Other variables and parameters have been defined previously.

A full SAR model differs from an AR model in that it considers the impact of the explanatory variables not directly through the expected realization of the response variable, but through its regression residuals. The specifications are as follows:

$$Y=(I-\rho W)X\beta+\rho WY+e, \tag{9}$$

which may be rewritten as

$$Y-X\beta=\rho W(Y-X\beta)+e. \tag{10}$$

This equation states that the residual from a linear regression of the response variable at location i is a function of the residuals at all locations connected to i, plus a truly random error term **e**.

One condition for equations (6), (7), (8), (9) and (10) that must hold is that **B**, which now may be defined as (I-$\rho$W), must be invertible. Suppose $\lambda_{min}$ and $\lambda_{max}$ are the smallest and largest eigenvalues of matrix **W**; then

$$(1/\lambda_{min})< \rho <(1/\lambda_{max}). \tag{11}$$

This restriction also ensures that the logarithmic argument of the Jacobian term is positive (see expression (26) in section IV). As $\lambda_{min}$ and $\lambda_{max}$ depend on *n* as well as the connectivity structure, so does the permissible range of $\rho$ .

With the general form of the function defined, the covariance matrix of the observations on the areal units becomes

$$
\begin{aligned}
\mathbf{V} &= E[(\mathbf{Y}\text{-}\boldsymbol{\mu})(\mathbf{Y}\text{-}\boldsymbol{\mu})^T] \\
&= E[\mathbf{B}^{-1}\mathbf{De}(\mathbf{B}^{-1}\mathbf{De})^T] \\
&= E[\mathbf{B}^{-1}\mathbf{Dee}^T\mathbf{D}^T(\mathbf{B}^{-1})^T] \\
&= E[\mathbf{B}^{-1}\mathbf{DV_e}\mathbf{D}^T(\mathbf{B}^{-1})^T] \\
&= \sigma^2[\mathbf{B}^{-1}\mathbf{DV_e}\mathbf{D}^T(\mathbf{B}^{-1})^T].
\end{aligned}
\tag{12}
$$

Substituting $\mathbf{D}=\mathbf{I}$ and $\mathbf{B}=(\mathbf{I}\text{-}\mathbf{S})$, we have

$$
\mathbf{V} = \sigma^2[(\mathbf{I}\text{-}\mathbf{S})^T(\mathbf{I}\text{-}\mathbf{S})]^{-1} .
\tag{13}
$$

Substituting $\mathbf{S}=\rho\mathbf{W}$, we have

$$
\mathbf{V} = \sigma^2[(\mathbf{I}\text{-}\rho\mathbf{W})^T(\mathbf{I}\text{-}\rho\mathbf{W})]^{-1} .
\tag{14}
$$

The covariance matrix defined by equation (14) is in fact one of the most elementary features distinguishing SAR and AR models from others. [Griffith, 1993, p. 43] This way of defining the model also permits the establishment of a joint density function for **Y**, which is the basis for parameter estimation and inference with the model.

## IV. EXPLICIT VARIABLES

### A. FERTILITY RATE

The linear regression model, the first of the aforementioned two, depicts the influence of various explicit determinants upon the variation of fertility rates, which is the sole response variable. Multiple fertility measures may be computed based on the data available, but the measure this paper relies on is the United Nations' Age-Sex Adjusted Birth Rate. It is calculated as follows:

$$
\frac{b}{\Sigma(w_a p_a^f)} \times 1,000 ,
$$

where,

$b$ is the total births within a year,

$p_a^f$ is the number of females in age group a, and

$w_a$ is the weight for age group a.

This weight $w_a$ merits further elaboration. In order to provide a basis for comparison among the wide array of indices from countries where the information is less complete, the United Nations developed a "standard population" based on the data from 52 nations for which the data are more complete. [Shryrock and Seigal, *et al.*, 1980] "Standard" age-specific birth rates are derived from this dataset, and standard weights, $w_a$, are computed accordingly. The actual birth rate estimates computed on the basis of such "standard" weights are considered quite robust. It is recognized that even a substantial modification of the weights would have only a slight effect on the relative levels of the adjusted birth rates being estimated. [Shryock and Siegel, *et al.*, 1980, p. 483] Its advantages over other measures may be summarized under three general headings. First of all, it utilizes all the information available, and at the same time avoids age specific birth rates that are not available in the Chinese dataset. Second, it allows one to exclude from an analysis fertility variation due to age and sex structure, as well as "effective" fertility, which is not immediately relevant to the relationship between fertility behavior and socio-economic determinants. Finally, it generates a single index of fertility for each areal unit (county), a great convenience for subsequent quantitative analyses.

## B.  GENERAL LIVING STANDARD

The most widely used index for living standard is per capita income. While this is a good index in general, it is unacceptable in the Chinese case, especially within the context of this study. The main reason is the large discrepancy between income in absolute U.S. dollar value and actual purchasing power.

Within the constraint of data availability, I choose the Indirect Age-sex Standardized Death Rate as a measure of the general living standard. It is seen as a measure of comprehensive quality of life rather than one particular aspect of it. This particular measure considers the age and sex structure of the underlying population, and at the same time generates a single index for each area. It is based on the indirect standardization formula as summarized by Shryock and Siegel [1980, p. 421] with the addition of $w_a$, the sex structure adjustment weight, and is calculated as follows:

$$m_2 = (\frac{d}{\Sigma(M_a W_a P_a)})M,$$

where $m_2$ is the indirect age-sex standardized death rate,

- $d$ is the total deaths within the year for the actual population,
- $P_a$ is the population for age group a for the actual population,
- $M_a$ is the age specific death rate for the "standard" population,
- $M$ is the crude death rate of the "standard" population, and is calculated as

$$\frac{\text{Total number of deaths with the year}}{\text{Total mid-year population}} \times 1000$$

and,

- $w_a$ is the sex structure adjustment weight for age group a, and is calculated for this age group as

$$\frac{\text{Standard population female, numbers}}{\text{Actual population female, numbers}} \times \frac{\text{Actual population male, numbers}}{\text{Standard population male, numbers}}$$

The population data of the United States in 1960 has been widely used in general demographic research, and it is adopted here as the "standard population" because of its completeness and historical stability.

## C.  ECONOMIC STRUCTURE

Economic structure may be appropriately characterized by employment structure of a region. The Chinese national census classifies regional economic activities into eleven broad categories. Generally there is very little discussion in the literature about a dominant influence of any particular industry on fertlity. Two possible exceptions are agriculture and mining. The former is regarded as fostering a life style that is essentially pronatalist. The latter contributes to fertility rates through the Chinese government's implementation of its population policies. Perhaps because of the dangerous nature of working conditions, miners consistently have been exempt from the restrictions of the one-child policy.

In this study, the first treatment of the employment figures is to compute the percentage of non-agricultural population and of mining employment. Of note is that the non-agricultural population percentage is based on total population rather than overall employment because of the more pervasive nature of the factor. Mining employment figures are subtracted from "employment in industries (light, heavy, and mining)" in which it is originally included. Subsequently "employment in industries" no longer contains that information. This alternation eventually results in 13 variables, all of which are expressed as a percentage of a region's total employment, except the non-agricultural population percentage (see Table 1). They roughly reflect the classification in the Chinese national census.

**Table 1**. A list of economic variables.

| Variables | Explanation |
|-----------|-------------|
| AGRI(%) | Percentage of non-agriculture population. [(non-agri.)/(tot.pop.)]*100 |
| PURE_IND(%) | Percentage of industrial employment, excluding mining industry employment. [(ind.emp.)/(tot.emp.)]*100 |
| CPGE(%) | Percentage of geological survey industry employment (*100). |
| CPCO(%) | Percentage of construction industry employment (*100). |
| CPTR(%) | Percentage of transportation and communication industry employment (*100). |
| CPRE(%) | Percentage of retail and wholesale industry employment (*100). |
| CPST(%) | Percentage of real estate and related service industry employment (*100). |
| CPMD(%) | Percentage of medical, sports human welfare service industry employment (*100). |
| CPED(%) | Percentage of education, entertainment, TV, and radio industry employment (*100). |
| CPSC(%) | Percentage of scientific industry employment (*100). |
| CPFI(%) | Percentage of finance and insurance industry employment (*100). |
| CPGO(%) | Percentage of government and socio-political organization industry employment (*100). |
| MING(%) | Percentage of mining industry employment (*100). |

These thirteen economic variables have been subjected to a factor analysis, in order to obtain a few synthetic indicators of a region's economic structure. It is needed because almost all classification schemes of industrial activities are based on either their market orientations or their characteristics of production processes, with no regard to their influence on fertility rates. That is apparently the case with the one employed here. However, although the measures individually have very little relevance to the determination of fertility behavior, a much smaller number of common factors from a factor analysis may become good indicators. If there is a significant relationship between a region's economic characteristic and its fertility rate, the relation should be captured by the scores of such common factors. Furthermore, this process will solve a potential lack of degrees of freedom problem, and at the same time reduce or eliminate predicted multicollinearity. Uniformity of measurement and equivalence of substantive meaning (all variables are percentages) make it possible to directly interpret the results from rotated factor loadings without any further manipulation.

Table 2 shows the first four eigenvalues from the factor analysis, together with their corresponding proportion of the total variance. Factor extraction is based on the correlation matrix of the variables. Hence the first three factors, which are the only ones with eigenvalues larger than unity, are retained. As the table shows, the first common factor accounts for about 70 percent of total variance, the second and third factors each account for less than 10 percent. Overall these three common factors account for 88 percent of the total variance among the original variables, indicating a fairly parsimonious description of the data.

**Table 2**. Eigenvalues of the Correlation Matrix

|  | Factor1 | Factor2 | Factor3 | Factor4 |
|---|---|---|---|---|
| Eigenvalue | 9.2705 | 1.0983 | 1.0110 | 0.5437 |
| Proportion | 0.7131 | 0.0845 | 0.0778 | 0.0418 |

The three retained factors then were subjected to a varimax rotation, a method that simultaneously rather than sequentially maximizes the variance of each factor's loadings for clearer interpretation. Table 3 lists the resulting rotated factor loadings as well as communalities associated with each original variable after standardization. It seems clear that AGRI, PURE_IND, CPTR, CPRE, CPST, CPMD, CPED, CPFI, and CPGO define the first common factor, with very high absolute values of loadings. CPGE defines the second common factor, and MING defines the third. The relationships between CPCO and CPSC and the first two common factors is somewhat unclear. They both have loadings slightly higher on the first than on the second. The difference, however, is not large enough to warrant a marked distinction. From the communality values it seems the variances of most original variables can be explained by the common factors, with the exceptions of CPCO and CPSC.

**Table 3**. Varimax Rotated Factor Loadings

| Variables | FACTOR 1 | FACTOR 2 | FACTOR 3 | Final Communality Estimates |
|---|---|---|---|---|
| AGRI | -0.88790 | -0.14820 | -0.26534 | 0.880734 |
| PURE_IND | 0.94222 | 0.05774 | 0.15080 | 0.913859 |
| CPGE | 0.06190 | 0.93807 | -0.00851 | 0.883886 |
| CPCO | 0.54507 | 0.47491 | 0.26138 | 0.590962 |
| CPTR | 0.91513 | 0.16293 | 0.08375 | 0.871018 |
| CPRE | 0.97955 | 0.06343 | 0.00223 | 0.963537 |
| CPST | 0.91711 | 0.32644 | 0.07546 | 0.953342 |
| CPMD | 0.96191 | 0.15374 | 0.13390 | 0.966836 |
| CPED | 0.89874 | 0.21970 | 0.08996 | 0.864096 |
| CPSC | 0.66009 | 0.48852 | 0.00884 | 0.674445 |
| CPFI | 0.94440 | 0.21188 | 0.03024 | 0.937710 |
| CPGO | 0.93677 | 0.14698 | 0.08348 | 0.906104 |
| MING | 0.10622 | 0.02716 | 0.98041 | 0.973225 |

These factors may be regarded as indices of the respective region's economic structure. Based on the loadings, it appears that the first factor measures the general economic strength. A region with a high score on this factor is expected to have less agricultural employment, but more industrial employment in almost all urban economic sectors. Regions having high scores on the second factor are supposed to have high geological survey employment, which in this province is mainly in oil exploration. This observation explains the undifferentiated loadings on CPCO and CPSC between the first and second factor: the geological exploration industry characteristically employs a high proportion of scientists, especially geologists and geophysicist, and demands constant construction because of the mobility of working sites. High scores on the third factor indicate an especially strong mining sector, which is mainly coal mining.

## D. EDUCATIONAL ATTAINMENT
The 1990 Chinese census provides six educational attainment variables. They are presented in Appendix I. In model building, they are lumped into three more comprehensive variables, as percentages of the total population holding diplomas from:

- [1] EDCO: college level institutions, including four year colleges, two-to-three year colleges or equivalents, and professional high schools,
- [2] EDHI: high school level institutions, including regular high schools and junior-high schools, and
- [3] EDEL: elementary schools.

Together these three variables measure a region's educational composition of its general population. These are the percentages of people who have graduated from, as opposed to currently enrolled in, various schools. The former, which includes all groups of child bearing age women, should be more closely related to current fertility rates than the latter, which excludes such groups. The separation of the three has to do with varying degrees of prevalence of different levels of education. Presently in most areas of the country, including He-Nan Province, elementary school education is universal, involving almost all children ages seven through twelve. College education, however, belongs largely to the elite group, with fewer than 20 percent of high school graduates having the opportunity to enter college each year. Opportunities for high school education lie between the two: more widespread in urban areas and less so in rural areas. It is suspected that different levels of education will have different impact on fertility behavior, which should be captured by the differentiation these three variables provide.

## E. STATE POPULATION POLICIES
We expect counties closer together to display similarities -- the presence of spatial autocorrelation -- in terms of the effects of state population policies. The basis of such an expectation is the spatial interaction among counties, which are probably the most important nodes in the Chinese territorial administrative hierarchy. They interact with each other through many channels, including information flows, human migration, commercial products transportation, financial transactions, and service deliverance. It is extremely difficult to construct a numerical index of interaction that covers all of these diverse categories. But distance decay theory suggests that a good proxy for such interaction is spatial contiguity, which indicates that counties next to each other interact more than those farther apart. Since regional

interaction brings about regional homogeneity, counties in one general area should have similar socio-economic attributes. Such similarities ensure that when faced with the same set of stimuli, counties tend to respond in similar ways. Thus, when the counties in one general area are pressured with the same set of fertility reduction policies, they should react with similar patterns in changes of fertility rates. This is the basis of our expectation of spatial autocorrelation in the effects of such policies.

## V.  A CLASSICAL LINEAR REGRESSION MODEL
## OF EXPLICIT VARIABLES

After a series of separate examinations of the individual determinant groups, I set out to disentangle their collective relationship with the fertility rates. At this stage, only explicit variables are considered. An obvious choice of method for such a model is linear regression.

The response variable in this model is the U.N. Age-Sex Specific Birth Rate. The explanatory variables include all those that are explicitly derived. Before further analysis, some descriptive statistics are computed and scatter-plots are drawn for the response and all explanatory variables. Several outliers have been identified based on the initial scatter-plots as well as Cook's D statistic. The outlying observations and variables are listed in Table 4. Possible explanations for some of the extreme values are offered in Table 5. Outliers concerning Factors II and III have an obvious explanation. Reasons, however, for those outliers, in relation to death rates and elementary school graduate percentages, are not very clear. A more substantive investigation seems needed in order to provide satisfactory explanations for them. Because none of the outliers are found as errors, all of the outlying observations are retained. However, caution needs to be used in interpretation.

**Table 4.** Outlying Variables and Observations

| ID | DETHR | EDEL | FACTOR2 | FACTOR3 |
|------|-------|--------|---------|---------|
| 1042 | - | - | - | 6.58 |
| 1058 | | - | 9.18 | - |
| 1063 | - | - | - | 6.97 |
| 1080 | 19.19 | - | - | - |
| 1082 | - | - | - | 3.78 |
| 1116 | - | 0..646 | - | - |

**Table 5.** Possible Explanations for Outliers

| ID | Out Lying Variable | Explanation |
|----|----|----|
| 1042 | FACTOR3(+) | Ping-ding-shan Shi, one of the largest coal mining areas in the country, and the largest in the province. |
| 1058 | FACTOR2(+) | Lin County, Yan-Jan area, which needs on going irrigation projects. |
| 1063 | FACTOR3(+) | He(4) Bi(4) Shi, a major coal mining area. |
| 1080 | DETHR(+) | Wei Hui County, unclear. |
| 1082 | FACTOR3(+) | Jiao Zou Shi, a major coal mining area. |
| 1116 | EDEL(+) | Ling Bao County, unclear. |

Next maps were drawn using Arc/Info GIS software for variables FERTR, DETHR, EDCO, EDHI, EDEL, and the three common factors extracted from the thirteen economic variables (see Appendix II, Maps 2 through 9). Areal units are classified according to the four quartiles of each variable. Inspection reveals that a group of areas that are small in geometrical size consistently stays in either the first or the fourth quartile, with essentially no relationship with their surrounding regions. It seems as if they are from a separate population. Further investigation shows that these areas actually are the major urban centers for which data for the districts are aggregated at city levels. Subsequently a binary indicator variable is introduced, with its value being one if the areal unit is a major urban center and zero otherwise, increasing the total number of explanatory variables to nine. Their measurement and content definitions are described in Table 6.

**Table 6.** Explanatory Variables in the Classical Linear Regression Model

| Variables | Measurement Unit | Definition |
|---|---|---|
| FERTR | UN Age-Sex Adjusted Birth Rate | Counts of births per thousand child-bearing age women, dependent variable |
| DETHR | Indirect Age-sex Standardized Death Rate | Used as a measure of comprehensive living standards. |
| EDCO | Percentage of total population holding diplomas from college level institutions. | Used as a measure of educational attainment structure, as well as education level of the general population. |
| EDHI | Percentage of total population holding diplomas from high school level institutions. | Used as a measure of educational attainment structure, as well as education level of the general population. |
| EDEL | Percentage of total population holding diplomas from elementary school level institutions | Used as a measure of educational attainment structure, as well as education level of the general opulation. |
| FACTOR1 | Common factor scores based on percentage. | Used as a measure of economic structure of the counties. It is positively correlated with general urban industries and negatively correlated with agricultural population proportion, with indifference to mining industries. |
| FACTOR2 | Common factor scores based on percentage. | Used as a measure of economic structure of the counties. It is positively correlated with geological survey industry, with a very week positive correlation with construction and scientific research industry. |
| FACTOR3 | Common factor scores based on percentage. | Used as a measure of economic structure of the counties. It is positively correlated with mining industries, with indifference to other industries. |
| DM | 1 if the areal unit is a major urban center and 0 otherwise. | May be a comprehensive socio-economic development indicator. |

To show the impact of the urban-regional dichotomy, a brief ANOVA report for the simple regressions on each explanatory variable is displayed in Table 7. It suggests that this effect made a difference that is statistically significant for *all* variables. Furthermore, it explains a large proportion of variance in fertility rate (FERTR), the three educational attainment variables (EDCO, EDHI and EDEL) as well as general economic strength (FACTOR1), thereby showing the intensity and the comprehensive nature of the effect. Correlation coefficients between DM and other variables may be found in Table 8. All of these point to one conclusion: the urban-regional dichotomy is indeed an important variable.

**Table 7.** A Brief ANOVA Table of Regressions on the Indicator Variable

| Explanatory Variables | Probability F | Adjusted R-Square |
|---|---|---|
| FERTR | 0.0001 | 0.4466 |
| DETHR | 0.0001 | 0.2403 |
| EDCO | 0.0001 | 0.8482 |
| EDHI | 0.0001 | 0.4550 |
| EDEL | 0.0001 | 0.4502 |
| FACTOR1 | 0.0001 | 0.6352 |
| FACTOR2 | 0.0024 | 0.0627 |
| FACTOR3 | 0.0007 | 0.0790 |

**Table 8.** Correlation coefficients between response and explanatory variables in linear regression model

(Pearson Correlation Coefficients / Prob > |R| under Ho: Rho=0)

| | FERTR | DETHR | EDCO | EDHI | EDEL | FACTOR1 | FACTOR2 | FACTOR3 |
|---|---|---|---|---|---|---|---|---|
| DETHR | 0.20956 | | | | | | | |
| | 0.0167 | | | | | | | |
| EDCO | -0.71308 | -0.39293 | | | | | | |
| | 0.0001 | 0.0001 | | | | | | |
| EDHI | -0.56507 | -0.27977 | 0.71263 | | | | | |
| | 0.0001 | 0.0013 | 0.0001 | | | | | |
| EDEL | 0.34221 | 0.36475 | -0.67912 | -0.68176 | | | | |
| | 0.0001 | 0.0001 | 0.0001 | 0.0001 | | | | |
| FACTOR1 | -0.65896 | -0.32534 | 0.89666 | 0.66025 | -0.60479 | | | |
| | 0.0001 | 0.0002 | 0.0001 | 0.0001 | 0.0001 | | | |
| FACTOR2 | -0.14405 | -0.16331 | 0.27865 | 0.03748 | -0.10124 | 0.00000 | | |
| | 0.1020 | 0.0634 | 0.0013 | 0.6720 | 0.2518 | 1.0000 | | |
| FACTOR3 | -0.13737 | -0.17158 | 0.19125 | 0.28549 | -0.18841 | 0.00000 | 0.00000 | |
| | 0.1191 | 0.0509 | 0.0293 | 0.0010 | 0.0318 | 1.0000 | 1.0000 | |
| DM | -0.67146 | -0.49618 | 0.92965 | 0.67763 | -0.60479 | 0.79879 | 0.26447 | 0.29352 |
| | 0.0007 | 0.0001 | 0.0001 | 0.0001 | 0.0001 | 0.0001 | 0.0024 | 0.0007 |

Next consider an examination of linearity. Pairwise relationships between each of the independent variables and the dependent variable may be examined through the plots in Appendix III. The relationships all seem either roughly linear or random. The only exception is variable

EDCO, which seems to have a quadratic relation with FERTR. A square root transformation is applied to EDCO, and the plot improves slightly (Plot 2). More precise measurements of correlation among the variables are revealed by the coefficients in Table 8. Most of them are high and statistically significant.

Finally, following the linearity examination, we may proceed to fit a linear regression model. The first step is to select the right explanatory variables. The PROC REG procedure in SAS is used with the MAXR option. This procedure examines all possible combinations of explanatory variables and, according to the highest R-square of each model, gives the best 1, 2, through 9 explanatory variable models. Next, R-squared, adjusted R-squared, and Mallow's C(p) statistic are used to select the best model among the ones given by the MAXR regression procedure. [Daniel and Wood, 1983, Chapter 6] A plot of R-square and adjusted R-square versus the number of variables in the best R-square model may be examined in Figures 1 and 2. The two figures show that both R-square and adjusted R-square rise quickly until there have been three variables entered into the model, which are EDCO, EDHI, and EDEL. Afterwards they continue to rise, only with a much flatter slope, until there have been five explanatory variables entered into the model. The three variables already in the model are kept, and two more, DM and DETHR, are added. After that, the adjusted R-square actually startes to fall. A similar pattern is displayed by the C(p) statistic, as is depicted in Figure 3. The C(p) statistic has an expected value that equals p, where p is the number of explanatory variables in the model, including the vector of ones for the intercept term. When C(p) values for all possible regression models are plotted against p, those models with little bias will tend to fall near the line C(p)=p. Models with substantial bias will tend to fall considerably above this line. C(p) values below the line C(p)=p are interpreted as showing no bias. In Figure 3, the C(p) value falls rapidly until there are three variables in the best R-square model, which are, as before EDCO, EDHI and EDEL. Then it starts to fluctuate slightly around the line C(p)=p. All of the three plots seem to point to one conclusion: that the three educational attainment variables should be included in the model. The inclusion of DM and DETHR, which only made a slight improvement in the R-square, are debatable. The decision concerning their inclusion has to be based on substantive grounds. Since all three firmly selected variables only provide information on educational development, it appears that inclusion of the two additional variables should add information that is different in nature. Thus the final linear regression model has five explanatory variables. They are EDCO, EDHI, EDEL, DM, and DETHR.

**Table 9.** Linear Regression Results

## Analysis of Variance

| Source | DF | Sum of Squares | Mean Square | F Value | Prob>F |
|--------|----|----------------|-------------|---------|--------|
| Model | 5 | 771.02577 | 154.20515 | 34.441 | 0.0001 |
| Error | 124 | 555.19695 | 4.47739 | | |
| C Total | 129 | 1326.22272 | | | |

| | | | |
|---|---|---|---|
| Root MSE | 2.11599 | R-square | 0.5814 |
| Dep Mean | 17.98400 | Adj R-sq | 0.5645 |
| C.V. | 11.76593 | | |

## Parameter Estimates

| Variable | DF | Parameter Estimate | Standard Error | T for H0: Parameter=0 | Prob > \|T\| |
|----------|----|--------------------|----------------|------------------------|-------------|
| INTERCEP | 1 | 31.715507 | 2.36320283 | 13.421 | 0.0001 |
| EDCO | 1 | -29.840538 | 8.43103179 | -3.539 | 0.0006 |
| EDHI | 1 | -9.993002 | 3.53977444 | -2.823 | 0.0055 |
| EDEL | 1 | -18.055830 | 4.36639516 | -4.135 | 0.0001 |
| DM | 1 | -2.098131 | 1.63157467 | -1.286 | 0.2009 |
| DETHR | 1 | -0.167109 | 0.15227221 | -1.097 | 0.2746 |

| Variable | DF | Standardized Estimate | Variance Inflation |
|----------|----|------------------------|---------------------|
| INTERCEP | 1 | 0.00000000 | 0.00000000 |
| EDCO | 1 | -0.60095460 | 8.53930537 |
| EDHI | 1 | -0.25377528 | 2.39358564 |
| EDEL | 1 | -0.36062950 | 2.25281089 |
| DM | 1 | -0.22147074 | 8.78562246 |
| DETHR | 1 | -0.07568409 | 1.40878091 |

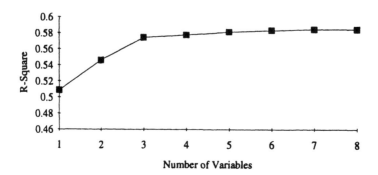

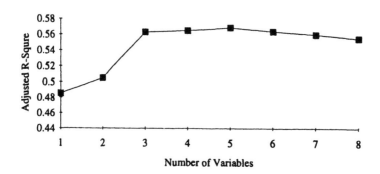

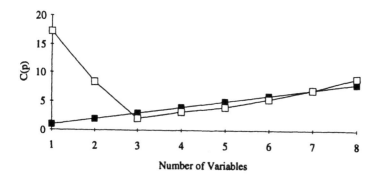

**Figure 1 (top).**  R-square model selection.
**Figure 2 (middle).**  Adjusted R-square model selection.
**Figure 3 (bottom).**  C(p) model selection

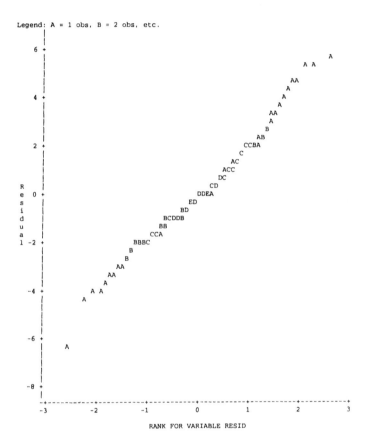

**Figure 4**. Plot of RESID*NORMRES. Legend: A=1, obs, B=2 obs, etc.

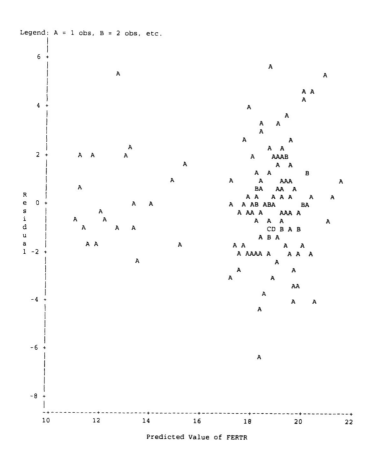

**Figure 5**. Plot of RESID*PRED. Legend: A=1, obs, B=2 obs, etc.

Linear regression results are displayed in Table 9. To ensure validity of model-based inferences, a test of normality as well as an inspection for heteroscedasticity of the error terms was conducted. The former is done first by a normal plot, which is a plot of actual residuals from the model against normal scores. The actual plot is displayed in Figure 4. It shows a relatively straight line. A more rigorous formal test is performed through the PROC UNIVARIATE procedure in SAS, yielding a Wilk-Shapiro statistic of 0.980733 and Pr<W=0.4933. Indeed, the hypothesis that the residuals are from a normal distribution may not be rejected. To inspect for nonconstant variance, the error terms are plotted against predicted values and the outcome is displayed in Figure 5. It does seem to suggest a minor problem of heteroscedasticity. However, since a transformation of the response variable would considerably complicate interpretation and the problem does not seem very serious, the results are taken without further action. As a result, interpretation of the parameter estimates will have to be approached with some caution.

Next the potential problem of multicollinearity is considered. This problem does not affect either the validity of inference or the goodness of fit of the model. However, when it is serious, the parameter estimates for the independent variables become unreliable, which may cause faulty interpretation. One way to examine the impact of multicollinearity is simply to add a VIF option to the PROC REG procedure. The Variance Inflation Factor is reported in the summary table (see the column under VIF in Table 9). More insights may be gained by examining the correlation matrix for parameter estimates (see Table 8). In Table 9, only the VIF values of EDCO and DM seem bothersome, but not troublesomely, high. Examination of Table 8 reveals that they actually have a correlation coefficient of 0.92159, which indicates that most information contained in DM also is contained in EDCO. Caution should be exercised in later interpretation of the standard errors of the parameter estimates of the two variables.

The univariate procedure results, normal and residual plots, as well as examination of residuals have provided the basis for adopting the regression outcomes. The overall model seems highly significant, with an Adjusted R-square of 0.5645, a very high value among social science studies. Variables DM and DETHR have regression parameter estimates that are not significantly different from zero, but those for all three educational attainment variables, EDCO, EDHI, and EDEL, are significant at the 0.05 level. Judging from the standardized parameter estimates, of the three variables that have non-zero impacts upon the response variable, EDCO is the most important factor, followed first by EDEL and then by EDHI. It needs to be noted at this point that faith put on the significance of parameter estimates should be discounted slightly because of heteroscedasticity. The eventual multiple linear regression model seems to be, with natural parameter estimates: FERTR=31.715507-29.840538 EDCO-18.055830EDEL-9.993002EDHI +e

## VI. IN SEARCH OF A SPATIAL PATTERN

The linear model developed in the last section demonstrates the co-variation of fertility rates in He-Nan province with its major socio-economic determinants. Residuals from the model represent the part of variation in fertility rates that does not go hand in hand with those of the included socio-economic variables. A natural ensuing question asks whether or not the residuals co-vary with other factors that have not been included in the model. This question may be answered by an examination of the residuals. Two possibilities arise. First, if no pattern still exists, then we may conclude that the explanatory variables already in the model have explained all variation in the response variable and therefore no further investigation is needed. However, as a second possibility, if significant pattern remains, we will have to go on to explain the part that has not been explained by the existing explanatory variables. Otherwise the model remains insufficient. At this point it seems clear that if we were to accept all assumptions of a linear regression, including that of independence among observations, then it would indeed seem that the residuals are random. However, given the strong spatial nature of the observations as well as the subject matter in general, it is only reasonable to ask if the residuals are also random spatially, or if there is still a latent spatial pattern. This question leads to a scrutiny of spatial autocorrelation among the residuals.

Among the most widely implemented measures of spatial autocorrelation in geo-referenced data are the Moran Coefficient (I) and the Geary Ratio (c). They have been briefly discussed earlier. In this section, focus is on their application to this particular data analysis.

Again, the essential task is to seek to uncover a potential spatial pattern. Computation is based on Griffith [1993, p. 24-25] and the results are as follows:

N   CSUM   I   ZAPP   PROBMC   c

130 656 0.39544 7.36852 8.6375E-14 0.58742

where N is the number of observations, CSUM is the total number of connections in the surface partioning, I is the calculated Moran

Coefficient, ZAPP is the approximate Z (standard normal) score for the Moran Coefficient, PROBMC is the probability to reject the null hypothesis that there is no spatial autocorrelation based on the Z score, and c is the calculated Geary Ratio.

Roughly speaking, the Moran Coefficient falls into the interval (-1, 1). Its expected value, signifying zero spatial autocorrelation, is $-1/(n-1)$, which in this particular case should be -0.00752. As one may see, the observed Moran Coefficient is highly significant and well above the expected value, showing that there is evidence of positive spatial autocorrelation. The Geary Ratio has a rough range of (0, 2) with an expected value of 1. The calculated Geary Ratio confirms the conclusion suggested by the Moran Coefficient that, indeed, there is spatial pattern not explained by the linear regression model. Our immediate task, then, is to uncover such pattern.

To understand the process of parameter estimation and statistical inference, there needs to be some introduction to the Jacobian term. A Jacobian term in this context serves as a normalizing constant. Specifically, suppose $X=(x_1, ..., x_n)^T$ is a vector of random variables that have a joint probability density function (p.d.f.) of $f(x_1, ..., x_n)$. Let $Z=(z_1, ..., z_n)^T$ be another vector of random variables, as well as a function of $X$ with the following form:

$$z_i=g_i(x_1, ..., x_n), i=1, 2, ..., n. \tag{15}$$

Also assume that a unique inverse of (15) always exists and is

$$x_i=w_i(z_1, ..., z_n), i=1, 2, ..., n. \tag{16}$$

If the partial derivatives $\partial w_i/\partial z_i$ exist for all i, then the Jacobian term for the derivation of the joint p.d.f. for $Z$ is

$$J = \det \begin{bmatrix} \partial w_1/\partial z_1 & ... & \partial w_1/\partial z_n \\ ... & & ... \\ \partial w_n/\partial z_1 & ... & \partial w_n/\partial z_n \end{bmatrix} \tag{17}$$

The derived joint p.d.f. of $(z_1, ..., z_n)$ then is

$$h(z_1, ..., z_n)=f(x_1, ..., x_n)|J| \tag{18}$$

for all values in the feasible parameter space, and 0 otherwise. That is, the joint p.d.f. of $Z$ is generated by substituting for each $x_i$ in its own p.d.f., $f(x_1, ..., x_n)$, $x_i=w_i(z_1, ..., z_n)$ and multiplying the result by the absolute value of the Jacobian term of the transformation.

The Jacobian term is important here because it enables us to derive a joint p.d.f. of the vector $Y$ in the original model specified by equations (1) and (2), on the basis of the joint p.d.f. of $e$, and this derived p.d.f. is the basis for a maximum likelihood estimation of the parameter $\rho$ as well as $\beta$ and $\sigma^2$ of $e$. Derivation of the joint p.d.f. of $Y$ is as follows:

Assume all random variables in $e=(e_1, ..., e_n)^T$ have identical probability density functions

$$f(e_i), i=1, 2, ..., n. \qquad (19)$$

Since $e$ is diagonal, the joint probability density function of $(e_1, ..., e_n)$ is

$$f_e(e_1, ..., e_n) = \prod_{j=1}^{n} f(e_j) \qquad (20)$$

Because, $Y$ is a function of $e$ (recall (2) and the assumptions that $C=I$, and $B=I-S$),

$$Y=\mu+B^{-1}e \qquad (21)$$

and,

$$e=B(Y-\mu) \qquad (22)$$

both hold. Then, without substituting $(I-S)$ for $B$ for simplicity and using algebraic notation, the p.d.f. of $Y$ based on the p.d.f. of $e$ is,

$$\phi(y_1, ..., y_n) = |J|f_e(e_1, ..., e_n)$$
$$= |J| \, f_e(\sum_{i=1}^{n} b_{1i}(y_i - \mu_i), ..., \sum_{i=1}^{n} b_{ni}(y_i - \mu_i)) \qquad (23)$$

where $|J|$ is the Jacobian term for transforming the joint p.d.f. from one of $e$ to one of $Y$. This process is also referred to as a transforming process between a spatially un-autocorrelated space and a spatially autocorrelated space.

Suppose $e$ has a standard normal distribution with no latent spatial autocorrelation. Then the joint p.d.f. of $e$ should be of the following form:

$$\prod_{i=1}^{n}(2\pi\sigma^2)^{-\frac{1}{2}}\exp[-\frac{1}{2\sigma^2}(e_i)^2] \qquad (24)$$

Based on equation (23), this results in the following joint p.d.f. of **Y**:

$$\phi(y_1, ..., y_n)$$

$$= |J|\prod_{j=1}^{n}(2\pi\sigma^2)^{-\frac{1}{2}}\exp[-\frac{1}{2\sigma^2}(b_j(y_i-\mu_i)^2] \qquad (25)$$

The above is also the effective part of the maximum likelihood function of **Y**, and some spatial autocorrelation information is contained in the Jacobian term |J|. Note that here we see each areal unit as a variable instead of an observation. In other words the single univariate sample of *n* is viewed here as a multivariate sample of size one with n variables.

The maximum likelihood method seeks to find the parameter estimates that specify a distribution that optimally represents the actual sample. This is accomplished through maximizing the likelihood function, which may be defined as the joint probability density function of the observed random sample. Mathematically, the estimates are obtained by setting the partial derivatives, of the maximum likelihood function with respect to the parameters, equal to zero. The values that satisfy this system of equations will be the sought estimates. This method is often deployed when the ordinary least squares method is not applicable or is inappropriate.

In this study, the ordinary least squares method is inappropriate because, as will be shown shortly, the estimation equation for p is non-linear by construction. Maximum likelihood seems therefore to be an appropriate method. However, as Ripley [1990] points out, precisely because of this non-linearity, analytical solutions to the equations are not implementable. The most problematic complication is the excessive numerical intensity involved, which is introduced mainly by the Jacobian term. For an SAR or AR model the Jacobian term is of the form

$$-J = \frac{2}{n}\sum_{i=1}^{n}\ln(1-\rho\lambda_i), \qquad (26)$$

where $\lambda_i$ is the i-th eigenvalue of the stochastic version of the connectivity matrix **W**. This specification requires the calculation of the eigenvalues of an n-by-n matrix. In addition, when non-linear optimization is used for maximum likelihood estimation, the n logarithm terms and their summation have to be recomputed for each iteration of the non-linear optimization. Both procedures are prohibitively numerically intensive for large data sets. Various scholars have attempted to simplify the procedure mainly by simplifying the Jacobian term. [Ord, 1975, Gasim, 1988] But the most practical method so far has been proposed by Griffith. [1988, 1992, 1993] For irregular lattices, he suggests that the Jacobian term may be approximated by the following equation:

$$J = \alpha_1 \ln(\delta_1) + \alpha_2 \ln(\delta_2) - \alpha_1 \ln(\delta_1 + \rho) - \alpha_2 \ln(\delta_2 - \rho) \quad (27)$$

where $\alpha_1$, $\alpha_2$, $\delta_1$ and $\delta_2$ are parameters that are functions of the eigenvalues of $W$, which is a stochastic version of $C$. After examining a number of empirical cases, Griffith concludes that these four parameters display remarkable consistency over geographical configurations and numbers of areal units, and the values should be generally in the neighborhoods of

$$\alpha_1 = 0.22,$$
$$\alpha_2 = 0.12,$$
$$\delta_1 = 1.75, \text{ and}$$
$$\delta_2 = 1.05. \quad (28)$$

This finding leads to the generalized Jacobian approximation equation

$$J = 0.22\ln(1.75) + 0.12\ln(1.05) - 0.22\ln(1.75+\rho) - 0.12\ln(1.05-\rho). \quad (29)$$

The above approximation reduces the estimation problem to one of solving a quadratic equation when only $\rho$ is unknown, or when both $\sigma$ and $\rho$ are unknown, thus considerably simplifying the estimation process. In particular applications, a more precise set of values of $\alpha_1$, $\alpha_2$, $\delta_1$, and $\delta_2$ may be required, and the generally recommended estimates in expressions (28) and (29) may provide good initial values for their maximum likelihood estimation. In the same articles Griffith also finds that this generalized approximation produces estimated standard errors that are consistent with exact asymptotic standard errors, with only a small amount of specification error when the test statistic is located close to a critical value. These results make it possible for widely available commercial statistical software packages, such as SAS and Minitab, to be used for spatial statistical parameter estimation.

In this study, an AR model is considered more appropriate because it is reasonable to assume that a region's fertility rate is influenced by its determinants and surrounding regions' fertility rates, rather than that a fertility residual is influenced by the surrounding regions' residuals. The necessary SAS programs for model estimation are provided in Griffith [1993, pp. 76-78]. There are three general steps of implementation. First of all, the eigenvalues of matrix **W** need to be computed. The importance of the eigenvalues is two-fold:

- First they are needed to calibrate equation (27), and
- second, the maximum and minimum eigenvalues are used to define the permissible range of $\rho$.

This strategy is used to ensure that the later Jacobian term (see expressions (28) and (29)) is evaluated correctly (see expression (11)). Griffith [1988] has more detailed description of the method to obtain the eigenvalues. At the second step, the eigenvalues generated at the first step are fed into a program to produce the appropriate Jacobian approximation. This substitution is based on the simplified expression (27). The eventual outcome of this procedure is a set of more precise estimates of coefficients $\alpha_1$, $\alpha_2$, $\delta_1$, and $\delta_2$ of (27). The values provided by Griffith [1993; also see expression (28)] are the initial values for the iterations. The eigenvalues are no longer needed after this step. Finally, the outcomes of Step 2 are fed into a third program for estimation of $\rho$, vector $\beta$, and $\sigma^2$. The core of this last SAS program is an NLIN procedure. Since OLS estimates of vector $\beta$ are unbiased, its elements are used as initial values of the elements in $\beta$ for the iterations. The initial value of $\rho$ is set to zero. The explanatory variables are the ones selected at earlier non-spatial linear regression analysis. The model being estimated is an AR model following expression (8). Three SAS programs are provided in Appendix VI, Programs 6, 7, and 8, corresponding to the three general implementation steps described here.

Again, as in earlier linear regression, the explanatory variables are EDCO, EDHI, EDEL, DM and DETHR. The response variable is FERTR. The eigenvalues associated with **I+C** and **I+W** are calculated with a program based on Griffith [1993, pp. 76-78]. Note that the true eigenvalues from matrices **C** and **W** should be the calculated values minus one.

The computed eigenvalues are then used to generate a Jacobian approximation by the same program. The non-linear least squares regression procedure is used for estimates, and the results are:

| Parameter | Estimate | Asymptotic Standard Error | Asymptotic 95 % Confidence Interval | |
|---|---|---|---|---|
| | | | Lower | Upper |
| A1 | 0.141363033 | 0.00679296728 | 0.1270916324 | 0.1556344332 |
| A2 | 0.115044004 | 0.00361734878 | 0.1074442861 | 0.1226437212 |
| D1 | 1.344742362 | 0.02327218579 | 1.2958496370 | 1.3936350870 |
| D2 | 1.063984257 | 0.00562773683 | 1.0521608986 | 1.0758076150 |

As Griffith [1993, pp. 79-80] points out, the asymptotic standard errors and their confidence intervals are biased because a systematic sample, rather than an unrestricted random sample, is drawn to produce a sample of the theoretical values of the Jacobian term. However experience says that the errors are irrelevant and the focus here is on the estimates themselves.

Final results of the estimation of the AR model are

| Source | DF | Sum of Squares | Mean Square |
|---|---|---|---|
| Regression | 7 | 46736.230166 | 6676.604309 |
| Residual | 123 | 377.338534 | 3.067793 |
| Uncorrected Total | 130 | 47113.568700 | |
| | | | |
| (Corrected Total) | 129 | 1440.652591 | |

| Parameter | Estimate | Asymptotic Std. Error | Asymptotic 95% Confidence Interval | |
|---|---|---|---|---|
| | | | Lower | Upper |
| RHO | 0.58900906 | 0.0851184910 | 0.420520878 | 0.757497232 |
| B0 | 34.68613633 | 4.6845827856 | 25.413218047 | 43.959054617 |
| B1 | -31.44874314 | 6.7308356110 | -44.772125706 | -18.125360577 |
| B2 | -1.96976518 | 3.0303019580 | -7.968109905 | 4.028579542 |
| B3 | -7.61638378 | 3.7393701808 | -15.018296759 | -0.214470801 |
| B4 | -1.43805071 | 1.2979453169 | -4.007274354 | 1.131172937 |
| B5 | -0.10732190 | 0.1211423307 | -0.347117631 | 0.132473822 |

As the results show, RHO, B0, B1, B3, are significantly different from zero. RHO is about 0.6, which indicates a moderate to strong degree of positive spatial autocorrelation, confirming previous conclusions. When spatial autocorrelation is considered, only two of the five original explanatory variables, plus the sample mean, provide significant influence on the realization of the response variable in an area. They are EDCO and DM. The eventual model is

$$(FERTR)=B0+(EDCO)*(B1)+(DM)*(B3)+\rho *W*(FERTR)+e \quad (30)$$

or,

$$(FERTR)=34.6931.45*(EDCO)$$
$$7.62*(DM)+0.589*W*(FERTR)+e \quad (31)$$

## VII. INTERPRETATIONS AND CONCLUSIONS

Two sets of calculation results require understanding in substantive terms: one from the linear regression model, the other from spatial autocorrelation analysis. The linear regression model is highly significant (Prob>F=0.0001), with a fit that is very good among social science studies (adjusted R-square=0.5645). Generally speaking, the results are in support of the conventional wisdom that the factors under the definition of modernization drive down fertility rates (at least in this particular province around 1990).

The most outstanding feature of the linear regression model is the overwhelming importance of educational attainment variables, especially the percentage of college graduates among a county's population. Each of the three educational variables exhibits a negative relationship with fertility rates, when the other two are included in the model but held constant. This finding is evidence that education is indeed a crucial factor in fertility reduction. In this case college education is more important than elementary school education, which is more important than high school education.

Of note here is what the variable EDCO represents. On the surface of course it represents the access a county's population possesses to higher education. However, a more careful examination reveals that EDCO really stands for many things beyond just education. Table 8 shows it has extremely high positive correlation with urbanization (DM, r=0.92965) and general economic strength (FACTORT1, 0.89666), moderately high positive correlation with high school graduate percentage (EDHI, 0.71263), moderately high negative correlation with elementary school graduate percentage (EDEL, -0.67912), and with negative correlation on the death rate (DETHR, -0.39293). In fact, judging from the variance inflation factors and correlation coefficients, multicollinearity with EDCO is probably the most important reason why DM and FACTOR1, both with high separate correlation coefficients with fertility rate, are not significant in the model. It seems high EDCO values actually represent comprehensive socio-economic development levels and living standards among the areas under study. It is probably such comprehensive development that drives down fertility rates, rather than just the higher education itself.

EDEL paints a rather different picture. It correlates negatively with most variables that represent better modernization, including FACTOR1 (-0.60479), EDCO (-0.67912), and EDHI (-0.68176) and positively with the death rate (0.36475). Standing alone it actually has a positive correlation with the fertility rate ($r=0.34221$); this relationship is reversed, however, in the multiple regression model. In that model, when the effects of EDCO and EDHI are included but held constant, the fertility rate goes down as elementary school graduate percentage goes up. This is a rather puzzling observation. One possible explanation lies in the degree of prevalence of elementary school education. Without considering high school and college graduate percentages, higher elementary school graduate percentage in the population means more people terminate their education at the most basic level, and fewer people go on to higher levels of education. Taking into account its negative correlations with the aforementioned modernization factors, this in general indicates a lower level of development, which as expected brings about negative correlation with the fertility rate. However, when college and high school graduate percentages are taken into account and fixed, a higher percentage of elementary school graduates signifies an increase in total number of people who have had some education, and a decrease in the number of those who have had no education at all. Given the fact that the average percentage of people who have not received any education at all is nearly thirty, such an expansion of the educated segment of the population can be very large. So the departing point between differentiated impact on fertility does not lie between elementary school education and high school and college education. Rather, it exists between no education and some education. This conclusion should have some policy implications.

The spatial statistical model defined by equations (30) and (31) shows a pattern that takes into account the spatial effects of the implicit variables, mainly governmental policies. That is, the new factor in this model is the autocorrelation among fertility rates of neighboring areas. On the surface, the spatial autocorrelation model suggests that when a county's college graduate percentage is held constant, and when it has been classified as one of the major urban areas, its fertility rate is higher when its surrounding counties have high fertility rates.

Population policies are issued by the central government and carried out through its administrative hierarchy, of which the counties are the most critical and active nodes. Since there has not been any evidence of discriminatory implementation in this part of the country, we will have to assume that one round of implementation effort takes effect throughout the region at about the same time. That is, we may expect the neighboring counties to show a simultaneous response in the form of a fertility rate change. For each individual county, this effect is originally mixed with those of explicit socio-economic factors. However, this latter part is filtered out by the linear regression, and what we detect in the spatial model should be basically the effect of the governmental policies.

In summary, the two models have provided insights into the mechanism of influence of both explicit and implicit variables upon regional fertility rates. A linear regression model supports the contention that modernization, urbanization and education do drive down the fertility rates. It also reveals a peculiar position of elementary school education in this part of the country, namely, that acting alone, it seems to be a force driving up the fertility rates, but given any development level, it is actually a factor driving down the fertility rates. The spatial model identifies the form of the additional pattern after the effects of the explicit variables have been accounted for, and therefore provides evidence of the effects of an important but not explicitly measurable determinant -- government population policy.

## REFERENCES

Becker, G. S.   An economic analysis of fertility, in A. J.Coale, (ed.), *Demographic and Economic Change in Developed Countries*, Princeton University Press, Princeton, NJ, 1960.
Becker, G. S.  A theory of the time allocation. *Economic Journal*, 75, 493-517, 1965.
Becker, G. S.  A theory of social interactions. *Journal of Political Economy*, .82, 1963-93, 1974.
Bulatao, R. A. and Lee, R. D.  (eds), *Determinants of Fertility in Developing Countries*, two volumes, Academic Press, New York, 1983.
Cleland, J. and Wilson, C.   Demand theories of the fertility transition: An iconoclastic view. *Population Studies*, 41, 5-30, 1987.
Cliff, A. and K. Ord.  *Spatial Processes*, Pion, London, 1981.
Coale, A. and Watkins, S.  (eds) *The Decline of Fertility in Europe*, Princeton University Press, Princeton, NJ, 1986.
Daniel, C., and Wood, F. S.  *Fitting Equations to Data: ComputerAnalysis of Multifactor Data*, John Wiley and Sons, New York, 1983.
Davis, K.  The theory of changes and response in modern demographic history. *Population Index*, 29, 345-66, 1963.
Eastline, R. A.  The economics and sociology of fertility: A synthesis, in C. Tilly (ed.), *Historical Studies of Changing Fertility*,  Princeton University Press, Princeton, NJ,  1978.
Feeney, G.; Wang, F.; Zhou, M.; and, Xiao, B.  Recent fertility dynamics in China: results from the 1987 one percent population survey. *Population and Development Review*, 15, no.1, 297-322, 1989.

**Feng, H. M.** The Contextual Determinants of Contemporary Chinese Fertility Transition, unpublished manuscript, Department of Geography, Syracuse University, Syracuse, NY, 1993.

**Freedman, R.** Theories of fertility decline: A reappraisal. *Social Forces*, 48, 1-17, 1979.

**Gasim, A.** First-order autoregressive models: a method for obtaining eigenvalues for weighting matrices. *Journal of Statistical Planning and Inference*, 18, 391-98, 1988.

**Goldscheider, C.** *Population, Modernization, and Social Structure*, Little Brown, Boston, 1971.

**Greenhalgh, S.** Toward a political economy of fertility: Anthropological contributions. *Population and Development Review*, 16, no.1, 85-106, 1990.

**Greenhalgh, S.; Zhu, C.; and, Li, N.** Restraining population growth in three Chinese villages, 1988-93. *Population and Development Review*, 20, no.2, 365-95, 1994.

**Griffith, D. A. and Amrhein, C.** *Statistical Analysis for Geographers*, Prentice Hall, Englewood Cliffs, NJ, 1991.

**Griffith, D. A.** *Spatial Autocorrelation: A Primer*, Association of American Geographers, Washington, D.C., 1987.

**Griffith, D. A.** Estimating Spatial Autoregressive Model Parameters with Commercial Statistical Packages. *Geographical Analysis*, 20, No.2, 176-86, 1988.

**Griffith, D. A.** Simplifying the normalizing factor in spatial autoregressions for irregular lattices. *Papers in Regional Science*, 71, 71-86, 1992.

**Griffith, D.A.** *Spatial Regression Analysis on the PC: Spatial Statistics Using SAS*, Association of American Geographers, Washington, D.C., 1993.

**Haining, R.** The use of added variable plots in regression modeling with spatial data. *The Professional Geographer*, 42, 336-44, 1990.

**Haining, R.** *Spatial Data Analysis for Social and Environmental Sciences*, Oxford University Press, London, 1991.

**He-Nan Province Census Bureau,** *He-Nan Province Population Census Statistics, 1990*, China Statistics Publishing Company, Beijing, China, 1992.

**Johnsson, S.R.** Implicit policy and fertility during development. Population and Development Review, 17, no.3, 377-414, 1991.

**Knodel, J. and van de Walle, E.** Lessons from the past: policy implications of historical fertility studies. *Development Review*, .5, no.2, 217-45, 1979.

**Lesthaeghe, R. and Surkyn, J.** Cultural dynamics and economic theories of fertility change. *Population and Development Review*, 14, no.1, 1-45, 1988.

**Luther, N.Y. Jr.; Feeney, G.; and, Zhang, W.** One-child families or a baby boom? Evidence from China's 1987 one-per-hundred survey. *Population Studies*, 44, no.2, 341-57, 1990.

**Nambookiri, N.K.** Some observations on the economic framework for fertility analysis. *Population Studies*, 26, 185-206, 1972.

**Notestein, F.** Economic problems of population change, in *Proceedings of the Eighth International Conference of Agri-Economics*, Oxford University Press, London, 1953.

**Notestein, F.** Population - the long view, in T.W.Schultz (ed.), *Food for the World*, University of Chicago Press, Chicago, 1954.

**Ord, K.** Estimation methods for models of spatial interaction. *Journal of American Statistical Association*, 70, 120-26, 1975.

**Peng, Peiyuan,** Accomplishments of China's family planning program: a statement by a Chinese official. *Population and Development Review*, 19, no.2, 399-403, 1993.

**Peng, X.** Major determinants of China's fertility transition. *The China Quarterly*, No.117 (March), 1-37, 1989.

**Pye, L.** *The Dynamics of Chinese Politics*, Delgeschlager, Gunn & Hain, Publishers, Cambirdge, MA, 1981.

**Ripley, B.** *Statistical Inference for Spatial Processes*, Cambridge University Press, Cambridge, 1988.

**SAS Institute Inc.** *SAS/STAT User's Guide* (two volumes), SAS Institute Inc., Cary, NC, 1989.

**Schultz, T. P.** *Economics of Population*, Addison-Wesley, MA, 1981.

**Shryock, H.S., and Siegel, J. S.** *et al. The Methods and Materials of Demography,* U.S.Bureau of Census, Washington, D.C., Vol.2, 1980.

**Smith, H.** Integrating theory and research on the institutional determinants of fertility. *Demography,* 26, 171-184, 1989.

**Willis, R. J.** A new approach to the economic theory of fertility behavior. *Journal of Political Economy,* 81, 14-69, 1973.

**Wolf, A. P.** The preeminent role of government intervention in China's family   revolution. *Population and Development Review,* 12, 106-16, 1986.

**Yi, Z.** Is the Chinese family planning program 'tightening up'? *Population and Development Review,* 15, No.2 (June), 1989.

## APPENDIX I: DESCRIPTION OF DATASET

### GROUP 1. Demographic variables

- (1) Total population, and total numbers of persons who are male, female, and non-agricultural
- (2) Total number of live births, broken down by gender, for three periods, January 1, 1989 to June 30, 1989, July 1, 1989 to December 30, 1989, and January 1 to June 30, 1990.
- (3) Total number of deaths, broken down by gender, for three periods, January 1, 1989 to June 30, 1989, July 1, 1989 to December 30, 1989, and January 1, 1990 to June 30, 1990.
- (4) Age structure (in persons): number of persons by gender, and age groups 0-4, 5-9, 10-14, 15-19, 20-24, 25-30, 31-34, 35-39, 40-44, 45-49, 50-54, 55-59, 60-64, 65-69, 70-74 and 75 years of age or older.
- (5) Total number of immigrants. An immigrant is defined as someone whose usual residence at the time of the census is different from his/her usual residence on July 1, 1985. The publications show the number of such persons whose usual residence on July 1, 1985 was, within the same province, other provinces, neither of the above two (mainly from other countries).

### GROUP 2. Economic structure variables

Total employees (persons) broken down by gender, in classified industries, including total employment, employment in agriculture, forestry, pastoralism, irrigation and drainage, and fishing, employment in industries (light, heavy and mining), employment in geology survey, employment in construction, employment in transportation and communication, employment in wholesale, retail and storage industry, employment in real estate management and public service, employment in medical, sports and social welfare service, employment in education, entertainment and TV-Radio industries, employment in scientific research and comprehensive technical services, employment in finance and insurance industry, and employment in government, political parties and social organizations.

### GROUP 3. Educational attainment variables

Educational attainment: Numbers of persons per 100,000 people who hold diplomas from a four year college, two to three year college or equivalent, professional high school, regular high school, junior-high school, and elementary school.

### GROUP 4. Cultural determinants (implicit variables)

They mainly include government population policies and historical factors. There is no explicit measure.

**APPENDIX II: MAPS**

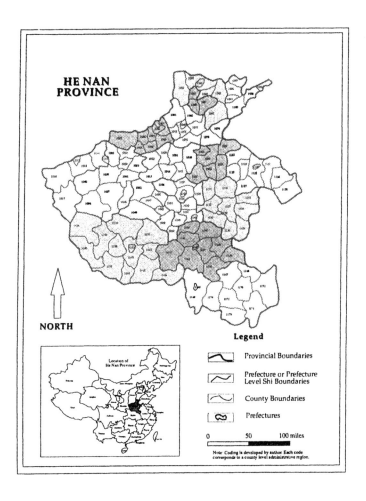

**Map 1.** He-Nan province, location map.

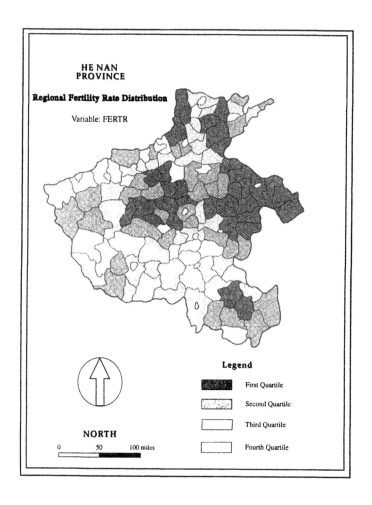

**Map 2.**  He-Nan province, FERTR variable.

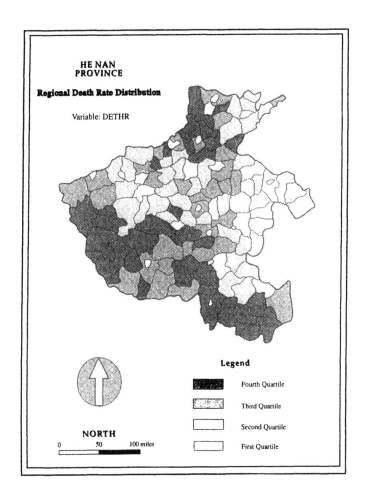

**Map 3.** He-Nan province, DETHR variable.

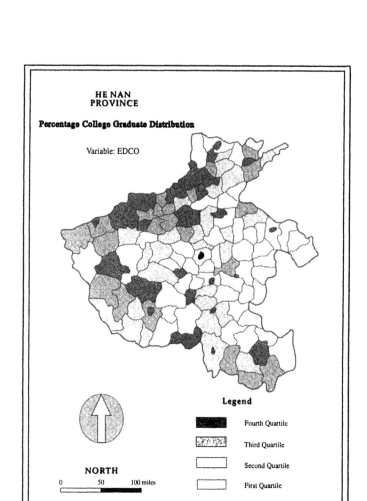

**Map 4.** He-Nan province, EDCO variable.

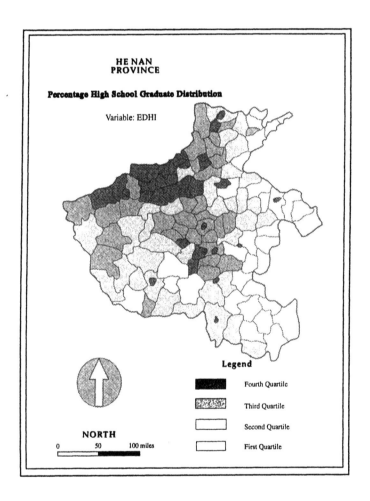

**Map 5.** He-Nan province, EDHI variable.

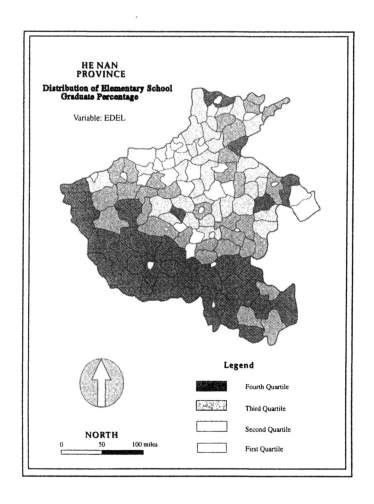

**Map 6.** He-Nan province, EDEL variable.

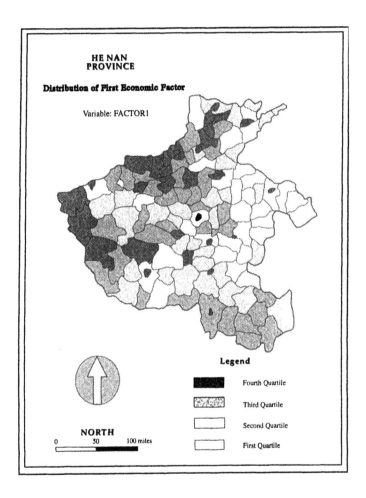

**Map 7.** He-Nan province, FACTOR1 variable.

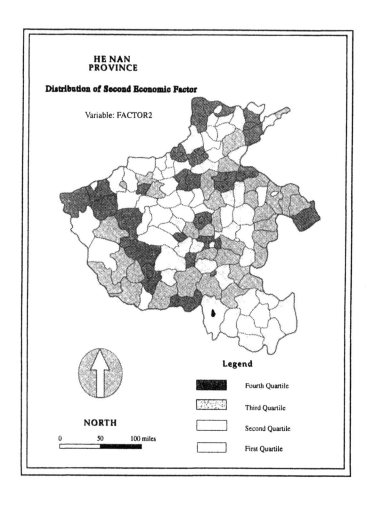

**Map 8.** He-Nan province, FACTOR2 variable.

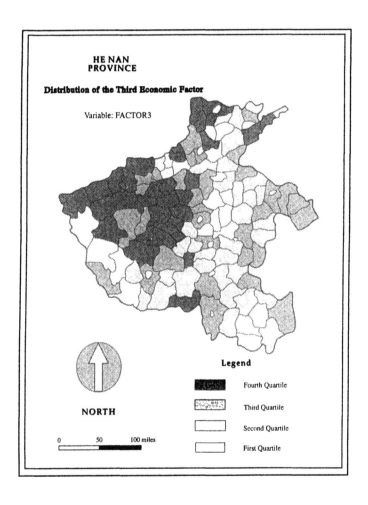

**Map 9.** He-Nan province, FACTOR3 variable.

## APPENDIX III:  SCATTER PLOTS

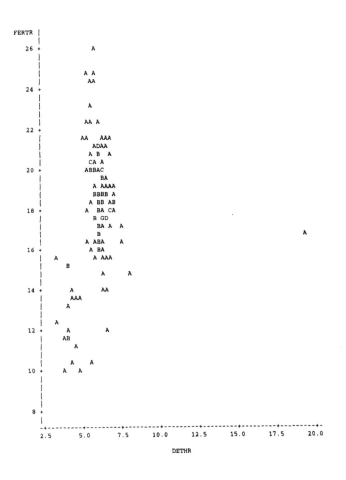

**Plot 1.**  FERTR*DETHR.  Legend:  A = 1 obs, B = 2 obs, etc.

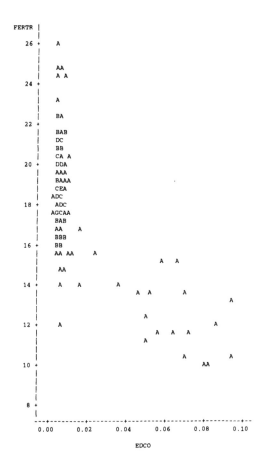

**Plot 2.** FERTR*EDCO. Legend: A = 1 obs, B = 2 obs, etc.

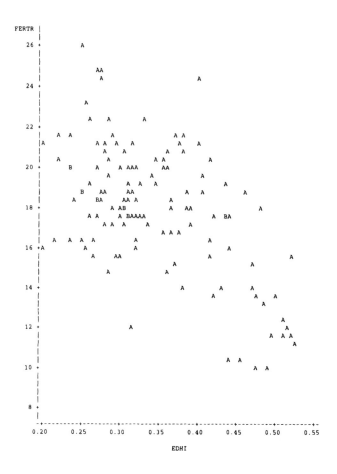

**Plot 3.** FERTR*EDHI. Legend: A = 1 obs, B = 2 obs, etc.

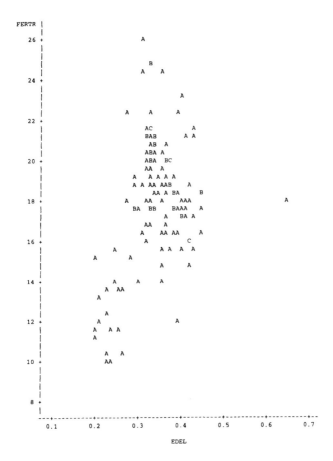

**Plot 4.**  FERTR*EDEL.  Legend:  A = 1 obs, B = 2 obs, etc.

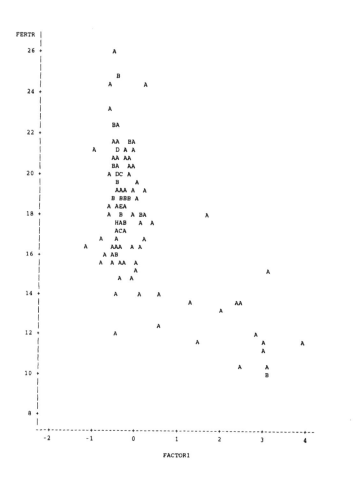

**Plot 5.** FERTR*FACTOR1. Legend: A = 1 obs, B = 2 obs, etc.

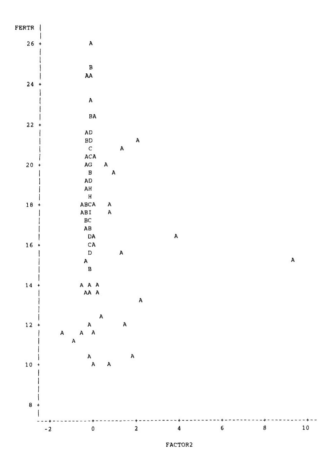

**Plot 6.** FERTR*FACTOR2. Legend: A = 1 obs, B = 2 obs, etc.

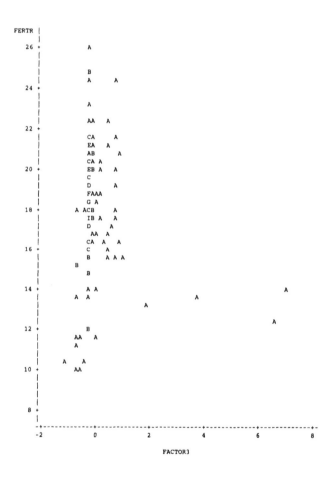

**Plot 7.** FERTR*FACTOR3.  Legend:  A = 1 obs, B = 2 obs, etc.

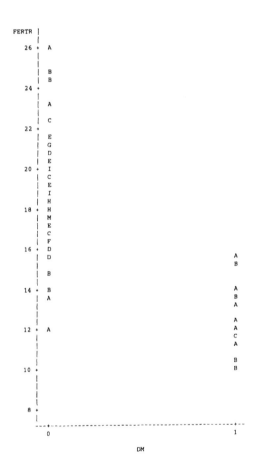

**Plot 8.** FERTR*DM. Legend: A = 1 obs, B = 2 obs, etc.

# Chapter 9

## SPATIAL STATISTICAL/ECONOMETRIC VERSIONS OF SIMPLE URBAN POPULATION DENSITY MODELS

Daniel A. Griffith[1] and Ayse Can

### I. INTRODUCTION AND BACKGROUND

The geographic organization of urban population within cities has been a topic of interest in urban studies for many decades, and has been investigated for a wide range of city sizes, types, cultural settings, and points in time. Virtually all attempts to model it have overlooked spatial dependence in their conceptualizations, for this approach has not yet diffused throughout the standard spatial modelling literature [see, for instance, Kau, Lee and Sirmans, 1986; see Batty and Kim, 1992, for a more recent example]. But housing prices are known to contain spatial dependence, the clustering of ethnic and socio-economic groups signify latent spatial dependence, and zoning policies dictate how population densities cluster in space. In fact, the notion of spatial externalities, which at least partially governs the geographic organization of urban population, is mathematically equivalent to spatial dependence. The primary objective of this chapter is to furnish an application, illustrating the implementation of spatial statistical/econometric modelling, that specifically examines spatial dependence concealed in the geographic distribution of urban population density, using 1986 census tract data for Toronto and Ottawa-Hull, and both 1980 and 1990 census tract data for Syracuse. Its results should help illuminate the role spatial dependence plays in geographically organizing urban population, and promote the dissemination of spatial statistical/econometric modelling procedures.

### A. THE RUDIMENTARY POPULATION DENSITY MODEL

Modelling the geographic distribution of population density in urban areas has been undertaken by many regional scientists during the past several decades (see Thrall, 1988, for one of many reviews of this literature). Most of these efforts have described this phenomenon with some variant of a negative exponential decline in population density with increasing distance from the city center, whose location often is

---

1

This research was supported, in part, by a 1993 U.S.-Canada Fulbright Research Fellowship.

represented by the central business district (CBD). This simple equational form will serve as the basis of this study, for purposes of comparison; it may be written as

$$D_i = \beta_0 \exp(-\beta_1 d_i), \tag{1}$$

where $D_i$ is the population density of census tract i,

$d_i$ is the distance separating census tract i from the city center, and

$\beta_0$ and $\beta_1$ are parameters ($\beta_0$ refers to the maximum possible population density, whereas $\beta_1$ refers to the rate of distance decay).

Error is latent in population density data, and may be introduced into equation (1) in two fundamentally different ways. If the error is multiplicative, then a logarithmic transformation of equation (1) can accommodate it; hence,

$$\ln(D_i) = \ln(\beta_0) - \beta_1 d_i + \xi_i, \tag{2}$$

where $\xi_i$ is the error associated with census tract I. If the error is additive, then

$$D_i = \beta_0 \exp(-\beta_1 d_i) + \xi_i . \tag{3}$$

Equation (2) is a very popular specification, in part because parameter estimation for equation (3) requires non-linear optimization techniques. In either case, the argument of the exponential operator 'exp' can be expanded by incorporating other predictor variables in order to accommodate more undulating surfaces.

## B.  SPATIAL AUTORRELATION AND POPULATION DENSITY MODELS[2]

If population density declines exponentially with increasing distance from a city center, then the idealized two-dimensional surface may be depicted with concentric isopleth rings of decreasing value. In such a situation, the level of spatial autocorrelation that can be measured for this geographic distribution will be positive (similar values of population density tend to be juxtaposed). The failure to have similar

---

[2]

A relevant geo-statistical treatment of population density appears in Griffith, Haining, and Arbia (1994).

values clustering on the surface in an optimally compact way will result in this positive spatial autocorrelation being moderate in degree, yielding intermediate to large (but never extremely large) spatial autocorrelation index values. But are such index values an artifact of the distance decay mechanism that induces spatial organization in urban population density? Or, is some type of autoregressive process at work that leads to detectable distance decay in population density surfaces? Presumably, as a partial answer to these two questions, population density surfaces result from both of these factors.

Indeed, few researchers have addressed the issue of spatial autocorrelation as associated with population density surfaces. Early attempts to represent these geographic distributions with trend surface models implicitly explored this issue. [Hill, 1973] Explicit autoregressive model specifications have been studied by Griffith [1981], and by Anselin and Can [1986]. Griffith [1981] examined a non-constant mean autoregressive response model, with an error structure similar to that for equation (3). The mean varied as a function of absolute location (which had significant coefficients, but accounted for only a negligible amount of the population density variance). A high level of positive spatial autocorrelation was detected, which accounted for much of the systematic variation uncovered in the geographic distribution of population density in Toronto.

Anselin and Can investigated a number of different specifications of the negative exponential component of population density gradients. They used the simultaneous autoregressive errors model to account for the presence of spatial autocorrelation, but with an error structure similar to that for equation (2). They uncovered evidence of residual spatial autocorrelation in 36% of the data sets they analyzed. Their findings include (p. 194)

> The degree of residual heteroskedasticity and spatial autocorrelation present in urban density functions should not be ignored, though it seems to depend on the spatial scale used in the application. In this respect, the use of a random sample rather than a contiguous sample does not preclude the presence of spatial autocorrelation of a general form.

This last comment emphasizes that spatial autocorrelation is not completely controllable through sampling design techniques; in part, though, it is linked to partitioning a surface into mutually exclusive and collectively exhaustive areal units.

## II. THE SELECTED METROPOLITAN LANDSCAPES

Three metropolitan databases have been selected for investigation here, in part because of the authors' familiarity with them. The first, Toronto, also has been chosen because of its relatively large size. The second, Ottawa-Hull, has been chosen because it is monocentric, enabling some comparisons to be made between results for it and those for the third city, Syracuse, which also is monocentric. Syracuse has been selected because it offers an opportunity to make an interesting contrast between two points in time, too.

### A. THE TORONTO METROPOLITAN AREA
The data to be analyzed here are from the 1986 Canadian Census of Population (a digital copy was retrieved from the University of Toronto library), and are for the 731 census tracts into which Toronto has been partitioned. The total population of this metropolitan area is 3,427,168, and its total area is 5599.33 square kilometers; hence the average density is 612.07 people per square kilometer, and the average census tract size is about 7.66 square kilometers. Census tract centroids are given in terms of Universal Transverse Mercator (UTM) co-ordinates; these co-ordinates have been rescaled for convenience. The CBD is considered to be contained in census tract #35.00, which houses the Eaton Center.

Three complicating features of Toronto are that it is a multi-centered metropolitan area, it has census tracts that appear to come from different populations (introducing mean response heterogeneity), and its radial influences are truncated by Lake Ontario. The presence of sub-centers will cause the single center model represented by equation (2) to provide a poorer description of the geographic distribution of population density across this metropolitan area. Toronto has five tracts whose densities are considerably lower than those of their surrounding tracts (hence being local anomalies). The city also has an artificially high density tract near the CBD (because of a social experiment that was conducted several decades ago) that displays aberrant behavior. And, it has a single tract with no population. Differentiations are incorporated into the analysis by including indicator variables to distinguish among these various groupings; these variables take on a value of 1 when a census tract is a member of a rogue group, and a value of -1 when a census tract is not (this parameterization allows for a direct difference of means test to be conducted). The abutment on the lake will introduce an edge effect into this geographic distribution of population density.

Ignoring spatial autocorrelation latent in the Toronto-centered region population densities, parameter estimation for equation (2) yields

| variable | Ordinary Least Squares OLS | | Weighted Least Squares WLS | |
|---|---|---|---|---|
| | b | $s_b$ | b | $s_b$ |
| intercept | 3.04336 | 0.77599 | 2.42321 | 0.36299 |
| distance from CBD | -0.05951 | 0.00323 | -0.03730 | 0.00287 |
| east-west coordinate | -0.01198 | 0.00302 | | |
| excessive density indicator | 1.04634 | 0.52290 | 0.99206 | 0.16615 |
| zero density indicator | -4.17348 | 0.52273 | -4.29926 | 0.26139 |
| local anomalies indicator | -2.75804 | 0.23427 | -3.33917 | 0.18985 |
| | | | | |
| (pseudo-)$R^2$ | 0.5019 | | 0.4340 | |
| residual heteroscedasticity[3] | 37.3192***[4] | | 13.6746* | |
| residual normality | 0.89443*** | | 0.83618*** | |

These estimations are the result of detailed analysis, in which various gradient (i.e., trend surface) terms were found actually not to contribute to the statistical explanation of the geographic distribution of population density across the Toronto centered region. Evidence supporting this contention ultimately argued for the removal of the east-west coordinate term; computed variance inflation factors (VIF) suggested that the removal of this single term would not dramatically alter the standard error estimates. The weighted least squares (WLS) has been employed in order to account for variance heterogeneity; the model specification uses the inverse of Area as the weight, which makes intuitive sense since one would expect variability in population density calculations to relate to the size of individual areal units. As expected, principal differences between these OLS and WLS estimates arise in the standard error estimates; these differences are almost strictly due to the taking into

---

3

The White test, based upon a $\chi^2$ statistic, which furnishes a test criterion concerning whether or not the first and second moments of the model are correctly specified, may be selected in SAS to evaluate the assumption of constant variance.

4

Statistical significance is denoted throughout this chapter as follows:
*   statistically significant at the 10% level
**  statistically significant at the 5% level
*** statistically significant at the 1% level

account of non-constant variance across the areal units. The pseudo-$R^2$ value [computed by squaring the correlation between $LN$(density) and $LN$(density)] for the WLS solution is less than its OLS counterpart because of the constraint introduced into the parameter estimates by the weight, Area-inverse.

## B. THE OTTAWA-HULL METROPOLITAN AREA

The data to be analyzed in this section are from the 1986 Canadian Census of Population (a digital copy was supplied by Statistics Canada), and are for the 192 census tracts into which Ottawa-Hull has been partitioned. The total population of this metropolitan area is 819,263, and its total area is 5138.33 square kilometers; hence the average density is 159.44 people per square kilometer, and the average census tract size is about 26.76 square kilometers. Census tract centroids again are given in terms of Universal Transverse Mercator (UTM) coordinates; these coordinates also have been rescaled for convenience. The CBD is considered to be embraced by adjacent census tracts #48.00 and #54.00; its location is denoted here by the centroid rendered by combining these two census tracts [rescaled UTM coordinate pair (44.5469, 50.301195)].

One complicating idiosyncrasy of Ottawa-Hull is that the metropolitan area is composed of two cultural regions. Roughly speaking, the boundary separating these two regions coincides with the Ottawa River, which also is a political boundary. Therefore, one cultural area is located in the Province of Ontario while the other is located in the Province of Quebec. This cultural differentiation has been incorporated into the analysis by including an indicator variable; this variable takes on a value of 1 when a census tract is located in Ontario, and a value of -1 when a census tract is in Quebec (as mentioned previously, this parameterization allows for a direct difference of means test to be conducted). A second complicating feature is that Ottawa-Hull has two tracts whose densities are conspicuously low--namely #140.01, and #160.03--and behave as though they are from a different density population; meanwhile, census tract #47.00 appears to be a local anomaly. A binary indicator variable has been included to accommodate the non-stationary mean response introduced into the geographic distribution of population density by these three anomalies.

Ignoring spatial autocorrelation latent in the Ottawa-Hull metropolitan area population densities, parameter estimation for equation (2) yields

|  | Ordinary Least Squares OLS | | Weighted Least Squares WLS | |
| --- | --- | --- | --- | --- |
| variable | b | $s_b$ | b | $s_b$ |
| intercept | 5.44773 | 0.38353 | 5.92992 | 0.23747 |
| distance from CBD | -1.57356 | 0.12268 | -0.66320 | 0.12626 |
| north-south coordinate |  |  | -0.35035 | 0.13169 |
| east-west coordinate | 3.56438 | 1.23973 | 5.58446 | 1.59157 |
| cultural indicator | 0.49322 | 0.12261 | 0.40942 | 0.08314 |
| low density indicator | -2.66096 | 0.36558 | -2.36447 | 0.21853 |
| (pseudo-)$R^2$ | 0.5554 | | 0.3904 | |
| residual heteroscedasticity | 47.6322*** | | 17.2087 | |
| residual normality | 0.92640*** | | 0.79048*** | |

These estimations are the result of considerable preliminary analysis (see Griffith, 1993). Again variance heterogeneity is accounted for with the inverse of Area. Not surprisingly, primary differences between these OLS and WLS estimates arise in the standard error estimates; marked differences are evident solely for the distance and Cartesian coordinate terms, partially because the north-south coordinate becomes significant once non-constant variance is adjusted for.

## C. THE SYRACUSE METROPOLITAN AREA

The data to be analyzed here are from the 1980 and 1990 United States Census of Population, and are for, respectively, the 145 and 143 census tracts into which the Syracuse area has been partitioned. The total populations of this metropolitan area are, respectively, 463,920 and 468,512, and its total area is 3252.52 square kilometer; hence the average densities are, respectively, 142.62 and 144.05 people per square kilometers, and the average census tract size is about 22.67 square kilometers. Census tract centroids again are given in terms of UTM coordinates; once more these coordinates have been rescaled for convenience. The CBD is considered to be embraced by adjacent census tracts #31.00 and #32.00; its location is denoted here by the centroid rendered by combining these two census tracts [rescaled UTM coordinate pair (40.632213, 76.656997)].

One complicating factor for Syracuse is that it contains several contiguous low density tracts in 1980, which have been merged into a single (local anomaly) tract for 1990, that behave differently from its other tracts. Additional rogue low density tracts in 1990 are #31, #37, and #41, all of which have populations less than 500 people. Again, then, differentiation will be incorporated into the analysis by including indicator variables; recall that each variable takes on a value of 1 when a census tract belongs to a designated group, and a value of -1 when it does not (allowing for a direct difference of means test to be conducted). One local anomaly is somewhat conspicuous in 1980, actually being a high density settlement in the southwestern corner of the metropolitan area. This deviant census tract disappears in 1990 because a number of the suburban county sub-regions have been repartitioned since 1980. Of note is that various attributes for Syracuse are similar to those reported for Ottawa-Hull, such as monocentricity, average population density, and average census tract size.

Ignoring spatial autocorrelation latent in the Syracuse metropolitan area population densities, for 1980, parameter estimation for equation (2) yields

| variable | Ordinary Least Squares OLS | | Weighted Least Squares WLS | |
|---|---|---|---|---|
| | $b$ | $s_b$ | $b$ | $s_b$ |
| intercept | 0.79521 | 0.53160 | -0.46872 | 0.20414 |
| distance from CBD | -2.06713 | 0.11558 | -1.83219 | 0.13591 |
| north-south coordinate | -0.24564 | 0.10752 | -0.47508 | 0.12712 |
| east-west coordinate | 0.26300 | 0.09676 | | |
| low density indicator | -2.26874 | 0.25517 | -2.09937 | 0.19952 |
| local anomalies indicator | 1.54227 | 0.45865 | | |
| | | | | |
| (pseudo-)$R^2$ | 0.7346 | | 0.6967 | |
| residual heteroscedasticity | 20.8026* | | 15.7080* | |
| residual normality | 0.9643** | | 0.9503*** | |

These estimations are the outcome of extensive exploratory analysis, too. This time variance heterogeneity is accounted for with the inverse of Area. As expected, principal differences between these OLS and WLS estimates arise in the standard error estimates, although here more of an impact on the parameter estimates, themselves, is detectable. Once more VIFs suggest that dropping both the east-west co-ordinate and local anomalies indicator variables have little impact on the estimated standard errors.

Meanwhile, ignoring spatial autocorrelation latent in the Syracuse metropolitan area population densities, for 1990, parameter estimation for equation (2) yields

| variable | Ordinary Least Squares OLS | | Weighted Least Squares WLS | |
|---|---|---|---|---|
| | b | $s_b$ | b | $s_b$ |
| intercept | -0.07785 | 0.43123 | 0.00128 | 0.18120 |
| distance from CBD | -2.08716 | 0.09936 | -2.09178 | 0.12052 |
| north-south coordinate | -0.17790 | 0.08878 | -0.25593 | 0.11140 |
| east-west coordinate | 0.34185 | 0.08152 | 0.33023 | 0.11532 |
| 1980 low density indicator | -0.51538 | 0.22081 | -0.56803 | 0.05829 |
| 1990 low density indicator | -1.20081 | 0.37561 | -1.34006 | 0.17269 |
| (pseudo-)$R^2$ | 0.7832 | | 0.7817 | |
| residual heteroscedasticity | 25.5582** | | 12.9283 | |
| residual normality | 0.93446*** | | 0.93237*** | |

These estimations are the outcome of extensive exploratory analysis, as well. This time variance heterogeneity is accounted for with the inverse of Area, rather than Area. Now the principal differences between these OLS and WLS estimates display a pattern of decreased standard error estimates for the intercept, the two low density indicator variables, and increased standard errors for distance and the Cartesian coordinate terms.

## III. PRELIMINARIES FOR ESTIMATING THE AUTOREGRESSIVE MODELS

Spatial statistics/econometrics models will be implemented both within SAS and with SpaceStat. The SAS computer code enables the features of a reputable commercial statistical package to be exploited when undertaking a spatial statistical analysis. [Griffith, 1993] Both exact and computationally simplified approximate solutions have been obtained and will be reported for the SAS code. SpaceStat implements many of the spatial econometrics results obtained by Anselin. [Anselin, 1992]

Three model properties are of concern here. The first is a recognition of the relationship between trend surface/gradient models and detected levels of spatial autocorrelation. The specific issue concerns measurement error associated with the designated coordinates of a given CBD. If these coordinates are shifted on a map, then resulting regression residuals will exhibit latent spatial autocorrelation. One way to circumvent this complication is to introduce coordinate terms in order to compensate for this trend in the residuals. A better solution would be to estimate an optimal CBD location as one of the parameters of the model; such a solution is best acquired within the context of equation (3), though. [Can and Griffith, 1993]

The second property is the spatial autoregressive model specification selected. The autoregressive response model has been chosen here, because it minimizes the number of spatial lag variables that must be dealt with, because it is consistent with models already appearing in the literature, and because casting population density at one location as a function of the densities in surrounding locations makes intuitive sense. Consider a set of P variables, whose data are entered into matrix $X$. In addition, a vector of ones, denoted 1, must be concatenated with matrix $X$ in order to include an intercept term. Let $Y$ be the response variable vector, and let $\xi$ be the error vector. The traditional linear statistical model that can be posited using these terms is expressed as $Y = X\beta + \xi$. Let $\rho$ be the spatial autocorrelation parameter. Therefore, the spatial autoregressive response model may be expressed as

$$(I - \rho W)Y = X\beta + \xi \quad \text{or} \quad Y = \rho WY + X\beta + \xi,$$

where $I$ is the identity matrix. Adjustments to equation (2) following this format render

$$Y/\exp(-J_W) = WY/\exp(-J_W) + X/\exp(-J_W) + \xi/\exp(-J_W). \quad (4)$$

[Griffith, 1988]

The third property is the necessary normalizing factor, or the Jacobian of the transformation from an autocorrelated domain to an unautocorrelated domain, for a given n. This term is denoted here by $J_W$, and $_W$ for its approximation [see Griffith, 1992], where matrix $W$ is the row standardized version of the standard binary geographic weights connectivity matrix $C$ (i. e., each row is scaled so that it sums to unity; $w_{ij} = c_{ij}/_{=1}c_{ij}$). Thus, equation (4) becomes

$$Y/\exp(-J_W) = WY/\exp(-J_W) + X/\exp(-J_W) + \xi/\exp(-J_W), \quad (5)$$

when the Jacobian approximation is substituted for the exact Jacobian term, $J_W$.

## A. THE TORONTO METROPOLITAN AREA

The Jacobian term, or normalizing factor required by equation (5), for Toronto involves a 731-by-731 matrix. When the autoregressive model is written in terms of the row-standardized version of matrix $C$, namely matrix $W$, then the eigenvalues of this matrix are such that

$\lambda_{max} = 1$ and $\lambda_{min} = -0.76976 \Rightarrow -1.299106 < \hat{\rho} < 1.$

Drawing a systematic sample of size 22 from across this feasible parameter space yields the following Jacobian approximation (SSE = $5.9 \times 10^{-6}$):

$$J_W = \exp[0.03561\rho + 0.14913$$
$$- 0.24941 \ln(1.73193 + \rho) - 0.12076 \ln(1.10584 - \rho)] \ .$$

## B. THE OTTAWA-HULL METROPOLITAN AREA

The Jacobian term for Ottawa-Hull involves a 192-by-192 matrix. When the autoregressive model is written in terms of the row-standardized matrix **W** for this case, then the eigenvalues of this matrix are such that

$$\lambda_{max} = 1 \text{ and } \lambda_{min} = -0.62371 \Rightarrow -1.601706 < \hat{\rho} < 1.$$

Again, drawing a systematic sample of size 22 from across this feasible parameter space yields the following Jacobian approximation (SSE = $1.2 \times 10^{-4}$):

$$J_W = \exp[0.01108\rho + 0.16630$$
$$- 0.24200 \ln(1.87613 + \rho) - 0.12878 \ln(1.11511 - \rho)] \ .$$

## C. THE SYRACUSE METROPOLITAN AREA

The Jacobian terms for Syracuse involve a 145-by-145[5] matrix for 1980 and a 143-by-143 matrix for 1990. When the autoregressive model is written in terms of the two different row-standardized **W** matrices here, then the eigenvalues of these matrices are such that

$$1980: \ \lambda_{max} = 1 \text{ and } \lambda_{min} = -0.64355 \Rightarrow -1.553880 < \hat{\rho} < 1.$$

$$1990: \ \lambda_{max} = 1 \text{ and } \lambda_{min} = -0.60038 \Rightarrow -1.665611 < \hat{\rho} < 1.$$

---

5

    Census tract #146.00 consisted of two geographically separated areal units. Each of these units is treated like an individual census tract here, since population counts and geo-referenced information are available for both.

Once more, drawing systematic samples of size 22 from across these feasible parameter spaces yields the following Jacobian approximations (SEE = $1.6 \times 10^{-4}$ and $1.7 \times 10^{-4}$, respectively):

$$1980: \; J_W = \exp[0.01543\rho + 0.15152$$
$$- 0.23006 \ln(1.84829 + \rho) - 0.11720 \ln(1.09092 - \rho)] \; .$$

$$1990: \; J_W = \exp[0.01347\rho + 0.17403$$
$$- 0.24540 \ln(1.94623 + \rho) - 0.12039 \ln(1.09221 - \rho)] \; .$$

Of note is that all four of the foregoing $J_W$s have a near-perfect correspondence with their associated $J_W$s. Each of these respective six sets of coefficients exhibits remarkably little within-set variation, too.

## IV. THE ESTIMATED POPULATION DENSITY MODELS

Generalized weighted least squares (GWLS) estimation results for the Toronto centered region are as follows:

| variable | SAS-approximate b | SAS-approximate $s_b$ | SAS-exact b | SAS-exact $s_b$ | SpaceStat[6] b | SpaceStat[6] $s_b$ |
|---|---|---|---|---|---|---|
| $\hat{\rho}$ spatial autocorrelation | 0.37929 | 0.03927 | 0.33622 | 0.02222 | 0.37653 | 0.01041 |
| intercept | -0.92647 | 0.48046 | -0.83086 | 0.38286 | -0.89414 | 0.33820 |
| distance from CBD | -0.01371 | 0.00359 | -0.01505 | 0.00296 | -0.01393 | 0.00266 |
| excessive density indicator | 1.02212 | 0.15224 | 0.93508 | 0.15306 | 1.02183 | 0.15184 |
| zero density indicator | -4.35952 | 0.23955 | -4.45803 | 0.24070 | -4.35894 | 0.23876 |
| local anomalies indicator | -3.23815 | 0.17424 | -3.33208 | 0.17504 | -3.23913 | 0.17327 |
| pseudo-$R^2$ | 0.5844 | | | | | |
| residual normality | 0.85522*** | | | | | |

These results differ from the foregoing ones in that

---

[6]

A pseudo-$R^2$ value is not reported here because SpaceStat performs a no-intercept, weighted regression analysis, and hence this summary value is not comparable to the one rendered by SAS.

- (1) weak-to-moderate spatial autocorrelation is uncovered,
- (2) parameter estimates for the intercept (changing from positive to negative) and distance decay (decreases in magnitude by nearly a third) variables change quite noticeably, and
- (3) all of the standard errors decrease.

Stability of the indicator parameters estimates is exactly what one should expect, since WLS estimators are unbiased in the presence of spatial autocorrelation. The AR spatial statistical model specification, itself, often leads to dramatic changes in the estimate of an intercept, in part because it relates to $(1 - \hat{\rho})b_{0,OLS}$. And, the very nature of spatial autocorrelation suggests that estimates of distance decay parameters may well become confounded with it. The standard error estimates for the spatial autocorrelation, intercept and distance decay parameter estimates based upon the Jacobian approximation are inflated by from 50% to more than 300%.

Generalized weighted least squares (GWLS) estimation results for the Ottawa-Hull metropolitan area are as follows:

| variable | SAS-approximate | | SAS-exact | | SpaceStat[7] | |
|---|---|---|---|---|---|---|
| | b | $s_b$ | b | $s_b$ | b | $s_b$ |
| $\hat{\rho}$ spatial autocorrelation | 0.19066 | 0.06934 | 0.21428 | 0.04451 | 0.18693 | 0.01973 |
| intercept | 4.48622 | 0.57400 | 4.28011 | 0.40788 | 4.51444 | 0.26927 |
| distance from CBD | -0.38547 | 0.15942 | -0.33469 | 0.13889 | -0.39089 | 0.12369 |
| north-south coordinate | -0.21214 | 0.13811 | -0.19215 | 0.13255 | -0.21485 | 0.12718 |
| east-west coordinate | 2.83130 | 1.84923 | 2.46878 | 1.68154 | 2.88511 | 1.54802 |
| cultural indicator | 0.26010 | 0.09770 | 0.24492 | 0.08844 | 0.26301 | 0.08107 |
| low density indicator | -2.33409 | 0.21375 | -2.33948 | 0.21355 | -2.33469 | 0.20962 |
| pseudo-$R^2$ | 0.5334 | | | | | |
| residual normality | 0.95141*** | | | | | |

In this case, these results differ from the foregoing WLS ones in that

---

[7]

    A pseudo-$R^2$ value is not reported here because SpaceStat performs a no-intercept, weighted regression analysis, and hence this summary value is not comparable to the one rendered by SAS.

- (1) weak spatial autocorrelation is uncovered,
- (2) parameter estimates for all but the cultural indicator variable discernibly decrease, and
- (3) all of the standard errors remain virtually unchanged.

The standard error estimates based upon the Jacobian approximation for the spatial autocorrelation and distance decay parameters are inflated by roughly 200-300%, too.

Generalized weighted least squares (GWLS) estimation results for the Syracuse metropolitan area, in 1980, are as follows:

| variable | SAS-approximate | | SAS-exact | | SpaceStat[8] | |
|---|---|---|---|---|---|---|
| | b | $s_b$ | b | $s_b$ | b | $s_b$ |
| $\hat{\rho}$ spatial | | | | | | |
| autocorrelation | 0.26236 | 0.09863 | 0.26065 | 0.05419 | 0.25640 | 0.09444 |
| intercept | -0.65208 | 0.20920 | -0.65056 | 0.20114 | -0.64791 | 0.21072 |
| distance from CBD | -1.31521 | 0.23465 | -1.31015 | 0.16901 | -1.32697 | 0.17915 |
| north-south | | | | | | |
| coordinate | -0.44657 | 0.12346 | -0.44990 | 0.12316 | -0.44722 | 0.12674 |
| low density indicator | -1.94162 | 0.20195 | -1.93945 | 0.19576 | -1.94521 | 0.19431 |
| | | | | | | |
| pseudo-$R^2$ | 0.7192 | | | | | |
| residual normality | 0.94733*** | | | | | |

Again, these results differ from the foregoing ones in that

- (1) weak-to-moderate spatial autocorrelation is uncovered,
- (2) parameter estimates for the intercept and distance decay variables dramatically change, and
- (3) the standard error for the distance decay parameter estimate increased a bit.

As before, stability of the indicator parameters estimates is exactly what one should expect, given that WLS estimators are unbiased in the presence of spatial autocorrelation; one should recall that the very nature of spatial autocorrelation suggests that estimates of distance decay parameters may well become confounded with it, too. Of note is that the distance decay and spatial autocorrelation standard error estimates based

---

8

A pseudo-$R^2$ value is not reported here because SpaceStat performs a no-intercept, weighted regression analysis, and hence this summary value is not comparable to the one rendered by SAS.

upon the Jacobian approximation essentially agree with those rendered by SpaceStat.

Generalized weighted least squares (GWLS) estimation results for the Syracuse metropolitan area, in 1990, are as follows:

| variable | SAS-approximate b $s_b$ | SAS-exact b $s_b$ | SpaceStat[9] b $s_b$ |
|---|---|---|---|
| $\hat{\rho}$ spatial autocorrelation | 0.24512 0.08536 | 0.23987 0.05305 | 0.24025 0.07329 |
| intercept | -0.35416 0.21421 | -0.34980 0.19097 | -0.34710 0.20286 |
| distance from CBD | -1.53888 0.22493 | -1.53999 0.16635 | -1.54985 0.14690 |
| north-south coordinate | -0.28370 0.10792 | -0.27988 0.10770 | -0.28315 0.10515 |
| east-west coordinate | 0.23194 0.11641 | 0.23006 0.11326 | 0.23389 0.10882 |
| 1990 low density indicator | -0.54497 0.05681 | -0.54029 0.05644 | -0.54543 0.05652 |
| 1980 low density indicator | -1.32791 0.16667 | -1.33019 0.16669 | -1.32816 0.16345 |
| pseudo-$R^2$ | 0.8057 | | |
| residual normality | 0.98164 | | |

In this situation these results differ from the foregoing ones in that

- (1) weak spatial autocorrelation is uncovered,
- (2) parameter estimates for the intercept, distance decay and east-west coordinate variables dramatically change, and
- (3) the standard errors tend to increase for parameter estimates attached to these two preceding variables and decrease for the remaining four, but only scarcely.

As before, stability of especially the indicator parameters estimates is exactly what one should expect, as is a confounding of spatial autocorrelation and distance decay parameter estimates. Of note is that the distance decay and spatial autocorrelation standard error estimates based upon the Jacobian approximation are inflated.

---

9

A pseudo-$R^2$ value is not reported here because SpaceStat performs a no-intercept, weighted regression analysis, and hence this summary value is not comparable to the one rendered by SAS.

## V. IMPLEMENTATION FINDINGS

The more comprehensive analyses summarized in this chapter support the contention of variance heterogeneity being present in urban population density values, here seemingly being related to the inverse of Area in some way. Furthermore, for a given metropolitan region, consistently census tracts with extremely low population counts appear to come from a statistical population other than the one to which the rest of the census tracts belong. Both of these non-stationarity features of the geographic distribution of urban population density should be recognized in refined model specifications of equation (2).

Implementation of a spatial statistical version of equation (2) reveals that spatial autocorrelation plays a prominent role in a description of the geographic distribution of population density, and is not merely another manifestation of distance-from-prominent-nodes factors or statistical non-stationarity. In all four cases reported here, weak-to-moderate levels of spatial autocorrelation were found to be latent in the geographic distribution of population density under study. The associated marginal increases in percent of variance accounted for are as follows: 15.0% for Toronto, 14.3% for Ottawa-Hull, 2.3% for Syracuse in 1980, and 2.4% for Syracuse in 1990. Additionally, numerical simplifications reported by Griffith [1992] enable the necessary spatial autoregressive specifications to be implemented with a minimum of effort. One potential drawback of this approximation approach is that the asymptotic standard error of $\hat{\rho}$ seems, on occasion, to be markedly inflated. If there is a concern that $s_\beta$ is excessively inflated, then it can be replaced with $\sqrt{2} / \sum_{i=1}^{n} \sum_{j=1}^{n} c_{ij}$, another asymptotic result. For the four preceding empirical applications this particular estimate compares as follows:

| metropolitan region | SAS- exact Jacobian $s_\beta$ | SAS-approximate Jacobian $s_\beta$ | SpaceStat | replacement $s_\beta$ |
|---|---|---|---|---|
| Toronto | 0.02222 | 0.03927 | 0.01041 | 0.02319 |
| Ottawa-Hull | 0.04451 | 0.06934 | 0.01973 | 0.04446 |
| Syracuse: 1980 | 0.05419 | 0.09863 | 0.09444 | 0.05064 |
| Syracuse: 1990 | 0.05305 | 0.08536 | 0.07329 | 0.05064[10] |

---

[10]

The equivalence of the replacement value for Syracuse in 1980 and 1990 is merely coincidental.

The SAS estimates, $s_\beta$, deviate from the SpaceStat ones largely because SAS is being tricked into estimating a spatial model, and hence SAS is unaware that the correct entry in the inverse asymptotic variance-covariance matrix of the maximum likelihood estimates (MLEs) , from which the standard error comes, is given by

$$\text{trace}[(\mathbf{I} - \hat{\rho}\,\mathbf{W}^T)^{-1}\mathbf{W}^T\mathbf{W}(\mathbf{I} - \hat{\rho}\,\mathbf{W}^T)^{-1}] + \sum_{j=1}^{n} \lambda_j^2 / (1 - \hat{\rho}_j)^2 +$$

$$\mathbf{b}^T\mathbf{X}^T(\mathbf{I} - \hat{\rho}\,\mathbf{W}^T)^{-1}\mathbf{W}^T\mathbf{W}(\mathbf{I} - \hat{\rho}\,\mathbf{W}^T)^{-1}\mathbf{X}\mathbf{b}/\hat{\sigma}^2 \ .$$

Meanwhile, slight differences appear between the mean response estimates, themselves, produced by SAS using the exact Jacobian and by SpaceStat.   These discrepancies can be attributed to the nonlinear analysis involved, particularly in terms of differences in nonlinear optimization algorithms employed and convergence criteria utilized.[11]

Statistical findings here include that biased distance decay results are obtained if spatial dependency is ignored when empirically describing urban population density patterns.  In fact, not surprisingly, a discernable relationship prevails among spatial autocorrelation, trend surface patterns, and distance decay effects from the CBD.   This collinearity noticeably impacts on the standard error estimates for parameter estimates associated with these variables, too.  Of note is that the standard errors for other mean response parameter estimates are little affected by the selection of implementation vehicle, whether it be SpaceStat, SAS with a Jacobian approximation, or SAS with an exact Jacobian term.  An additional statistical feature is that errors persist in appearing to come from a non-normal population, compromising to some degree the inferential basis of this type of spatial statistical analysis.

Generally speaking, these findings are consistent with those previously reported by Griffith [1981] for Toronto, and by Anselin and Can [1986] for Columbus, Ohio.  Overall, findings summarized in this chapter indicate that Syracuse has a steeper density gradient for both

---

11

    In both cases the estimation of the autoregressive response model parameters is by maximum likelihood (ML) using a non-linear optimization procedure.  SpaceStat uses a bisection search over values of $\rho$ in the interval of $1/\lambda_{min}$ to $1/\lambda_{max}$, where $\lambda_{min}$ and $\lambda_{max}$ are the extreme eigenvalues of the row-standardized geographic weights matrix, $\mathbf{W}$. SAS uses the Marquardt approach, which requires a spatial model to be written in non-linear regression terms (see Griffith, 1981).  It, too, restricts attention to values of $\rho$ in the interval $(1/\lambda_{min}, 1/\lambda_{max})$.

1980 and 1990 than does either Toronto or Ottawa-Hull for 1986.  A relatively higher level of positive spatial autocorrelation revealed by the geographic distribution of Toronto's population may be indicative of higher levels of residential segregation associated with its metropolitan size, and/or the presence of multiple centers, too.

In conclusion, then, indeed spatial autocorrelation appears to be an important component of the geographic distribution of urban population density, and it should not continue to be overlooked by urban scientists; rather, as called for by Anselin and Can [1986], it needs to be incorporated into modelling efforts.   This component of urban population spatial structure can be incorporated into an analysis using SAS, MINITAB, SpaceStat, or perhaps even other software packages. The mean response coefficient estimates, as well as their standard errors, obtained using the Jacobian approximation procedure with SAS code tend to be very similar to those obtained using SpaceStat.  By using SAS, rather than a specialized package such as SpaceStat, though, the practitioner has a considerably larger set of statistical tools directly accessible, and does not have to worry about transporting data and/or output between software packages.  This convenience was encountered in the analyses performed for this chapter when executing weighted least squares regression, which is easily implemented in SAS with weights statements while SpaceStat requires the construction of specially transformation variables.   Given the associated benefits in computer resource demands and software availability, then, this finding is encouraging for practitioners interested in more user-friendly implementation of spatial statistical/econometric models.

# REFERENCES

Anselin, L.  *SPACESTAT TUTORIAL: A Workbook for Using SpaceStat in the Analysis of Spatial Data.* Technical Software Series S-92-1, NCGIA, University of California, Santa Barbara,1992.
Anselin, L., and Can, A.  Model comparison and model validation issues in empirical work on urban density functions. *Geographical Analysis*, 18: 179-197, 1986.
Batty, M., and Kim, K.  Form follows function:  reformulating urban population density functions. *Urban Studies*, 29:  1043-1070, 1992.
Baumont, C.  Preferences spatiales et éspaces urbaines multicentriques. *Document de Travail No. 9308*, Laboratoire d'Analyse et de Techniques Economiques, Université de Bourgogne, Dijon, France, 1993.
Can, A., and Griffith, D.  Spatial dependence in urban density functions.  Paper presented at the 40th North American meetings of RSAI, Houston, TX, November 12-14, 1993.
Griffith, D.  Modelling urban population density in a multi-centered city. *Journal of Urban Economics*, 9:  298-310, 1981.

**Griffith, D.** Estimating spatial autoregressive model parameters with commercial statistical packages. *Geographical Analysis*, 20: 176-186, 1988.

**Griffith, D.** Simplifying the normalizing factor in spatial autoregressions for irregular lattices, *Papers in Regional Science*, 71: 71-86, 1992.

**Griffith, D.** *Spatial Regression Analysis on the PC: Spatial Statistics Using SAS.* The Association of American Geographers, Washington D.C., 1993.

**Griffith, D.; Haining, R.; and Arbia, G.** Heterogeneity of attribute sampling error in spatial data sets. *Geographical Analysis*, 26: 300-320, 1994.

**Hill, F.** Spatio-temporal trends in urban population density: a trend surface analysis, in *The Form of Cities in Central Canada*, edited by L. Bourne; MacKinnon, R.; and Simmons, J. University of Toronto Press, Toronto, pp. 103-119, 1973.

**Kau, J., C. Lee and Sirmans, C.** *Urban Econometrics.* Vol. 6, Research in Urban Economics Series, JAI Press, Greenwich, CT, 1986.

**Thrall, G.** Statistical and theoretical issues in verifying the population density function, *Urban Geography*, 9: 518-537, 1988.

# Chapter 10

## SPATIAL STATISTICS FOR ANALYSIS OF VARIANCE OF AGRONOMIC FIELD TRIALS

### D. S. Long

The analysis of variance (ANOVA) is used by agronomists to compare the means of treatment groups in field trials. To perform an ANOVA, the treatment groups are assumed to be randomly sampled, normally distributed, and have equal variances. In practice, the method works well even with data being non-normal and having unequal variances, provided the groups are of equal size. However, the ANOVA is seriously affected by data that violate the randomization assumption, or are spatially autocorrelated.

Positive spatial autocorrelation, a term used to describe the tendency for similarly valued observations to cluster together in space [Haining, 1980], distorts inferential tests and interval estimates. [Legendre *et al.*, 1990] Inferences based on classical distribution theory are valid only if the observations are independently and identically distributed. [Glass *et al.*, 1972] Autocorrelated data do not adhere to this theory. Furthermore, such data are redundant and carry less total information than do independent, unautocorrelated data. [Cliff and Ord, 1975] In this situation, measured observations behave as partially repeated measures of a single observation rather than as single observations. [Griffith, 1992]

Randomization is used to satisfy the condition of independence of the observations. It is supposed to ensure that no treatment is consistently favored by being placed under the best field conditions. [Petersen, 1985] Autocorrelation can be imparted to crop data because non-treatment variables such as soil and topography are autocorrelated [Mulla *et al.*, 1990; van Es and van Es, 1993] and they influence crop response. Unfortunately, randomization may not completely neutralize this autocorrelation when it is too non-random to promote equal likelihood of treatments being placed under all possible field conditions. [Olson *et al.*, 1985]

Presently, there is great interest in methods of analysis of field experiments that take into account spatial autocorrelation. [Bartlett, 1978; Besag and Kempton, 1986; Green *et al.*, 1984; Zimmerman and Harville, 1991; Brownie *et al.*, 1993] Citing evidence of up to 30 percent greater efficiency, Grondona and Cressie [1991] found that spatial analyses that approximate the error variance-covariance structure are more precise than are conventional analyses that assume this structure to be independent. They recommend simple modeling techniques based on time series analysis for spatial data from designed experiments.

In this chapter I propose such a procedure involving a spatial statistical technique known as autoregressive response (AR) modeling. I advocate this method for ANOVA of agronomic field trials because, through its specification of the geographic configuration of areal units, the AR model accommodates plots that are located on the nodes of a regular lattice. In addition, by partialling out of the error term the autocorrelated component of variance, AR-based ANOVA increases the precision of detecting real treatment differences. I instruct how to conduct AR modeling of agronomic field data using SAS computer code, illustrating it with an example data set consisting of a 102-plot variety trial.

## I. THE EXAMPLE DATA SET

The focus of this chapter is on a data set that was derived from a 1983 variety trial of spring wheat (*Triticum aestivum L.*) from Montana State University, Northern Agricultural Research Center, Havre, MT. This field experiment consists of three blocks of 34 cultivars planted in accordance with a randomized complete block (RCB) design (Figure 1a). Individual plots measure 0.914- by 4.88-m equalling 4.46-m$^2$ (48-ft$^2$) and when planted consist of three rows of wheat spaced 0.305-m (12-in) apart. However, only the center row of each plot, representing 1.49-m$^2$ (16-ft$^2$), is actually harvested for yield. The datafile VARIETY.DAT, described in Section 4, contains values of crop yield in units of kilograms per hectare (kg ha$^{-1}$) for 102 plots located on a 6 column by 17 row rectangular grid. Plotting the expected normal distribution versus sample distribution of errors obtained from a linear regression of yield on cultivars will show these data to approximately satisfy the assumption of normality. Plus, plotting the errors versus predicted yield will reveal a near-uniform band, thus indicating no serious departure from the assumption of equal variances.

**(A)**

| | BLOCK 1 | BLOCK 2 | | BLOCK 3 | |
|---|---|---|---|---|---|
| 1 | 18 | 8 | 6 | 17 | 34 |
| 2 | 19 | 9 | 12 | 25 | 6 |
| 3 | 20 | 7 | 21 | 21 | 32 |
| 4 | 21 | 30 | 9 | 3 | 24 |
| 5 | 22 | 4 | 32 | 14 | 31 |
| 6 | 23 | 23 | 28 | 26 | 18 |
| 7 | 24 | 3 | 25 | 28 | 30 |
| 8 | 25 | 13 | 20 | 16 | 5 |
| 9 | 26 | 10 | 3 | 13 | 1 |
| 10 | 27 | 22 | 18 | 8 | 20 |
| 11 | 28 | 31 | 2 | 15 | 2 |
| 12 | 29 | 19 | 5 | 23 | 19 |
| 13 | 30 | 27 | 14 | 4 | 33 |
| 14 | 31 | 24 | 16 | 27 | 7 |
| 15 | 32 | 1 | 26 | 9 | 10 |
| 16 | 33 | 11 | 33 | 12 | 29 |
| 17 | 34 | 15 | 17 | 22 | 11 |

**(B)  ROW**

| | COLUMN 1 | 2 | 3 | 4 | 5 | 6 |
|---|---|---|---|---|---|---|
| 2 | 2989.955 | 3272.153 | 3608.103 | 3950.772 | 3184.806 | 3836.549 |
| 3 | 3231.839 | 3151.211 | 3567.789 | 3997.805 | 3601.384 | 4595.796 |
| 4 | 3231.839 | 3157.93 | 3823.111 | 3359.5 | 2969.798 | 3829.83 |
| 5 | 3406.533 | 3272.153 | 4125.466 | 3890.301 | 3587.946 | 3970.929 |
| 6 | 3292.31 | 3782.797 | 3903.739 | 3366.219 | 3352.781 | 3809.673 |
| 7 | 3339.343 | 3648.417 | 3413.252 | 4132.185 | 3406.533 | 3991.086 |
| 8 | 2539.782 | 3359.8 | 3997.805 | 3043.707 | 3587.946 | 4179.218 |
| 9 | 3500.699 | 3366.219 | 4306.879 | 3594.665 | 3708.888 | 3312.467 |
| 10 | 4521.887 | 3231.839 | 3937.334 | 3453.566 | 1894.758 | 4125.466 |
| 11 | 4152.342 | 3090.74 | 4447.978 | 3278.872 | 3587.946 | 3917.177 |
| 12 | 4884.713 | 3614.822 | 3648.417 | 3231.839 | 4058.276 | 4300.16 |
| 13 | 4320.317 | 3776.078 | 3823.111 | 3742.483 | 3964.21 | 3977.648 |
| 14 | 3917.177 | 4071.714 | 3829.83 | 3090.74 | 3325.905 | 3984.367 |
| 15 | 3883.582 | 4481.873 | 3164.649 | 3682.012 | 3003.393 | 4421.102 |
| 16 | 3446.847 | 3903.739 | 3823.111 | 3594.665 | 4118.747 | 4038.119 |
| 17 | 4259.846 | 3923.896 | 3816.392 | 3984.367 | 3674.508 | 3722.326 |

**(C)  ROW**

| | COLUMN 1 | 2 | 3 | 4 | 5 | 6 | 7 |
|---|---|---|---|---|---|---|---|
| 1 | 1 | 2 | 3 | 4 | 5 | 6 | 7 |
| 2 | 8 | 9 | 10 | 11 | 12 | 13 | 14 |
| 3 | 15 | 16 | 17 | 18 | 19 | 20 | 21 |
| 4 | 22 | 23 | 24 | 25 | 26 | 27 | 28 |
| 5 | 29 | 30 | 31 | 32 | 33 | 34 | 35 |
| 6 | 36 | 37 | 38 | 39 | 40 | 41 | 42 |
| 7 | 43 | 44 | 45 | 46 | 47 | 48 | 49 |
| 8 | 50 | 51 | 52 | 53 | 54 | 55 | 56 |
| 9 | 57 | 58 | 59 | 60 | 61 | 62 | 63 |
| 10 | 64 | 65 | 66 | 67 | 68 | 69 | 70 |
| 11 | 71 | 72 | 73 | 74 | 75 | 76 | 77 |
| 12 | 78 | 79 | 80 | 81 | 82 | 83 | 84 |
| 13 | 85 | 86 | 87 | 88 | 89 | 90 | 91 |
| 14 | 92 | 93 | 94 | 95 | 96 | 97 | 98 |
| 15 | 99 | 100 | 101 | 102 | 103 | 104 | 105 |
| 16 | 106 | 107 | 108 | 109 | 110 | 111 | 112 |
| 17 | 113 | 114 | 115 | 116 | 117 | 118 | 119 |

**Figure 1.** Field plot layout of the spring wheat variety trial planted in accordance with a randomized complete block design of 34 varieties in three blocks (A), 6 column by 17 row rectangular grid of plots with yield observations in kg ha$^{-1}$ (B), and order of observations in SAS datafile VARIETY.DAT including column 7 for missing plots on right side of lattice (C).

## II. GOALS OF THE CASE STUDY

In this chapter I provide guidance for the use of AR-based ANOVA and accompanying SAS computer code for the analysis of agronomic field data. I describe the AR model specification in Section III. In Sections IV through VII I explain the following steps using SAS computer code to perform AR-based ANOVA of the 102-plot variety trial:

- •      a. indexing of autocorrelation with the Moran Coefficient,
- •      b. calculating the eigenvalues of the geographic weights matrix,
- •      c. estimating the Jacobian term, and
- •      d. estimating an autoregressive response (AR) model.

Finally, in Section VIII, I show clearly the basic difference between AR-based ANOVA and conventional ANOVA with regard to the results from the 102- plot variety trial.

## III. THE AUTOREGRESSIVE RESPONSE MODEL

The following AR model specification describes the spatial variation in the (n-by-1) crop response vector Y:

$$Y = \rho WY + X_p \beta_p + \epsilon \qquad (1)$$

The term $\rho WY$ is a crop response vector whose values in adjacent locations are specified by the (n-by-n) configuration matrix W, and are autocorrelated by an amount $\rho$, $\beta$ is a vector of p slope parameters, X is an (n-by-p) matrix of classificatory indicator variables, and $\epsilon$ is the error vector.

In regression the independence assumption centers on the errors rather than predictor variable X and response variable Y. Therefore, equation (1) is rewritten with respect to the error term $\epsilon$ as

$$\epsilon = (I - \rho W)Y - X\beta \qquad (2)$$

where I is the identity matrix. Equation (2) specifies the basic form of the error variance-covariance matrix. Accordingly, the actual error sum of squares, D, for an AR model with nonconstant mean is

$$D = [I - \rho W)Y - X\beta]' [(I - \rho W)Y - X\beta]. \qquad (3)$$

Estimation of the AR model requires the maximum likelihood (ML) technique. The ML is based on the principle of selecting values for the parameters that maximize the probability of the observed data. It is complicated here by the nonlinear mathematical structure, requiring an iterative numerical approach to obtain a solution for the autocorrelation parameter, $\rho$. Nevertheless, ML estimates of $\rho$ consistently converge to the true parameter with increasing n [Cliff and Ord, 1975; Haining, 1978], are restricted to a feasible parameter space, and are associated with a probability density function that integrates to unity [Griffith, 1988a].

The ML estimate of $\rho$ is found by maximizing with respect to $\rho$ the following nonlinear equation:

$$L = (2\pi)^{-n/2} \det[(I - \rho W)^t(I - \rho W)]^{1/2}(\sigma^2)^{-n/2} \exp[-D/(2\sigma^2)] \qquad (4)$$

where $\det[(I - \rho W)^t (I - \rho W)]$ is the Jacobian (J) that is written in terms of the eigenvalues of matrix W. Equation (4) expresses the probability density function of the normally distributed errors of the response variable Y. For matrix W, n eigenvalues need to be calculated.

Equation (4) is iteratively estimated until the mean sum of squared errors D, weighted by $\exp((J)^{1/2})$, reaches a minimum, and the value of $\rho$ is selected that is associated with this minimum. Conceptually, this procedure is similar to computing orthogonal components in principle components analysis whereby data are transformed from a correlated to an uncorrelated mathematical space. The ML function accomplishes this transformation by being constrained by the Jacobian term, a normalizing factor ensuring that this transformation yields a probability density function whose complete integration equals unity. [Griffith 1990]

## IV. CALCULATING THE MORAN COEFFICIENT

Moran [1948] first proposed a coefficient to measure the nature and degree of autocorrelation in geo-referenced variables. The Moran Coefficient (MC) can be used to test the null hypothesis of zero spatial autocorrelation in agronomic field data that are geo-referenced by a regular lattice arrangement of plots. It is calculated by comparing neighboring pairs of observations with their deviation from the mean of all observations according to

$$MC = \frac{n}{\Sigma\Sigma c_y} \times \frac{\Sigma\Sigma c_y (y_i - y_j)^2}{\Sigma(y_i - \bar{y})^2}, \qquad (5)$$

where MC is the Moran Coefficient, n is the number of plots, $c_{ij}$ is the ij-th entry in a binary configuration matrix with scores of 1 and 0, $y_i$ is the ith observation, $y_j$ is the jth observation, and $\bar{y}$ is the mean of the data series. Matrix entries $c_{ij}$ equal unity if plots i and j are adjacent, and zero otherwise.

Values of the MC roughly range between 1 and -1 for strongly positive and strongly negative autocorrelation. Its calculated value given by equation (5) should equal its expected value, which is $-[1/(n-1)]$, if the data of a series of observations are truly independent. Values that exceed $-[1/(n-1)]$, within the limits of statistical significance, indicate positive autocorrelation and values below $-[1/(n-1)]$ indicate negative autocorrelation.

The 102-plot variety trial data have been collected for a regular rectangular grid as illustrated in Figure 1b. The uniform structuring of this two-dimensional surface partitioning allows the MC to be calculated without having to construct the geographic configuration matrix C. Instead, the SAS code for calculating the MC uses the LAG function to automatically construct this matrix. A user can easily modify this code for analysis of any row by column rectangular lattice. Some SAS code, with filename MC102.SAS, for computing the MC of the variety trial data appears below. A commercially available software package such as GS+ (Gama Design Inc., Plainwell, MI) could also be used for computing the MC.

### FILE MC102.SAS

```
*THIS CODE IS COPYRIGHTED BY SYRACUSE UNIVERSITY, 1991*;        1
FILENAME YIELDDAT 'C:VARIETY.DAT';                              2
OPTIONS LINESIZE=72;                                            3
TITLE 'MORAN COEFFICIENT FOR YIELD VALUES';                     4
*------------------------------------------------------------*;
*DATA ARE FOR A SQRT(N)-BY-SQRT(N) REGULAR LATTICE, WHICH HAS*
*BEEN AUGMENTED TO A SQRT(N)-BY-(SQRT(N)+1) REGULAR LATTICE.*
*THE ADDITIONAL LATTICE COLUMN IS ADDED TO THE RIGHT HAND*
*SIDE, AND CONTAINS MISSING VALUES.  DATA ARE ENTERED INTO*
*"YIELD.DAT" BEGINNING WITH THE UPPER LEFT HAND CORNER OF THE*
*LATTICE, AND MOVING FROM LEFT TO RIGHT.  THIS RESULTS IN A*
*CONCATENATION OF SUCCESSIVE LATTICE ROW ENTRIES, WITH EACH*
*SUCCESSIVE PAIR BEING SEPARATED BY A MISSING VALUE AREAL UNIT*
*VALUE.*
*------------------------------------------------------------*;
DATA STEP1;                                                     16
     INFILE YIELDDAT;                                           17
     INPUT NUMBER YIELD;                                        18
RUN;                                                            19
PROC STANDARD MEAN=0 OUT=TEMP1;                                 20
     VAR YIELD;                                                 21
RUN;                                                            22
*------------------------------------------------------------*;
*TWO DIRECTIONS OF SPATIAL LAG VALUES ARE DETERMINED. N4 MUST*
*BE LAGGED BY SQRT(N)+1, RESULTING IN A CHANGE IN "LAG__".*
*------------------------------------------------------------*;
DATA STEP2;                                                     27
     SET TEMP1;                                                 28
XSQ=X*X;                                                        29
N2=LAG1(X);                                                     30
N4=LAG7(X);                                                     31
PROC SORT DATA=STEP2;                                           32
     BY DESCENDING ORDER;                                       33
```

```
RUN;                                                                           34
*-------------------------------------------------------------------*;
*THE REMAINING TWO DIRECTIONS OF SPATIAL LAG VALUES ARE*
*DETERMINED. N3 MUST BE LAGGED BY SQRT(N)+1, RESULTING IN A*
*CHANGE IN "LAG__".*
*-------------------------------------------------------------------*;
DATA STEP3;                                                                    40
    SET STEP2;                                                                 41
N1=LAG1(X);                                                                    42
N3=LAG7(X);                                                                    43
RUN;                                                                           44
*-------------------------------------------------------------------*;
*DETERMINATION OF THE NUMBER OF NEIGHBORS AND THE*
*AVERAGE NEIGHBORING VALUES FOR EACH AREAL UNIT.*
*-------------------------------------------------------------------*;
DATA STEP4;                                                                    49
    SET STEP3;                                                                 50
IF N1 EQ '.' THEN NBR1=0; ELSE NBR1=1;                                         51
IF N2 EQ '.' THEN NBR2=0; ELSE NBR2=1;                                         52
IF N3 EQ '.' THEN NBR3=0; ELSE NBR3=1;                                         53
IF N4 EQ '.' THEN NBR4=0; ELSE NBR4=1;                                         54
NEIGH=NBR1+NBR2+NBR3=NBR4;                                                     55
IF X EQ '.' THEN NEIGH=0;                                                      56
IF N1 EQ '.' THEN N1=0;                                                        57
IF N2 EQ '.' THEN N2=0;                                                        58
IF N3 EQ '.' THEN N3=0;                                                        59
IF N4 EQ '.' THEN N4=0;                                                        60
CX = (N1+N2+N3+N4);                                                            61
IF X EQ '.' THEN CX='.';                                                       62
RUN;                                                                           63
*-------------------------------------------------------------------*;
*COMPUTATION OF TERMS NEEDED FOR COMPUTING THE MC*
*-------------------------------------------------------------------*;
DATA STEP5;                                                                    67
    SET STEP4;                                                                 68
    SMCX=X*CX;                                                                 69
RUN;                                                                           70
PROC MEANS DATA=STEP5 NOPRINT;                                                 71
    VAR NEIGH XSQ SMCX;                                                        72
        OUTPUT OUT=TEMP2 SUM=CSUM XSQ SMCX;                                    73
RUN;                                                                           74
DATA STEP6;                                                                    75
    SET TEMP2;                                                                 76
N=102;                                                                         77
MC = N*SMCX/(CSUM*XSQ);                                                        78
ZAPP=(MC+1/(N-1))/SQRT(2/CSUM);                                                79
PROBMC=1-PROBNORM(ZAPP);                                                       80
PROC PRINT;                                                                    81
    VAR N CSUM MC ZAPP PROBMC;                                                 82
RUN;                                                                           83
```

Line 2 of the SAS code defines the path for accessing the 102-plot variety trial data named VARIETY.DAT. The initial file path declaration is written for the personal computer (PC). The experimental lattice of 17 rows and 6 columns has rows numbered from 1 to 17 and columns numbered from 1 to 6 (Fig. 1b). The 1,1 entry of this lattice is the plot in the top left-hand corner. While the logical structure of the lattice is a grid, the actual structure of the SAS datafile is 2 columns of 119 records. Below is a segment of this data file. Lines 16-19 read the sequential ordering variable denoted NUMBER and the crop response variable denoted YIELD. Lines 20-22 standardize the observations of YIELD so that they are centered around the mean of the data series (i.e., $x_i$ - ).

**FILE VARIETY.DAT**

| 1 | 386 |
|---|---|
| 2 | 452 |
| 3 | 371 |
| 4 | 498 |
| 5 | 543 |
| 6 | 541 |
| 7 | . |
| 8 | 445 |
| 9 | 487 |
| 10 | 537 |
| 11 | 588 |
| 12 | 474 |
| 13 | 571 |
| 14 | . |
| 15 | 481 |
| 16 | 469 |
| 17 | 531 |
| 18 | 595 |
| 19 | 536 |
| 20 | 684 |
| 21 | . |

The map in Figure 1c represents how the plots are ordered in the data file by the variable NUMBER. Accordingly, the lattice entries are read from the columns of the data file in the order of columns 1-6 by rows 1-17. Each row of six plots is separated by a decimal (.) to denote in SAS the presence of missing values of a variable. This special value is encountered in the datafile once after each six records and hence 17 of the 119 entries are missing value entries. Its purpose is to attach a phantom column of field plots in order to allow the LAG function of lines 30-31 and lines 42-43 to properly identify those plots lying immediately right of, below, left of, and above each individual plot.

The regular two-dimensional arrangement of the data allows the time series LAG function to be used. The LAG function operates on a data series in the direction that the data are read from the data file. Accordingly, Lines 30-31, for DATA STEP2, identify the YIELD variable values lying immediately to the right of (LAG1) and below (LAG7) for each plot. A missing-values column has been added to the map (Figure 1c) since the extreme right plots have no values immediately to their right. Lines 32-34 reverse the original geographic distribution of the data. Now when the LAG1 function extracts values lying to the right, it is identifying those values originally lying to the left, and now when the LAG7 function extracts those values lying below, it is identifying those values originally lying above. Lines 51-54, 56-60, and 62 dispose of values identified for the appended ghost field plots.

Results obtained with the SAS code are as follows:

| | |
|---|---|
| N | 102. |
| CSUM | 362. |
| MC | 0.159 |
| ZAPP | 2.28 |
| PROBMC | 0.0113 |

where N is the number of observations, CSUM is the sum of the ones contained in matrix C, MC is the Moran Coefficient, ZAPP is the z-score for the normal distribution, and PROBMC is the normal probability value of the MC for zero autocorrelation. The MC for the yield observations is 0.16 which exceeds an expected value of -0.0099 for zero autocorrelation. Hence, weak positive autocorrelation is being detected among the observations and AR-based ANOVA is deemed warranted.

In regression, the MC can be used to test for autocorrelation in the residuals. This autocorrelation can arise from a spatially varying variable that is missing from a model's specification [Cressie, 1991]. According to Miron [1984], if the missing variable is partly related to the dependent and independent variables of a regression model, then the remaining effect may be manifested as autocorrelation in the residuals.

## V. CALCULATING THE NECESSARY EIGENVALUES

Having computed the MC, the second step in AR modeling is computing the eigenvalues for the geographic configuration matrix. The purpose of this matrix is to describe the spatial connectivity of the sample locations of the crop response variable represented as a row-by-column grid of plots. Eigenvalues computed from this matrix reflect the systematic behavior in the arrangement, interconnectivity, and interdependence of the yield observations. Furthermore, the eigenvalues facilitate computation of the Jacobian of the transformation from an autocorrelated to an unautocorrelated mathematical space. Computing the Jacobian is introduced in Section 6 as the third step in AR modeling.

The SAS code, filename EIGEN102.SAS, for computing the eigenvalues is presented below. Line 2 of this code accesses the datafile CON102.DAT that contains the entries of a 102- by-102 binary configuration matrix C. The initial file path declaration is written for the PC. The name CONN is the SAS reference file defined for CON102.DAT The datafile was constructed with text editor software, but computer programs can be written for automatic construction of such a matrix for regular lattices.

```
FILE EIGEN102.SAS
*THIS CODE IS COPYRIGHTED BY SYRACUSE UNIVERSITY, 1991*;          1
FILENAME CONN 'C:\CON102.DAT';                                    2
OPTIONS LINESIZE=72;                                              3
TITLE "EIGENVALUES OF MATRIX W FOR VARIETY TRIAL DATA";           4
SET C;                                                            5
INFILE CONN;                                                      6
INPUT OBS C1-C102;                                                7
ARRAY CMTX{102} C1-C102;                                          8
     RSUM = 0;                                                    9
     DO I=1 TO 102;                                              10
          RSUM = RSUM + CMTX{I};                                 11
          IF OBS=I THEN CMTX{I} = 1;                             12
     END;                                                        13
RUN;                                                             14
DATA IC;                                                         15
     SET C;                                                      16
DROP OBS RSUM;                                                   17
RUN;                                                             18
DATA W;                                                          19
     SET C;                                                      20
     ARRAY CMTX{102} C1-C102;                                    21
     ARRAY WMTX{102} C1-C102;                                    22
     DO I=1 TO 102;                                              23
          WMTX{I} = CMTX{I}/SQRT(RSUM);                          24
     END;                                                        25
DROP C1-C102;                                                    26
RUN;                                                             27
PROC TRANSPOSE DATA=W PREFIX=TRSM OUT=TEMP1;                     28
     VAR RSUM;                                                   29
RUN;                                                             30
DATA IW;                                                         31
     SET W;                                                      32
     IF _N_ = 1 THEN SET TEMP1;                                  33
     ARRAY RMTX{102} TRSM1-TRSM102;                              34
     ARRAY WMTX{102} W1-W102;                                    35
```

```
    DO I=1 TO 102;                                          36
        WMTX{I} = WMTX{I}/SQRT(RMTX{I});                    37
    END;                                                    38
    DO I=1 TO 102;                                          39
        IF OBS=1 THEN WMTX{I} = 1;                          40
    END;                                                    41
DROP _NAME_ RSUM OBS TRSM1-TRSM102;                         42
RUN;                                                        43
DATA WCORREL(TYPE=CORR);                                    44
    SET IW;                                                 45
    _TYPE_='CORR';                                          46
RUN;                                                        47
PROC FACTOR DATA=WCORREL METHOD=PRIN N=0;                   48
    VAR W1-W102;                                            49
RUN;                                                        50
```

The first 3 records of the datafile CON102.DAT containing matrix C are presented below. Single records of this file wrap several times as indicated; SAS recognizes the end of a record when the number of values that are read matches the number of variables indicated in Line 7. This line of the code reads the ordering variable NUMBER (column 1) and the variables C1-C102 (columns 2-103). The variable NUMBER has sequential values that correspond to the order in which the plots are read from the lattice. Again, the plot numbering is in order of lattice columns 1-6 by rows 1-17. Cells C1-C102 contain the binary entries of matrix C. Entries equal one (1) for plots that share common boundaries and are adjacent. Otherwise, entries equal zero (0) for plots that do not share common boundaries. Figure 2 shows 102 plots of the example data set arranged in 17 rows and 6 columns, and the associated (102-by-102) binary configuration matrix C.

### FILE CON102.DAT

```
1   0 1 0 0 0 0 1 0 0 0 0 0 0 0 0 0 0 0 0 0 0 0 0 0 0 0 0 0 0 0 0 0 0 0 0 0 0 0 0 0 0 0 0 0 0
0 0 0 0 0 0 0 0 0 0 0 0 0 0 0 0 0 0 0 0 0 0 0 0 0 0 0 0 0 0 0 0 0 0 0 0 0 0 0 0 0 0 0 0 0 0 0 0
0 0 0 0 0 0 0
2   1 0 1 0 0 0 0 1 0 0 0 0 0 0 0 0 0 0 0 0 0 0 0 0 0 0 0 0 0 0 0 0 0 0 0 0 0 0 0 0 0 0 0 0 0
0 0 0 0 0 0 0 0 0 0 0 0 0 0 0 0 0 0 0 0 0 0 0 0 0 0 0 0 0 0 0 0 0 0 0 0 0 0 0 0 0 0 0 0 0 0 0 0
0 0 0 0 0 0 0
3   0 1 0 1 0 0 0 0 1 0 0 0 0 0 0 0 0 0 0 0 0 0 0 0 0 0 0 0 0 0 0 0 0 0 0 0 0 0 0 0 0 0 0 0 0
0 0 0 0 0 0 0 0 0 0 0 0 0 0 0 0 0 0 0 0 0 0 0 0 0 0 0 0 0 0 0 0 0 0 0 0 0 0 0 0 0 0 0 0 0 0 0 0
0 0 0 0 0 0 0
```

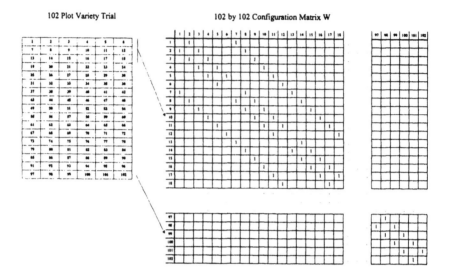

**Figure 2**. Relationship between geographic arrangement of 6-by-17 plot variety trial and 102-by-102 entry configuration matrix (with a portion shown for plots 1-18 and 97-102).

The INPUT and ARRAY statements of lines 7-8 read each row of this matrix into a one-dimensional array and compute an integer row sum. This process is repeated until a row sum is calculated for all rows resulting in the creation of the variable RSUM. In practice, AR models frequently are specified in terms of matrix W, the stochastic form of binary matrix C. Matrix W is derived by dividing each entry of the binary matrix by the corresponding row sum (RSUM) resulting in each row summing to unity. [Cliff and Ord, 1975] For example, replacing binary values with row-standardized values would result in the nonzero entries of Figure 2 equaling 0.5 for the first row, 0.33 for the second through fifth rows, 0.5 for the sixth row, 0.33 for the seventh row, 0.25 for the eighth through eleventh rows, and so forth.

The variable RSUM is used to create the W matrix in lines 19-27 for the first DATA statement. This DATA statement calls for the creation of a new data file W. It reads the observations in data file C into a new one-dimensional array WMTX{102}, and divides the square root of RSUM into each element of array WMTX{102}. The square root is used in order to simplify computation of the eigenvalues of W. Next, in lines 28-30, the variable RSUM is transposed with the PROC

TRANSPOSE procedure. Lines 31-43 create the matrix (I+W) by adding the identity matrix I to matrix W. This step ensures that diagonal entries of matrix (I+W) have values of one (1), as is necessary for any correlation matrix from which eigenvalues are to be computed with PROC FACTOR (lines 48-50).

The DATA WCORREL(TYPE=CORR) statement in line 44 specifies that the data in matrix (I+W) created from the binary matrix C is a correlation matrix. The aforementioned PROC FACTOR then computes the eigenvalues of newly created matrix (I+W). Current technology dictates that PC memory required to compute these eigenvalues will be limited to a matrix of about 300-by-300 entries for the DOS operating system. PCs using UNIX and OS/2 operating systems access large memory, and hence can handle much larger matrices.

The data file EIGEN102.DAT, containing 102 eigenvalues and their corresponding matrix row indexing numbers that have been obtained with this SAS code, is presented below. They were obtained with this code under Release 6.08 of the SAS statistical software for the OS/2 operating system (SAS Institute Inc., 1988). A 486-based personal computer was used having a 66 mHz, 32 bit microprocessor. These eigenvalues and their row numbers need to be organized with text editing software as two columns in a standard text datafile. The eigenvalues are larger by a value of one because they are based on matrix (I+W) and hence need to be corrected by subtracting this value. The principle eigenvalue should be equal to unity for matrix W. In Section 7 we shall note that this principle eigenvalue in AR models for regular square lattices constrains values of the autocorrelation parameter, $\rho$, to within negative and positive one.

### EIGEN102.DAT

| 1 | 2.0000 | 27 | 1.3771 | 53 | 0.9818 | 79 | 0.5624 |
| 2 | 1.9901 | 28 | 1.3682 | 54 | 0.9800 | 80 | 0.5542 |
| 3 | 1.9608 | 29 | 1.3499 | 55 | 0.9698 | 81 | 0.5268 |
| 4 | 1.9186 | 30 | 1.3226 | 56 | 0.9569 | 82 | 0.5059 |
| 5 | 1.9134 | 31 | 1.3025 | 57 | 0.9290 | 83 | 0.4699 |
| 6 | 1.9051 | 32 | 1.2637 | 58 | 0.9290 | 84 | 0.4251 |
| 7 | 1.8729 | 33 | 1.2526 | 59 | 0.9267 | 85 | 0.4083 |

| 8 | 1.8500 | 34 | 1.2459 | 60 | 0.9200 | 86 | 0.4003 |
|---|--------|----|--------|----|--------|-----|--------|
| 9 | 1.8215 | 35 | 1.2151 | 61 | 0.8655 | 87 | 0.3477 |
| 10 | 1.7731 | 36 | 1.1671 | 62 | 0.8628 | 88 | 0.3307 |
| 11 | 1.7527 | 37 | 1.1737 | 63 | 0.8621 | 89 | 0.3141 |
| 12 | 1.7061 | 38 | 1.1686 | 64 | 0.8365 | 90 | 0.3126 |
| 13 | 1.6874 | 39 | 1.1635 | 65 | 0.8314 | 91 | 0.2939 |
| 14 | 1.6859 | 40 | 1.1379 | 66 | 0.8263 | 92 | 0.2473 |
| 15 | 1.6693 | 41 | 1.1372 | 67 | 0.8029 | 93 | 0.2269 |
| 16 | 1.6523 | 42 | 1.1345 | 68 | 0.7849 | 94 | 0.1785 |
| 17 | 1.5997 | 43 | 1.0800 | 69 | 0.7541 | 95 | 0.1500 |
| 18 | 1.5917 | 44 | 1.0733 | 70 | 0.7474 | 96 | 0.1271 |
| 19 | 1.5749 | 45 | 1.0710 | 71 | 0.7363 | 97 | 0.0949 |
| 20 | 1.5301 | 46 | 1.0710 | 72 | 0.6975 | 98 | 0.0866 |
| 21 | 1.4941 | 47 | 1.0431 | 73 | 0.6774 | 99 | 0.0814 |
| 22 | 1.4732 | 48 | 1.0302 | 74 | 0.6501 | 100 | 0.0392 |
| 23 | 1.4458 | 49 | 1.0200 | 75 | 0.6318 | 101 | 0.0099 |
| 24 | 1.4376 | 50 | 1.0182 | 76 | 0.6229 | 102 | 0.0000 |
| 25 | 1.4153 | 51 | 1.0110 | 77 | 0.6035 | | |
| 26 | 1.3965 | 52 | 0.9890 | 78 | 0.5847 | | |

## VI.  ESTIMATING THE JACOBIAN TERM

Having computed eigenvalues from the stochastic configuration matrix W, the third step in AR modeling is computation of the Jacobian term.  This term provides a means of transforming the data from an autocorrelated to an unautocorrelated mathematical space.

Griffith [1990] simplified computation of the Jacobian by identifying the following function that accurately approximates this term based on matrix W for regular lattice data:

$$J^{1/2} = 2\alpha_n \ln(\delta_n) - \alpha_n \ln(\delta_n + \rho) - \alpha_n \ln(\delta_n - \rho) \qquad (6)$$

where $\alpha$ and $\delta$ are synthetic parameters for a rectangular gridded region, and $\rho$ is the autocorrelation parameter. The synthetic parameters are obtained by doing a nonlinear least squares regression of generated values of J on sampled values of $\rho$. Values of J may be generated for values of $\rho$ systematically sampled from the interval between the smallest and largest eigenvalues using

$$J^{1/2} = \sum_{i=1}^{n} (-\ln(1 - \rho\lambda)) \qquad (7)$$

where the $\lambda$s are the eigenvalues extracted from matrix W, n is the number of observations, and ln is the natural logarithm [Griffith, 1988b]. The value J is a logarithmic sum of n eigenvalues for the locations of n observations that are uniquely depicted by a matrix W. The eigenvalues constrain the absolute value of $\rho$ to be less than unity. The upper limit of this space, $\rho = 1$, is determined by the principle eigenvalue of matrix W, which equals unity.

The necessary steps for approximating the Jacobian include: (1) computing the Jacobian for selected values of RHO, $\rho$, drawn from the interval within the inverse of the minimum and maximum eigenvalues, and (2) calibrating this Jacobian equation. The SAS code for computing the Jacobian is presented below.

```
*THIS CODE IS COPYRIGHTED BY SYRACUSE UNIVERSITY, 1991       *;    1
FILENAME EIGEN 'C:\EIGEN.DAT';                                     2
TITLE 'JACOBIAN APPROXIMATION FOR MATRIX W';                       3
*-----------------------------------------------------------*;
*JACOBIAN APPROXIMATION FOR MATRIX W, VARIETY TRIAL DATA     *;
*-----------------------------------------------------------*;
*CHOOSE A STARTING VALUE THAT IS .999/LAMBDA-MIN.            *;
*CHOOSE AN INCREMENT THAT EQUALS                             *;
*       (.999/LAMBDA-MAX - .999/LAMBDA-MIN)/21.              *;
*-----------------------------------------------------------*;
DATA STEP1;                                                       11
    START = -0.999;                                               12
    FINISH = 1;                                                   13
    INC = 0.0951428;                                              14
RUN;                                                              15
DATA STEP2;                                                       16
    IF _N_=1 THEN SET STEP1;                                      17
INFILE EIGEN;                                                     18
INPUT OBS EIGENW;                                                 19
ARRAY JACOB{22} JAC1-JAC22;                                       20
RHO=START;                                                        21
DO I = 1 TO 22;                                                   22
    JACOB{I} = -LOG(1 - RHO*EIGENW);                              23
```

```
        RHO = RHO+INC;                                          24
END;                                                            25
RUN;                                                            26
PROC MEANS NOPRINT;                                             27
        VAR JAC1-JAC22;                                         28
        OUTPUT OUT=JACOB1 MEAN=;                                29
RUN;                                                            30
PROC TRANSPOSE OUT=JACOB2;                                      31
        VAR JAC1-JAC22;                                         32
RUN;                                                            33
DATA STEP3;                                                     34
        SET STEP2;                                              35
        DO RHO = START TO FINISH BY INC;                        36
                OUTPUT;                                         37
        END;                                                    38
DROP START FINISH INC;                                          39
PROC PRINT;                                                     40
RUN;                                                            41
DATA STEP4;                                                     42
        SET JACOB2;                                             43
        SET STEP3;                                              44
J = COL1;                                                       45
DROP COL1;                                                      46
RUN;                                                            47
PROC PLOT;                                                      48
        PLOT J*RHO;                                             49
RUN;                                                            50
PROC NLIN MAXITER=500 METHOD=MARQUARDT;                         51
        PARMS A1 = 0.25                                         52
              A2 = 0.25                                         53
              D1 = 2.0                                          54
              D2 = 2.0;                                         55
        BOUNDS D1 > 1.5, D2 > 1.00;                             56
        MODEL J = A1*LOG(D1) + A2*LOG(D2) -                     57
                    A1*LOG(D1 + RHO) - A2*LOG(D2 - RHO);        58
        DER.A1 = LOG(D1) - LOG(D1 + RHO);                       59
        DER.A2 = LOG(D2) - LOG(D2 - RHO);                       60
        DER.D1 = A1/D2 - A1/(D1 + RHO);                         61
        DER.D2 = A2/D2 - A2/(D2 - RHO);                         62
        DER.A1.A1 = 0;                                          63
        DER.A2.A2 = 0;                                          64
        DER.A1.A2 = 0;                                          65
        DER.D1.D1 = -A1/D1**2 + A1/(D1 + RHO)**2;               66
        DER.D2.D2 = -A2/D2**2 + A2/(D2 - RHO)**2;               67
        DER.D1.D2 = 0;                                          68
        DER.A1.D2 = 1/D1 - 1/(D1 + RHO);                        69
        DER.A2.D2 = 1/D2 - 1/(D2 - RHO);                        70
        DER.A1.D2 = 0;                                          71
        DER.A2.D1 = 0;                                          72
RUN;                                                            73
```

Line 2 defines EIGEN as the reference file for the external data file, EIGEN.DAT. A value for START (line 12) is based on 0.999 divided by the value of the minimum eigenvalue based on matrix W. A value for INC (line 14) is based on 0.999 minus the START value and dividing this term by 21.

Lines 16-18, for DATA STEP2, instruct SAS to read the values entered in STEP1 and the 102 eigenvalues from the datafile EIGEN.DAT. The DO loop in line 22 is iterated 22 times. Each iteration reads the datafile EIGEN.DAT and calculates 102 values of JACOB (the Jacobian) and RHO. A mean Jacobian value is generated for each of the 22 columns of 102 values with the PROC MEANS procedure in lines 27-30. Lines 31-33, involving PROC TRANSPOSE, transpose these mean values into one column of 22 values. Lines 48-50 permit a plot of the Jacobian, generated in DATA STEP2, versus values of RHO, generated in DATA STEP3.

The Jacobian approximation is accomplished in lines 51-73 by means of a nonlinear optimization procedure involving the Marquardt algorithm. Prior to running this SAS code, values for parameters A1, A2, D1, and D2 (lines 52-55) are initialized at values of 0.25, 0.25, 2.0, and 2.0, respectively. Parameters D1 and D2 in the BOUNDS statement (line 56) are set to values that are slightly greater in absolute value than the corresponding START and FINISH values. This BOUNDS statement restricts the iterative nonlinear procedure to within the feasible parameter space of estimation.

The nonlinear estimation results obtained with this SAS code for approximating the J term are as follows:

Non-Linear Least Squares Summary Statistics    Dependent Variable J

| Source | DF | Sum of Squares | Mean Square |
|---|---|---|---|
| Regression | 4 | 0.29980159745 | 0.07495039936 |
| Residual | 18 | 0.00022873971 | 0.00001270776 |
| Uncorrected Total | 22 | 0.30003033715 | |
| | | | |
| (Corrected Total) | 21 | 0.16971181840 | |

| Parameter Estimate | Asymptotic Std. Error | Asymptotic 95 % Confidence Interval | |
|---|---|---|---|
| | | Lower | Upper |
| A1  0.143533465 | 0.00341746696 | 0.1363536809 | 0.1507132494 |
| A2  0.143535653 | 0.00341769230 | 0.1363553957 | 0.1507159109 |
| D1  1.067927543 | 0.00500209540 | 1.0574186002 | 1.0784364863 |
| D2  1.067936727 | 0.00500295298 | 1.0574259826 | 1.0784474721 |

Except for slight rounding error, the symmetric eigenvalues that have been derived from the regular square lattice result in A1=A2 and D1=D2. Consequently, the forthcoming SAS code for estimating an AR model can be simplified by reducing the J specification from four parameters (A1, A2, D1, and D2) to two parameters (A1 and D1). The value given to A1 is the mean of A1 and A2, or (0.143533465+0.143535653)/2=0.143535, and the value given to D1 is the mean of D1 and D2, or (1.067927543 + 1.067936727) / 2 =1.067932.

## VII. ESTIMATING AN AUTOREGRESSIVE RESPONSE MODEL

The fourth step in AR modeling is estimation, which is undertaken once the J term has been approximated. The SAS code with filename AR102W.SAS for estimating an AR model appears below. Comments are included so that a user can modify it to accept any P-by-Q rectangular lattice. This SAS code obtains an estimate of the autocorrelation parameter when the class indicator variables are present in the AR model. The resulting estimate is then used to adjust the yield observations for autocorrelation; ANOVA is performed with PROC GLM using these yield observations that have been adjusted for autocorrelation.

```
        AR102W.SAS FILE
*THIS CODE IS COPYRIGHTED BY SYRACUSE UNIVERSITY, 1991  *;          1
FILENAME VARIETY 'C:\VARIETY.DAT';                                  2
*---------------------------------------------------------------*
* "VARIETY.DAT" IS THE NAME OF THE FILE ON THE SYSTEM.*
*---------------------------------------------------------------*;
OPTIONS LINESIZE=72;                                               6
TITLE 'AR MODEL OF VARIETY TRIAL BASED ON W MATRIX';              7
*---------------------------------------------------------------*
* DATA ARE FOR A SQRT(N)-BY-SQRT(N) REGULAR LATTICE, WHICH*
* HAS BEEN AUGMENTED TO A SQRT(N)-BY-(SQRT(N)+1) REGULAR *
* LATTICE. THE ADDITIONAL LATTICE COLUMN IS ADDED TO THE*
* RIGHT HAND SIDE, AND CONTAINS MISSING VALUES. DATA ARE*
* ENTERED INTO "SQUARE DATA" BEGINNING WITH THE UPPER*
* LEFTHAND CORNER OF THE LATTICE, AND MOVING FROM LEFT TO*
* RIGHT. THIS RESULTS IN A CONCATENATION OF SUCCESSIVE*
* LATTICE ROW ENTRIES, WITH EACH SUCCESSIVE PAIR BEING*
* SEPARATED BY A MISSING VALUE AREAL UNIT VALUE*.
*---------------------------------------------------------------*;
DATA STEP1;                                                        19
   INFILE VARIETY;                                                 20
   INPUT NUMBER YIELD BLOCK R2 R3 CULTIVAR V2 V3 V4 V5             21
          V6 V7 V8 V9 V10 V11 V12 V13 V14 V15 V16                  22
          V17 V18 V19 V20 V21 V22 V23 V24 V25 V26                  23
          V27 V28 V29 V30 V31 V32 V33 V34;                         24
   Y = (YIELD*0.1)*67.19;                                          25
RUN;                                                               26
*---------------------------------------------------------------*
* TWO DIRECTIONS OF SPATIAL LAG VALUES ARE DETERMINED.*
* N4 MUST BE LAGGED BY SQRT(N) + 1, RESULTING IN A*
* CHANGE IN "LAG__". *
```

```
*-----------------------------------------------------------------*;
DATA STEP2;                                                          32
    SET STEP1;                                                       33
N2 = LAG1(Y);                                                        34
N4 = LAG7(Y);                                                        35
PROC SORT DATA=STEP2;                                                36
    BY DESCENDING NUMBER;                                            37
RUN;                                                                 38
*-----------------------------------------------------------------*
* THE REMAINING TWO DIRECTIONS OF SPATIAL LAG VALUES ARE*
* DETERMINED.  N3 MUST BE LAGGED BY SQRT(N)+1, RESULTING*
* IN A CHANGE IN "LAG__".*
*-----------------------------------------------------------------*;
DATA STEP3;                                                          44
    SET STEP2;                                                       45
N1 = LAG1(Y);                                                        46
N3 = LAG7(Y);                                                        47
RUN;                                                                 48
*-----------------------------------------------------------------*
* DETERMINATION OF THE NUMBER OF NEIGHBORS AND THE AVERAGE*
* NEIGHBORING VALUES FOR EACH AREAL UNIT.*
*-----------------------------------------------------------------*;
DATA STEP4;                                                          53
    SET STEP3;                                                       54
IF N1 EQ '.' THEN NBR1 = 0; ELSE NBR1 = 1;                           55
IF N2 EQ '.' THEN NBR2 = 0; ELSE NBR2 = 1;                           56
IF N3 EQ '.' THEN NBR3 = 0; ELSE NBR3 = 1;                           57
IF N4 EQ '.' THEN NBR4 = 0; ELSE NBR4 = 1;                           58
NEIGH = NBR1 + NBR2 + NBR3 + NBR4;                                   59
IF N1 EQ '.' THEN N1 = 0;                                            60
IF N2 EQ '.' THEN N2 = 0;                                            61
IF N3 EQ '.' THEN N3 = 0;                                            62
IF N4 EQ '.' THEN N4 = 0;                                            63
WX = (N1 + N2 + N3 + N4)/NEIGH;                                      64
IF Y EQ '.' THEN WY = '.';                                           65
RUN;                                                                 66


*-----------------------------------------------------------------*
* ESTIMATION OF THE SAR SPATIAL AUTOREGRESSIVE PARAMETER*
* (RHO).  "ALPHA" AND "DELTA" ARE SPECIFIC TO SQRT(N),*
* AND COME FROM THE JACOBIAN APPROXIMATION TABULATED*
* RESULTS.*
*-----------------------------------------------------------------*;
PROC NLIN METHOD=MARQUARDT MAXITER=500;                              73
    PARMS RHO=0.5, B0=0, B1=0, B2=0, B3=0,B4=0, B5=0,               74
        B6=0, B7=0, B8=0, B9=0, B10=0, B11=0, B12=0,                75
        B13=0, B14=0, B15=0, B16=0, B17=0, B18=0,                   76
        B19=0, B20=0, B21=0, B22=0, B23=0, B24=0,                   77
        B25=0, B26=0, B27=0, B28=0, B29=0, B30=0,                   78
        B31=0, B32=0, B33=0, B34=0, B35=0;                          79
    BOUNDS -1.0 < RHO < 1.0;                                        80
    A1 = 0.143535;                                                  81
    D1 = 1.067932;                                                  82
    JHAT = EXP(A1*(2*LOG(D1) - LOG(D1+RHO) - LOG(D1-RHO)));         83
        ZY = Y*JHAT;                                                84
    MODEL ZY = (RHO*WY + (1-RHO)*B0 + B1*R2 +B2*R3 + B3*V2 +        85
        B4*V3 + B5*V4 + B6*V5 + B7*V6 + B8*V7 +                     86
```

```
         B9*V8 + B10*V9 + B11*V10 + B12*V11 +                                   87
         B13*V12 + B14*V13 + B15*V14 + B16*V15 +                                88
         B17*V16 + B18*V17 + B19*V18 + B20*V19 +                                89
         B21*V20 + B22*V21 + B23*V22 + B24*V23 +                                90
         B25*V24 + B26*V25 + B27*V26 + B28*V27 +                                91
         B29*V28 + B30*V29 + B31*V30 + B32*V31 +                                92
         B33*V32 + B34*V33 + B35*V34)*JHAT;                                     93
   DER.B0 = (1-RHO)*JHAT;                                                       94
   DER.B1 = R2*JHAT;                                                            95
   DER.B2 = R3*JHAT;                                                           96
   DER.B3 = V2*JHAT;                                                            97
   DER.B4 = V3*JHAT;                                                            98
   DER.B5 = V4*JHAT;                                                            99
   DER.B6 = V5*JHAT;                                                           100
   DER.B7 = V6*JHAT;                                                           101
   DER.B8 = V7*JHAT;                                                           102
   DER.B9 = V8*JHAT;                                                           103
   DER.B10 = V9*JHAT;                                                          104
   DER.B11 = V10*JHAT;                                                         105
   DER.B12 = V11*JHAT;                                                         106
   DER.B13 = V12*JHAT;                                                         107
   DER.B14 = V13*JHAT;                                                         108
   DER.B15 = V14*JHAT;                                                         109
   DER.B16 = V15*JHAT;                                                         110
   DER.B17 = V16*JHAT;                                                         111
   DER.B18 = V17*JHAT;                                                         112
   DER.B19 = V18*JHAT;                                                         113
   DER.B20 = V19*JHAT;                                                         114
   DER.B21 = V20*JHAT;                                                         115
   DER.B22 = V21*JHAT;                                                         116
   DER.B23 = V22*JHAT;                                                         117
   DER.B24 = V23*JHAT;                                                         118
   DER.B25 = V24*JHAT;                                                         119
   DER.B26 = V25*JHAT;                                                         120
   DER.B27 = V26*JHAT;                                                         121
   DER.B28 = V27*JHAT;                                                         122
   DER.B29 = V28*JHAT;                                                         123
   DER.B30 = V29*JHAT;                                                         124
   DER.B31 = V30*JHAT;                                                         125
   DER.B32 = V31*JHAT;                                                         126
   DER.B33 = V32*JHAT;                                                         127
   DER.B34 = V33*JHAT;                                                         128
   DER.B35 = V34*JHAT;                                                         129
   DER.RHO = ((RHO*WY + (1-RHO)*B0 + B1*R2 + B2*R3 + B3*V2                      130
        + B4*V3 + B5*V4 + B6*V5 + B7*V6 + B8*V7 + B9*V8                         131
        + B10*V9 + B11*V10 + B12*V11 + B13*V12 + B14*V13                        132
        + B15*V14 + B16*V15 + B17*V16 + B18*V17 + B19*V18                       133
        + B20*V19 + B21*V20 + B22*V21 + B23*V22 + B24*V23                       134
        + B25*V24 +B26*V25 + B27*V26 + B28*V27 + B29*V28                        135
        + B30*V29 + B31*V30 + B32*V31 + B33*V32 + B34*V33                       136
        + B35*V34-Y)*(2*A1*RHO/(D1**2-RHO**2))+WX - B0)*JHAT;                   137
  ID JHAT;                                                                     138
  OUTPUT OUT=TEMP P=YHAT;                                                      139
RUN;                                                                          140
```

Line 2 defines the path for accessing the data file VARIETY.DAT. Lines 19-26, for DATA STEP1, read the 119 records of data from this file. Below is a segment of this data file that now includes 37 additional columns of data. As before, column 1 is ordering variable NUMBER and column 2 is response variable YIELD. Column 3 is the variable BLOCK whose values indicate blocks 1, 2, or 3. Columns 4-5 are binary indicator variables R2-R3 that take on a value of 1 for plots belonging to a particular block and 0 otherwise. Column 6 represents the variable CULTIVAR whose values indicate cultivar 1 through 34. Binary indicator variables V2-V33 in columns 7-39 take on a value of 1 for plots belonging to a particular cultivar and 0 otherwise.

```
1  386 1 0 0  1 0 0 0 0 0 0 0 0 0 0 0 0 0 0 0 0 0 0 0 0 0 0 0 0 0 0 0 0 0 0 0
2  452 1 0 0 18 0 0 0 0 0 0 0 0 0 0 0 0 0 0 0 0 1 0 0 0 0 0 0 0 0 0 0 0 0 0 0
3  371 2 1 0  8 0 0 0 0 0 1 0 0 0 0 0 0 0 0 0 0 0 0 0 0 0 0 0 0 0 0 0 0 0 0 0
4  498 2 1 0  6 0 0 0 1 0 0 0 0 0 0 0 0 0 0 0 0 0 0 0 0 0 0 0 0 0 0 0 0 0 0 0
5  543 3 0 1 17 0 0 0 0 0 0 0 0 0 0 0 0 0 1 0 0 0 0 0 0 0 0 0 0 0 0 0 0 0 0 0
6  541 3 0 1 34 0 0 0 0 0 0 0 0 0 0 0 0 0 0 0 0 0 0 0 0 0 0 0 0 0 0 0 0 0 0 1
7  . . . . . . . . . . . . . . . . . . . . . . . . . . . . . . . . . . .
8  445 1 0 0  2 1 0 0 0 0 0 0 0 0 0 0 0 0 0 0 0 0 0 0 0 0 0 0 0 0 0 0 0 0 0 0
```

The classificatory variables BLOCK and CULTIVAR have been included for ease of performing an ANOVA with PROC GLM of SAS. These variables will be used once the yield observations have been adjusted for autocorrelation by means of AR modeling. The binary indicator variables allow the AR method to cast the ANOVA as a regression of grain yield on the three blocks and 34 cultivars. Accordingly, two indicator variables define all blocks except one and 33 indicator variables define all cultivars except one. The resulting intercept term expresses the combined mean of the excluded cultivar and block. Each slope term expresses a difference in mean yield between the cultivar represented by this slope term and the cultivar and block represented by the intercept.

The statistical form of the AR model statement (lines 85-93), for PROC NLIN (lines 73-140), is

$$Y \exp(J^{1/2}) = (X_k\beta_k + \rho WY - \rho\beta_0) \exp(J^{1/2}) + \epsilon \exp(J^{1/2}), \qquad (8)$$

where $\rho$ is the autocorrelation parameter, and $X_k$ specifies binary indicator variables for k different classes of cultivars and blocks, $\beta_0$ expresses mean grain yield in kg ha$^{-1}$ and represents an excluded class, and $\epsilon$ is a random error term. The value $\beta_k$ expresses a difference in mean grain yield in kg ha$^{-1}$ for the kth class and $X_k$ takes on values of 1 for observations belonging to the kth class and 0 otherwise. Both sides of this model statement are multiplied by $\exp(J^{1/2})$ as represented by

JHAT from lines 83-84. The yield response vector WY is a weighted average of neighboring plots that is computed in line 64 by division of variable CY by variable NEIGH. Computation of NEIGH in line 59 specifies the number of neighboring plots immediately adjacent to each yield observation. Note that, except for division by NEIGH, lines 1-66 are the same as in MC102.SAS for computing the MC of section 4.

The Marquardt algorithm for PROC NLIN is employed for estimating the AR model parameters. Parameter values for A1 and D1, previously acquired in estimating the Jacobian term, are required to be entered on lines 81-82. These values, 0.143535 for A1 and 1.067932 for D1, are specific to the 6-by-17 regular lattice of the variety trial. In addition, the BOUNDS statement of line 80 is assigned a minimum value of -1.0 and a maximum value of 1.0 corresponding to the same minimum and maximum eigenvalues used in estimating the J term. The Marquardt algorithm requires the analytical first derivatives for the 37 terms in the model as specified in lines 94-137.

Results obtained with this part of the SAS code are

| Non-Linear Least Squares Summary Statistics | | | Dependent Variable ZY | |
|---|---|---|---|---|
| Parameter | Estimate | Asymptotic Std. Error | Asymptotic 95 % Confidence Interval | |
| | | | Lower | Upper |
| RHO | 0.345856 | 0.12695436 | 0.0923103 | 0.5994017 |
| B0 | 2610.037767 | 336.17591714 | 1938.6472722 | 3281.4282614 |
| B1 | 65.940025 | 72.97446997 | -79.8002344 | 211.6802851 |
| B2 | 87.912481 | 74.03868958 | -59.9531749 | 235.7781365 |
| B3 | 401.154971 | 245.83335850 | -89.8087532 | 892.1186945 |
| B4 | 204.005535 | 245.73682030 | -286.7653885 | 694.7764585 |
| B5 | 910.920639 | 245.75496472 | 420.1134785 | 1401.7277995 |
| B6 | 612.006388 | 245.83858264 | 121.0322310 | 1102.9805454 |
| B7 | 501.914600 | 245.91048300 | 10.7968474 | 993.0323520 |
| B8 | 636.771822 | 245.84068825 | 145.7934595 | 1127.7501843 |
| B9 | -654.163826 | 246.59193069 | -1146.6425253 | -161.6851272 |
| B10 | 257.244669 | 245.61042214 | -233.2738199 | 747.7631575 |
| B11 | 1386.682171 | 245.60804482 | 896.1684298 | 1877.1959114 |
| B12 | 804.252589 | 246.99356070 | 310.9717782 | 1297.5333992 |
| B13 | 1277.937909 | 245.67232501 | 787.2957914 | 1768.5800261 |
| B14 | 1002.881854 | 245.68015319 | 512.2241025 | 1493.5396052 |
| B15 | 703.382160 | 245.73285058 | 212.6191641 | 1194.1451550 |
| B16 | 719.014866 | 245.70736987 | 228.3027588 | 1209.7269726 |
| B17 | 320.127150 | 245.86889073 | -170.9075363 | 811.1618371 |
| B18 | 967.900148 | 245.80030850 | 477.0024296 | 1458.7978664 |
| B19 | 552.579421 | 251.12227299 | 51.0529924 | 1054.1058494 |
| B20 | 718.048859 | 245.59734658 | 227.5564840 | 1208.5412339 |
| B21 | 508.239228 | 247.89087652 | 13.1663520 | 1003.3121043 |
| B22 | 584.426117 | 245.73285060 | 93.6631219 | 1075.1891129 |
| B23 | 519.474827 | 246.32727713 | 27.5246787 | 1011.4249761 |
| B24 | 891.387305 | 245.59684252 | 400.8959369 | 1381.8786734 |
| B25 | 759.106309 | 245.64534214 | 268.5180802 | 1249.6945378 |
| B26 | 594.924672 | 246.57077852 | 102.4882168 | 1087.3611271 |
| B27 | 514.523721 | 246.97850346 | 21.2729815 | 1007.7744596 |

| B28 | 434.798466 | 245.59787122 | -55.6949563 | 925.2918891 |
| B29 | 209.477535 | 246.41934171 | -282.6564794 | 701.6115498 |
| B30 | 697.146163 | 245.89512301 | 206.0590866 | 1188.2332392 |
| B31 | 801.080313 | 246.00934929 | 309.7651112 | 1292.3955155 |
| B32 | 1109.974471 | 245.84386374 | 618.9897669 | 1600.9591755 |
| B33 | 1281.771740 | 245.67831242 | 791.1176652 | 1772.4258154 |
| B34 | 694.389138 | 248.74156110 | 197.6173254 | 1191.1609508 |
| B35 | 649.111512 | 246.15627598 | 157.5028767 | 1140.7201475 |

The SAS NLIN procedure for estimating the AR model converged after 11 iterations. The autocorrelation parameter, $\rho$, is 0.345856 and slope-intercept, $\beta_0$, is 2610.037767 which are significant (probability<0.05). That the yield data are autocorrelated is consistent with the finding based on the MC. The procedures described here allow one to derive $\rho$ and $\beta$ estimates simultaneously. This information is used in solving the next problem, which is to adjust the data values to correct for the effect of autocorrelation on conventional ANOVA. The adjusted ANOVA results are derived by appending lines 144-180 of the following SAS code to the rest of the code, setting RO and MN to their estimated values of 0.345856 and 2610.037767, and running this SAS code a second time.

```
***********************************************************
*OBTAIN THE MEAN SQUARE ERROR FOR THE AR MODEL*
***********************************************************;
DATA STEP5;                                                   144
   SET STEP1;                                                 145
   SET STEP4;                                                 146
   SET TEMP;                                                  147
   CORY = YHAT/JHAT;                                          148
   ME = ((Y - CORY)**2);                                      149
RUN;                                                          150
PROC MEANS SUM;                                               151
   VAR ME;                                                    152
   OUTPUT OUT=MS SUM=MSE;                                     153
PROC PRINT DATA=MS;                                           154
   VAR MSE;                                                   155
RUN;                                                          156
***************************************************
*OBTAIN THE PSEUDO R2 FOR THE AR MODEL *
***************************************************;
PROC CORR;                                                    160
   VAR Y CORY;                                                161
RUN;                                                          162
*****************************************************************
*PERFORM ANOVA WITH YIELD UNADJUSTED FOR AUTOCORRELATION*
*AND YIELD ADJUSTED FOR AUTOCORRELATION. *
*****************************************************************;
DATA STEP6;                                                   167
   SET STEP5;                                                 168
   RO = 0.345856;                                             169
   MN = 2610.037767;                                          170
   ADJY = Y - (RO*WY - RO*MN);                                171
```

```
RUN;                                              172
PROC GLM;                                         173
   CLASSES BLOCK CULTIVAR;                         174
   MODEL Y = BLOCK CULTIVAR;                       175
RUN;                                              176
PROC GLM;                                         177
   CLASSES BLOCK CULTIVAR;                         178
   MODEL ADJY = BLOCK CULTIVAR;                    179
RUN;                                              180
```

Lines 144-150, for DATA STEP5, compute a value for the mean squared error based on the squared difference between observed yield and predicted yield.    Values for predicted yield, denoted YHAT, are returned to DATA STEP5 by means of the output statement in line 138. In line 148 they have the Jacobian term removed (it was introduced with multiplication in lines 83-93) by dividing by JHAT.  In addition, a value for the coefficient of determination ($R^2$) is computed in line 161 by correlating observed yield, Y, with corrected predicted yield, CORY. Meanwhile, the mean square error computed in lines 148-152 is 5880986.38 and the correlation value computed in lines 160-162 is 0.86366 ($R^2 = 0.746$).

Lines 167-172, for DATASTEP6, adjust the yield observations for autocorrelation according to the following equation

$$Y_{adj} = Y - (\rho WY - \rho \beta_0) \tag{9}$$

where $Y - (\rho WY - \beta_0)$ equals the adjusted values of grain yield and $\rho$ is the autocorrelation parameter obtained from its nonlinear estimation in Equation (10).

PROC GLM in lines 173-180 is used to separately determine the standard ANOVA tables for the AR and conventional methods.  The AR method is based on observed yield that has been adjusted for autocorrelation.  The conventional method is based on observed yield that has not been adjusted for autocorrelation.  We now compare the results from AR-based and conventional ANOVA in the following section.

## VIII.  COMPARISON OF AR-BASED ANOVA AND CONVENTIONAL ANOVA

Table 1 compares the results of AR-based and conventional ANOVA.  Listed are the coefficient of determination ($R^2$), error sum of squares (SSE), mean square error (MSE), and F values for the error, treatment, and total sources of variation in grain yield.

The correct MSE for the AR method has been determined from

$$MSE = \frac{SSE}{(n - p - 1)} \tag{10}$$

where the SSE returned from the GLM procedure is divided by n observations minus p regression terms (including the intercept) and one degree of freedom for the autocorrelation parameter estimate, $\hat{\rho}$. The correct total sum of squares equals the component of total variation in grain yield, Y. The correct conditional sum of squares for the autocorrelation parameter is the difference between the total sum of squares received from PROC GLM for the OLS method and that received from PROC GLM for the AR method.

**Table 1.** Analysis of variance of cultivar yields from conventional and AR-based methods.

| Source of Variation | Sum of Squares | Degrees of Freedom | Mean Squares | F Ratio |
|---|---|---|---|---|
| **Conventional ANOVA, $R^2 = 0.699$** | | | | |
| Cultivars | 15925020 | 33 | 482576 | 4.58* |
| Blocks | 256461 | 2 | 128231 | 1.22 |
| Error | 6954633 | 66 | 105373 | |
| Total | 23136116 | 101 | | |
| **AR-based ANOVA, $R^2 = 0.746$** | | | | |
| ρ | 2263059 | 1 | 2263059 | 25.4* |
| Cultivars | 14849730 | 33 | 449992 | 5.05* |
| Blocks | 142341 | 2 | 71171 | 0.80 |
| Error | 5880986 | 65 | 89106 | |
| Total | 23136116 | 101 | | |

\* Significant at the five percent level of probability.

The F statistic value is significant for cultivars and is nonsignificant for blocks in both types of ANOVA. The AR-based ANOVA shows that the F value for cultivars is more extreme than one would find for unautocorrelated data while the F value for blocks is less extreme. The amount of explained variation ($R^2$) in grain yield increases from about 70 % for OLS-based ANOVA to about 75 % for AR-based ANOVA. This is accompanied by a reduction in the MSE from 105,373 for OLS-based ANOVA to 89,106 for AR-based ANOVA. This difference represents a 15.4 % reduction in experimental error, or improvement in precision, for AR-based ANOVA over OLS-based ANOVA.

The autocorrelation parameter, $\rho$, accounts for 9.8% of the variation in grain yield. Meanwhile, the sum of squares for cultivars is 14,849,730 rather than 15,925,020 and for blocks it is 142,341 rather than 256,461. The reduction in sum of squares for cultivars in AR-versus OLS-based ANOVA corresponds to a decrease in explained yield variation from 68.8 to 64.2 % relative to the total sum of squares.

This variety trial presents a total of 561 possible one-way contrasts between the 34 cultivars. Based on the MSE values and Tukey's test for significant differences [Tukey, 1949], two contrasted means would be significantly different if their difference exceed 1058 kg ha$^{-1}$ for OLS-based ANOVA and 973 kg ha$^{-1}$ for AR-based ANOVA. Accordingly, the OLS method detects 38 contrasts as significant whereas the AR method detects 45 contrasts as significant.

These results from this variety trial suggest that randomization does not completely provide for independence of observations in data that are spatially autocorrelated. Modest amounts of autocorrelation can inflate levels of explained variance in yield, thereby leading to incorrect inferences concerning the significance of F values. The outcome of AR-based ANOVA differs from its conventional counterpart because extraneous variance in yield due to autocorrelation has been partialed out by $\rho WY$ in Equation (1). Consequently, the $R^2$ value is larger and the sum of squares for error and explanatory variables is smaller in AR-based ANOVA versus conventional ANOVA. Furthermore, the decrease in the MSE has increased capacity for detecting real treatment differences.

## IX. CONCLUSIONS

Agronomists should begin to utilize spatial statistics in analyzing autocorrelated experimental field data. We recommend a two-step approach. First, use the MC to see if a crop response variable contains spatial autocorrelation. Second, apply spatial statistics if significant autocorrelation is detected. The SAS computer code for computing the MC and estimating an AR model can be modified to accept any regular square lattice structure. Potential users can consult Griffith (1993) for further information on this code.

In conventional ANOVA, positive spatial autocorrelation tends to inflate the sum of squares for errors and deflate those for treatments relative to the total sum of squares. The AR method gives improved performance because autocorrelation can be partialed out in a regression equation. Consequently, this method increases capacity in detecting treatment differences and making correct statements concerning the significance of treatment effects. Results from the variety trial corroborate these contentions.

The AR method is ideally suited for agronomic field data because it easily accommodates the data of a single response variable that have been collected on a regular lattice, and it provides for adjusting the data to correct for the negative effect of autocorrelation on conventional ANOVA. Estimating an AR model can be performed in a minimum number of steps with a personal computer and the package statistical software SAS. The method is adaptable to agronomic field data because the results of regression are identical to an ANOVA of a continuous response variable on any number of classificatory indicator variables; ANOVA with regression is essentially the same approach used in generalized linear models. Therefore, many statistical procedures such as analysis of covariance, weighted least squares, or ANOVA of unbalanced designs can be implemented with the method.

Finally, the use of spatial statistics is not a panacea for local control of extraneous variance by means of blocking but rather can complement it. [Grondona and Cressie 1991] On the other hand, a drawback of blocking is that use of blocks assumes the existence of relatively uniform regions in the field, which is often not the case.

## Acknowledgments
Statistical results are from a thesis presented by the author in partial fulfillment of the requirements for the Ph.D degree at Cornell University (S.D. DeGloria, Chair). The 1983 variety trial data were provided by G.R. Carlson, Montana State University, Northern Agricultural Research Center.

# REFERENCES

Barlett, M.S.   Nearest neighbour models in the analysis of field experiments (with discussion). *J. Roy. Statist. Soc. Ser. B.* 40:147-174, 1978.

Besag, J.E. and  Kempton, R  Statistical analysis of field experiments using neighbouring plots. *Biometrics* 42:231-251, 1986.

Brownie, C.; Bowman, D. T.; and Burton, J. W..  Estimating spatial variation in analysis of data from field trials: A comparison of methods. *Agron. J.* 85:1244-1253, 1993.

Cliff, A.D. and Ord, J. K.   The comparison of means when samples consist of spatially autocorrelated observations. *Env. Plan. A.* 7:725-734, 1975.

Cressie, N. 1991. *Statistics for Spatial Data.* Wiley, New York. p. 900, 1991.

Glass, G.V.; Peckham, P.D.; and Sanders, J. R.  Consequences of failure to meet assumptions underlying the fixed effects analysis of variance and covariance. *Rev. Educ. Res.* 42(3):237-288, 1972.

Green, P.J.; Jennison, C.;  and Seheult, A. H..  Analysis of field experiments by least squares smoothing. *J. Roy. Statist. Soc. Ser. B.* 47(2):299-315, 1984.

Griffith, D.A.  Estimating spatial autoregressive model parameters with commercial statistical packages. *Geogr. Anal.* 20(2):176-186, 1988b.

Griffith, D.A.  A numerical simplification for estimating parameters of spatial autoregressive models. p. 185-195. *In* D.A. Griffith (ed.) *Spatial Statistics: Past, Present and Future.* Institute of Mathematical Geography, Ann Arbor, MI, 1990.

Griffith, D.A.  What is spatial autocorrelation? Reflections on the past 25 years of spatial statistics. *L'Espace geographique* 3:265-280, 1992.

Griffith, D.A.  *Spatial Regression Analysis on the PC: Spatial Statistics using SAS.* Assoc. Am. Geog. Washingtion D.C. p. 130, 1993.

Grondona, M.O. and Cressie, N.  Using spatial considerations in the analysis of experiments. *Technomet.* 33(4):381-392, 1991.

Haining, R.P.  A spatial model for high plains agriculture. *Ann. Ass. Am. Geog.* 68:(4):493-504, 1978.

Haining, R.P. 1980. Spatial autocorrelation problems. *In* D.T. Herbert and R.J. Johnson (eds.), *Geography and the urban environment, progress in research and applications.* 3:1-44, Wiley, New York, 1980.

Legendre, P.; Oden, N. L.; Sokal, R. R.; Vaudor, A.; and Kim J.  Approximate analysis of variance of spatially autocorrelated regional data. *J. Class.* 7:53-75, 1990.

Miron, J.  Spatial autocorrelation in regression analysis: a beginners guide. *In* G.L. Gaile and C.J. Willmot (ed.) *Spatial Statistics and Models.* Dordrectht. p. 201-222, 1984.

Moran, P.  The interpretation of statistical maps. *J. Roy. Statist. Soc. Ser. B* 37:243-251, 1948.

Mulla, D.J.; Bhatti, A. J.; and Kunkel, R.  Methods for removing spatial variability from field research trials. *Advances in Soil Science.* Springer-Verlag New York Inc., 13:201-213, 1990.

Olson, K.R.; Carmer, S. G.; and Olson, G. W.  *Assessment of effects of soil variability on maximum alfalfa yields,* GEDMAB, New York, 36:1-14, 1985.

Petersen, R.G.  *Design and Analysis of Experiments.* Marcel Dekker, Inc. p. 429, 1985.

SAS Institute Inc.  *SAS Procedures Guide,* release 6.03 edition. Cary, NC. 210 p., 1988.

Tukey, J.W.  Comparing individual means in the analysis of variance. *BIOKA.* 5:99-114, 1949.

van Es, H.M. and  van Es, C. L. The spatial nature of randomization and its effect on the outcome of field experiments. *Agron. J.* 85:420-428, 1993.

Zimmerman, D.L. and Harville, D. A.  A random field approach to the analysis of field-plot experiments and other spatial experiments. *Biometrics,* 47:223-239, 1991.

# References, by Chapter

The following listing offers student and researcher alike the opportunity to gain an understanding, at a glance, of the emphases of the various chapters, in relation to one another.

## Foreword

**Bookstein, Fred L.** *The Measurement of Biological Shape and Shape Change.* Berlin, Springer-Verlag, 1978.

**Dacey, Michael F.** Imperfections in the uniform plane. *Michigan Inter-University Community of Mathematical Geographers,* John D. Nystuen, Editor, May, 1964. Reprinted, *Solstice: An Electronic Journal of Geography and Mathematics,* Summer, 1994.

**Ness, Gayl D.; Drake, William D.; Brechin, Steven R., eds.** *Population-Environment Dynamics.* Ann Arbor, University of Michigan Press, 1993.

**Nystuen, Jeffrey A.; McGlothin, Charles C.; and, Cook, Michael S.** The underwater sound generated by heavy rainfall. *Journal of the Acoustical Society of America,* 93 (6), 3169-3177, 1993.

**Sadowski, Frank G. and Covington, Stephen J.** *Processing and Analysis of Commercial Satellite Image Data of the Nuclear Accident near Chernobyl, U.S.S.R.* U. S. Geological Survey, Bulletin 1785. U.S. Government Printing Office, Washington, D. C., 1987.

**Thompson, D'Arcy W.** *On Growth and Form.* Cambridge University Press, 1917.

**Tobler, Waldo R.** *Map Transformations of Geographic Space.* Ph.D. Thesis, University of Washington, 1961.

**Tobler, Waldo R.** Preliminary representation of world population by spherical harmonics. *Proceedings of the National Academy of Sciences of the United States of America,* 89:6262, 4 Jul. 15 1992

**Warntz, William.** *Macrogeography and Income Fronts.* Philadelphia, Regional Science Research Institute, Monograph #3, 1965.

## Chapter 1

**Ahuja, N., and Schachter, B.** *Pattern Models.* Wiley, New York, 1983.

**Amrhein, C.** Searching for the elusive aggregation effect: evidence from statistical simulations. *Environment and Planning A,* 26, 1994.

**Anselin, L.** *Spatial Econometrics.* Kluwer, Dordrecht, 1988.

**Anselin, L.** SPACESTAT TUTORIAL: A workbook for using SpaceStat in the analysis of spatial data. *Technical Software Series S-92-1,* NCGIA, Santa Barbara CA, 1992.

**Anselin, L., and Can, A.** Model comparison and model validation issues in empirical work on urban density functions. *Geographical Analysis,* 18, 179-197, 1986.

**Anselin, L., and Griffith, D.** Do spatial effects really matter in regression analysis? *Papers of the Regional Science Association,* 65, 11-34, 1987.

**Anselin, L., and Hudak.** Spatial econometrics in practice: a review of software options. *Regional Science and Urban Economics,* 22, 509-536, 1992.

**Bartlett, M.** *Statistical Analysis of Spatial Pattern.* Chapman and Hall, London, 1975.

**Brown, D., Bian, L., and Walsh, S.** Response of a distributed watershed erosion model to variations in input data aggregation levels. *Computers and Geosciences,* 19(4), 499- 509, 1993.

**Brown, D. and Walsh, S.** Spatial autocorrelation in remotely sensed and GIS data. *Proceedings* of the ACSM/ASPRS Annual Convention, New Orleans, LA, Vol. 3, 13- 39, 1993.

**Brown, D. and Bara, T.** Recognition and reduction of systematic error in elevation and derivative surfaces from 7 1/2-minute DEMs. *Photogrammetric Engineering and Remote Sensing*, 60, 189-194, 1994.

**Chorley, R.** [chairman]. *Handling Geographic Information*. Her Majesty's Stationary Office, Report to the Select Committee on GIS, London, 1987.

**Cliff, A., and Ord, K.** *Spatial Processes*. Pion, London, 1981.

**Cordy, C., and Griffith, D.** Efficiency of least squares estimators in the presence of spatial autocorrelation. *Communications in Statistics--Simulation and Computation*, 22, 1161-1179, 1993.

**Cressie, N.** *Statistics for Spatial Data*. Wiley, New York, 1991.

**Durrett, R.** Stochastic spatial models, *Forefronts* (newsletter of the Cornell Theory Center), 9 (#4, Spring), 4-6, 1994.

**Goodchild, M.** *Spatial Autocorrelation*. CATMOG, Norwich, England, 1986.

**Goodchild, M., Haining, R., and Wise, S.** Integrating GIS and spatial data analysis: problems and possibilities. *International Journal of Geographical Information Systems*, 6, 407-423, 1992.

**Green, M.** Ecological fallacies and the modifiable areal unit problem. *Research Report No. 27*, North West Regional Research Laboratory, Lancaster University, UK, 1993.

**Griffith, D.** *Spatial Autocorrelation*. Association of American Geographers, Washington, D. C., 1987.

**Griffith, D.** *Advanced Spatial Statistics*. Kluwer, Dordrecht, 1988.

**Griffith, D.** Spatial regression Analysis on the PC: spatial statistics using MINITAB. *Discussion Paper #1*, Institute of Mathematical Geography, Ann Arbor, MI, 1989.

**Griffith, D.** (ed.). *Spatial Statistics: Past, Present, and Future*. Institute of Mathematical Geography, Ann Arbor, MI, 1990.

**Griffith, D.** Which spatial statistics techniques should be converted to GIS functions? in *Geographic Information Systems, Spatial Modelling and Policy Evaluation*, edited by M. Fischer and P. Nijkamp. Springer-Verlag, 103-114, Berlin, 1993a.

**Griffith, D.** Advanced spatial statistics for analysing and visualizing geo-referenced data. *International Journal of Geographical Information Systems*, 7, 107-123, 1993b.

**Griffith, D.** *Spatial Regression Analysis on the PC: Spatial Statistics Using SAS*. Association of American Geographers, Washington, D.C., 1993c.

**Griffith, D., and Csillag, F.** Exploring relationships between semi-variogram and spatial autoregressive models. *Papers in Regional Science*, 72, 283-295, 1993.

**Griffith, D., and Sone, A.** Trade-offs associated with computational simplifications for estimating spatial statistical models. *Working Paper*, l'Institut de Mathématiques Economiques, Université de Bourgogne, Dijon, France (with French Resumé), 1992.

**Griffith, D., and Sone, A.** Some trade-offs associated with computational simplifications for estimating spatial statistical/econometric models: preliminary results. *Discussion Paper* No. 103, Department of Geography, Syracuse University, 1993.

**Grondona, M., and Cressie, N.** Using spatial considerations in the analysis of experiments. *Technometrics*, 33, 381-392, 1991.

**Haining, R.** *Spatial Data Analysis in the Social and Environmental Sciences*. Cambridge University Press, Cambridge, England, 1990.

**IBM** Exploring new worlds with GIS. *Directions*, Summer/Fall, 12-19, 1991.

**Isaaks, E. and Srivastava, R.** *An Introduction to Applied Geostatistics*. Oxford University Press, Oxford, England, 1989.

**Lindgren, B.** *Statistical Theory*, 3rd ed. Macmillan, New York, 1976.

**Mardia, K., and Marshall, R.** Maximum likelihood estimation of models for residual covariance in spatial regression. *Biometrika*, 71, 135-146, 1984.

**Matérn, B.** *Spatial Variation*, 2nd ed. Springer-Verlag, Berlin, 1986.

**National Research Council (Mapping Science Committee; Commission on Physical Sciences, Mathematics, and Resources).** *Spatial Data Needs: The Future of the National Mapping Program*, National Academy Press, Washington, D.C., 1990a.

**National Research Council (Board on Mathematical Sciences).** *Renewing U.S. Mathematics: A Plan for the 1990s.* National Academy Press, Washington, D.C., 1990b.

**National Research Council (Panel on Spatial Statistics and Image Processing).** *Spatial Statistics and Digital Image Analysis.* National Academy Press, Washington, D.C., 1991.

**National Science Foundation.** *Solicitation: National Center for Geographic Information and Analysis.* Biological, Behavioral, and Social Sciences Directorate, Washington, D.C., 1987.

**Odland, J.** *Spatial Autocorrelation.* Sage, Beverly Hills, CA, 1988.

**Okabe, A., Boots, B., and Sugihara, K.** *Spatial Tessellations: Concepts and Applications of Voronoi Diagrams.* Wiley, New York, 1992.

**Olsen, J.** Autocorrelation and visual map complexity. *Annals*, Association of American Geographers, 65, 189-204, 1975.

**Ord, K.** Estimation methods for models of spatial interaction. *Journal of the American Statistical Association*, 70, 120-126, 1975.

**Overton S., and Stehman, S.** Statistical properties of designs for sampling continuous functions in two dimensions using a triangular grid. *Technical Report* No. 143, Department of Statistics, Oregon State University, 1990.

**Paelinck, J. and Klaassen, L.** *Spatial Econometrics.* Saxon House, Farnborough, England, 1979.

**Ripley, B.** *Statistical Inference for Spatial Processes.* Cambridge University Press, Cambridge, England, 1988.

**Stehman, S., and Overton, W.** Variance estimation for fixed-configuration, systematic sampling. *Technical Report* No. 134, Department of Statistics, Oregon State University, 1989.

**Stetzer, F.** Specifying weights in spatial forecasting models: the results of some experiments. *Environment and Planning A*, 14, 571-584, 1982.

**Upton, G., and Fingleton, B.** *Spatial Data Analysis by Example*, vol. 1. Wiley, New York, 1985.

**Warnecke, L.** GIS in the states: applications abound. *GIS World*, 3, (# 3), 54-58, 1990.

**Warnecke, L.** *State Geographic Information Activities Compendium.* Council of State Governments, Lexington, KY, 1991.

## Chapter 2

**Boots, B., and Dufouraud, D.** A programming approach to minimizing and maximizing spatial autocorrelation statistics. *Geographical Analysis* 26(1):54-66, 1994.

**Cliff, A. and Ord, J.** *Spatial Processes, Models, and Applications.* Pion, London, 1981.

**Dacey, Michael F.** Analysis of central place and point patterns by a nearest neighbour method. Lund Studies in Geography, B, Human Geography, 24, 55-75, 1962.

**Dykes, J.** Area-value data: new visual emphases and representations, in *Visualization in Geographical Information Systems*, edited by H. Hearnshaw and D. Unwin. Wiley, New York, pp. 103-114, 1994.

**Gatrell, Anthony C.** Complexity and redundancy in binary maps. *Geographical Analysis*, Vol. IX, No. 1, 29-41, 1977.

**Getis, A. and Boots, B.** *Models of Spatial Processes: An Approach to the Study of Point, Line, and Area Patterns.* Cambridge University Press, Cambridge, 1978.

**Griffith, Daniel A.** *Spatial Regression Analysis on the PC: Spatial Statistics Using SAS.* Association of American Geographers, Washington, D.C., 1993.

**Griffith, Daniel A.** *Spatial Autocorrelation. A Primer.* Washington, D.C.: Association of American Geographers, 1987.

**Griffith, Daniel A.** *Advanced Spatial Statistics.* Kluwer, Dordrecht, 1988.

**Haggett, Peter; Cliff, Andrew D.; and Frey, Allan.** *Locational Analysis in Human Geography* (in two volumes), 2nd edition. Wiley, New York, 1977.
**Moellering, Harold.** Real maps, virtual maps and interactive cartography, pp. 109- 132 in Gaile and Willmott (ed.), *Spatial Statistics and Models*, Reidel, 1984.
**Olson, Judith M.** Autocorrelation and Visual Map Complexity. *Annals of the Association of American Geographers* 65(2):189 - 204, 1975.
**Plane, David A. and Rogerson, Peter A.** *The Geographical Analysis of Population.* Wiley, New York, 1994.

# Chapter 3

**Arbia, G.** The use of GIS in spatial statistical surveys. *International Statistical Review*, 61, 339-359, 1993.
**Bellhouse, D. R.** Some optimal designs for sampling in two dimensions. *Biometrika*, 64, 605-611, 1977.
**Bellhouse, D. R.** Spatial sampling in the presence of a trend. *Journal of Statistical Planning and Inference*, 5, 365-375, 1981.
**Bellhouse, D. R.** Systematic sampling of periodic functions. *Canadian Journal of Statistics*, 13, 17-28, 1985.
**Bellhouse, D. R.** *Systematic sampling.* Handbook of Statistics, Vol. 6 (P. R. Krishnaiah and C. R. Rao, Eds.). Elsevier Science Publishers, Amsterdam, 1988.
**Brus, D.J., and De Gruijter, J.J.** Design-based versus model-based estimates of spatial means: Theory and applications in environmental soil science. *Environmetrics*, 4, 123-152, 1993.
**Buckland, S.T., Anderson, D.R., Burnham, K.P., and Laake, J.L.** *Distance Sampling: Estimating Abundance of Biological Populations.* Chapman & Hall, New York, 1993.
**Cochran, W.G.** Relative accuracy of systematic and stratified random samples for a certain class of population. *Annals of Mathematical Statistics*, 17, 164-177, 1946.
**Cochran, W.G.** Sampling Methods (3$^{rd}$ ed.). Wiley, New York, 1977.
**Cordy, C.B.** An extension of the Horvitz-Thompson theorem to point sampling from a continuous universe. *Statistics & Probability Letters*, 18, 353-362, 1993.
**Cordy, C.B., and Thompson, C.M.** An application of the deterministic variogram to design-based variance estimation. *Mathematical Geology*, 27, 173-205, 1995.
**Cressie, N. A. C.** *Statistics for Spatial Data.* Wiley, New York, 1991.
**Das, A.C.** Two-dimensional systematic sampling and the associated stratified and random sampling. *Sankhya*, 10, 95-108, 1950.
**De Gruijter, J.J., and Ter Braak, C.J.F.** Model-free estimation from spatial samples: A reappraisal.of classical sampling theory. *Mathematical Geology*, 22, 407-415, 1990.
**Deming, W.E.** *Some Theory of Sampling.* Wiley, New York, 1950.
**De Vries, P.G.** Sampling Theory for Forest Inventory. Springer-Verlag, New York, 1986.
**Dunn, R., and Harrison, A.R.** Two-dimensional systematic sampling of land use. *Applied Statistics*, 42, 585-601, 1993.
**Gilbert, R.O.** *Statistical Methods for Environmental Pollution Monitoring.* Van Nostrand Reinhold, New York, 1987.
**Griffith, D.A.** A spatially adjusted ANOVA model. *Geographical Analysis*, 10, 296-301, 1978.
**Haining, R.P.** *Spatial Data Analysis in the Social and Environmental Sciences.* Cambridge University Press, 1990.
**Jager, H.I., and Overton, W.S.** Explanatory models for ecological response surfaces. In *Environmental Modeling with GIS*, M.R. Goodchild, B.O. Parks, and L.T. Steyaert (ed.). Oxford University Press, pp. 422-431, 1993.
**Jessen, R.J.** *Statistical Survey Techniques.* Wiley, New York, 1978.
**Kish, L.** *Survey Sampling.* Wiley, New York, 1965.
**Kish, L.** *Statistical Design for Research.* Wiley, New York, 1987.
**Linthurst, R.A., Landers, D.H., Eilers, J.M., Brakke, D.F., Overton, W.S., Meier, E.P., and Crowe, R.E.** Characteristics of lakes in the eastern United States. Volume I: Population

descriptions and physico-chemical relationships. U.S. Environmental Protection Agency, 401 M Street SW, Washington, DC 20460. (EPA-600/4-86/007a), 1986.

**Mandallaz, D.** (1993). *Geostatistical Methods for Double Sampling Schemes: Application to Combined Forest Inventories. Chair of Forest Inventory and Planning*, Department of Forest and Wood Sciences, ETH-Zentrum, CH-8092, Zürich, 1993.

**Matérn, B.** *Spatial Variation* (2nd ed.). Springer-Verlag, New York, 1986.

**Olea, R.A.** Sampling design optimization for spatial functions. *Mathematical Geology*, 16, 369-392, 1984.

**Overton, W.S., Kanciruk, P., Hook, L.A., Eilers, J.M., Landers, D.H., Brakke, D.F., Blick, D.J., Linthurst, R.A., DeHaan, M.D.** *Characteristics of Lakes in the Eastern United States. Volume II: Lakes Sampled and Descriptive Statistics for Physical and Chemical Variables*, EPA600/486007b, Washington, DC, U. S. Environmental Protection Agency, 1986.

**Overton, W.S., and Stehman, S.V.** Properties of designs for sampling continuous spatial resources from a triangular grid, *Communications in Statistics* - Theory and Methods, 21, 2641-2660, 1993.

**Overton, W.S., and Stehman, S.V.** Variance estimation in the EMAP strategy for sampling discrete ecological resources, *Environmental and Ecological Statistics*, 1, 133-152, 1994.

**Overton, W.S., and Stehman, S.V.** The Horvitz-Thompson theorem as a unifying perspective for probability sampling: with examples from natural resource sampling. *American Statistician* (forthcoming), 1995.

**Overton, W.S., White, D., and Stevens, D.L.** *Design Report for EMAP: Environmental Monitoring and Assessment Program.* Washington, DC, U. S. Environmental Protection Agency (EPA/600/3-91/053), 1991.

**Quenouille, M.H.** Problems in plane sampling. *Annals of Mathematical Statistics*, 20, 355-375, 1949.

**Ripley, B.D.** *Spatial Statistics.* Wiley, New York, 1981.

**Royall, R. M.** On finite population sampling theory under certain linear regression models. *Biometrika*, 57, 377-387, 1970.

**Särndal, C.E., Swensson, B., and Wretman, J.** (1992). *Model Assisted Survey Sampling.* Springer-Verlag, New York, 1992.

**Stehman, S.V., and Overton, W.S.** (1989). Variance Estimation for Fixed-Configuration, Systematic Sampling. *Biometrics Unit Manuscript* Bu-1010-M, Cornell University, Ithaca, NY, 1989.

**Stehman, S.V., and Overton, W.S.** Comparison of variance estimators of the Horvitz-Thompson estimator for randomized variable probability systematic sampling. *Journal of the American Statistical Association*, 89, 30-43, 1994.

**Stuart, A.** *The Basic Ideas of Scientific Sampling* (1984 ed.), Charles Griffin and Company, London, 1962.

**Upton, G.J.G., and Fingleton, B.** *Spatial Data Analysis by Example, Volume I:  Point Pattern and Quantitative Data.* Wiley, New York, 1985.

**Upton, G. J. G., and Fingleton, B.** *Spatial Data Analysis by Example, Volume II: Categorical and Directional Data.* Wiley, New York, 1989.

**Webster, R., and Oliver, M.A.** *Statistical Methods in Soil and Land Resource Survey.* Oxford University Press, New York, 1990.

**Wolter, K.** Introduction to Variance Estimation. Springer-Verlag, New York, 1985.

**Yates, F.** *Sampling Methods for Censuses and Surveys* (4th ed.). Charles Griffin and Company, London, 1981.

**Zubrzycki, S.** Remarks on random, stratified and systematic sampling in a plane. *Colloquium Mathematicum*, 6, 251-264, 1958.

### Chapter 4

**Ahuja, N., and Schachter, B.** *Pattern Models.* Wiley, New York, 1983.

**Bartlett, M.** *The Statistical Analysis of Spatial Pattern.* Chapman and Hall, London, 1975.

**Cliff, A., and Ord, J.** *Spatial Processes.* London: Pion, London, 1981.

**Cordey, C., and Griffith, D.** Efficiency of least squares estimators in the presence of spatial autocorrelation. *Communications in Statistics--Simulation,* 22: 1161-1179, 1993.

**Florax, R., and Rey, S.** The impacts of misspecified spatial interaction in linear regression models, in *New Directions in Spatial Econometrics,* edited by L. Anselin and R. Florax, Springer-Verlag, Berlin, pp. 111-135, 1995.

**Griffith, D.** *Advanced Spatial Statistics.* Kluwer, Dordrecht, 1988.

**Griffith, D., and Sone, A.** Trade-offs associated with normalizing constant computational simplifications for estimating spatial statistical models. *Journal of Statistical Computation and Simulation,* 51: 165-183, 1995.

**Hordijk, L.** Problems in estimating econometric relations in space. *Papers of the Regional Science Association,* 42: 99-115, 1979.

**Stetzer, F.** Specifying weights in spatial forecasting models: the results of some experiments. *Environment and Planning A,* 14: 571-584, 1982.

**Upton, G.** Information from regional data, in *Spatial Statistics: Past, Present, and Future,* edited by D. Griffith, Institute of Mathematical Geography, Ann Arbor, pp. 315-359, 1990.

## Chapter 5

**Amrhein, C. G.** Searching for the elusive aggregation effect: evidence from statistical simulation. *Environment and Planning A* (forthcoming), 1994.

**Amrhein, C. G. and Flowerdew, R.** The effect of data aggregation on a Poisson regression model of Canadian migration. *Environment and Planning A* 24, 1381-1391, 1992.

**Anselin, L.** *Spatial Econometrics: Methods and Models.* Kluwer Academic Publishers, Dordrecht,1988.

**Arbia, G.** *Spatial Data Configuration in Statistical Analysis of Regional Economic and Related Problems.* Kluwer Academic Publishers, Dordrecht, 1989.

**Bian, L. and Walsh, S.** Scale dependencies of vegetation and topography in a mountainous environment of Montana. *The Professional Geographer* 45, 1-11, 1993.

**Blair, P. and Miller, R. E.** Spatial aggregation in multiregional input-output models. *Environment and Planning A* 15, 187-206, 1983.

**Blalock, H. M.** *Causal Inferences in Nonexperiental Research.* University of North Carolina Press, Chapel Hill, 1964.

**Bureau of the Census.** *Census of Population and Housing, 1990: Summary Tape File 1 on CD-ROM (Connecticut).* The Bureau, Washington, 1992a.

**Bureau of the Census.** *Census of Population and Housing, 1990: Summary Tape File 3 on CD-ROM (Connecticut).* The Bureau, Washington, 1992b.

**Cliff, A.D. and Ord, J. K.** *Spatial Autocorrelation,* Pion, London, 1973.

**Current, J. R. and Schilling, D. A.** Analysis of errors due to demand data aggregation in the set covering and maximal covering location problems. *Geographical Analysis* 22, 116-26, 1990.

**Fotheringham, A. S. and Wong, D. W. S.** The Modifiable Areal Unit Problem in multivariate statistical analysis. *Environment and Planning A* 23, 1025-44, 1991.

**Fotheringham, A. S.** Scale-independent spatial analysis. In *Accuracy of Spatial Databases,* edited by M. F. Goodchild and S. Gopal, pp.221-28, and Francis, London, 1991.

**Gehlke, C. E. and Biehl, K.** Certain effects of grouping upon the size of the correlation coefficient in Census tract material. *Journal of the American Statistical Association Supplement* 29, 169-70, 1934.

**Green, M.** Ecological fallacies and the Modifiable Areal Unit Problem. North West Regional Research Laboratory, Lancaster University, *Research Report No. 27,* 1993.

**Griffith, D. A.** *Advanced Spatial Statistics.* Kluwer Academic Publishers, Boston, 1988..

**Griffith, D. A. and Amrhein, C. G.** *Statistical Analysis for Geographers.* Prentice-Hall, Englewood Cliffs, 1991..

**Hubert, L. J.; Golledge, R. G.; and, Costanzo, C. M.** Generalized procedures for evaluating spatial autocorrelation. *Geographical Analysis* 13, 224-33, 1981.

**Moellering, H. and Tobler, W.** Geographic variances. *Geographical Analysis* 4, 34-50, 1972.

**Openshaw, S.** Optimal zoning systems for spatial interaction models. *Environment and Planning A* 9, 169-84, 1977a.

**Openshaw, S.** A geographical solution to scale and aggregation problems in region-building, partitioning and spatial modelling. *Transactions of Institute of British Geographers* 2, 459-72, 1977b.

**Openshaw, S.** *Concepts and Techniques in Modern Geography, Number 38. The Modifiable Areal Unit Problem.* Geo Books, Norwich, 1984.

**Openshaw, S. and Taylor, P. J.** A million or so correlation coefficients: three experiments on the modifiable areal unit problem. Pages 127-144 in N. Wrigley (ed.) *Statistical Applications in the Spatial Sciences,* Pion Limited, London, 1979.

**Putman, S. H. and Chung, S-H.** Effects of spatial systems design on spatial interaction models. 1: the spatial definition problem. *Environment and Planning A* 21, 27-46, 1989.

**Ripley, B. D.** *Spatial Statistics.* Wiley, New York, 1981.

**Robinson, A. H.** The necessity of weighing values in correlation analysis of areal data. *Annals, Association of American Geographers* 46: 233-236, 1956.

**Robinson, W. S.** Ecological correlations and the behavior of individuals. *American Sociological Review* 15: 351-357, 1950.

**Steel, D.G.; Holt, D.; and, Tranmer, M.** Modelling and adjusting aggregation effects. Paper presented at the Annual Conference of the US Bureau of the Census, 1993.

**Tobler, W.** Cellular geography. In *Philosophy in Geography,* edited by S. Gale and G. Olsson, pp. 379-86, Reidel, Dordrecht, 1979.

**Tobler, W.** Frame independent spatial analysis. In *Accuracy of Spatial Databases,* edited by M. F. Goodchild and S. Gopal, pp. 115-22. Taylor and Francis, London, 1991.

**Wartenberg, D.** Multivariate spatial correlation: a method for exploratory geographical analysis *Geographical Analysis* 17, 263-83, 1985.

**Wong, D.W.S.** Spatial dependency of segregation index. Paper presented at 41st North American Meeting of Regional Science Association International, Niagara Falls, Ontario, Canada, Nov.17-20, 1994.

**Wrigley, N.** Revisiting the Modifiable Areal Unit Problem and the ecological fallacy. In *Festschrift for Peter Haggett,* edited by Cliff, A.D., Gould, P.R., Hoare, A.G. and Thrift, N.J., Blackwell, Oxford, 1994.

### Chapter 6

**Adams, J., et al.** *Fortran90 Handbook, Complete ANSI/ISO Reference,* McGraw-Hill Book Company, New York, 1992.

**Anselin, L.** *Spatial Econometrics: Methods and Models,* Kluwer Academic Publisher, Dordrecht, The Netherlands, 1988.

**Anselin, L., and Hudak S.** Spatial econometrics in practice: a review of software options, *Regional Science and Urban Economics,* Vol. 22, 509-536, 1992.

**Blelloch, G.** *Vector Models for Data-Parallel Computing,* MIT Press, Cambridge, MA, 1991.

**Cliff, A. D., and Ord, J. K.** *Spatial Processes, Models & Applications,* Pion, London, 1981.

**ElGindy, H.** Optimal parallel algorithms for updating planar triangulations, *Proceedings of the Fourth International Symposiums on Spatial Data Handling,* Zurich, Switzerland, 200-208, 1990.

**ESRI.** *ARC Command References,* Redlands, CA, 1991.

**Fang, T. P., and Piegl, L.** Delaunay Triangulation Using a Uniform Grid, *IEEE Computer Graphics and Applications,* Vol. 13, 36-47, 1993.

**Flynn, M. J.** Very high-speed computing systems, *Proceedings of the IEEE,* Vol. 54, 1901-1909, 1966.

**Fox, G.** *Parallel Computers and Complex Systems,* SCCS-370, Syracuse University, Syracuse, NY, 1992.

**Griffith, D.** *Spatial Autocorrelation, A Primer,* Resource Publications in Geography, American Association of American Geographers, Washington, DC, 1987.

**Griffith, D.** *Advanced Spatial Statistics, Advanced Studies in Theoretical and Applied Econometrics,* Kluwer Academic Publishers, Dordrecht, The Netherlands, 1988a.

**Griffith, D.** Estimating spatial autoregressive model parameters with commercial statistical packages, *Geographical Analysis*, Vol. 20, 176-186, 1988b.

**Griffith, D.** *Spatial Regression Analysis on the PC: Spatial Statistics Using Minitab*, Discussion Paper No. 1. Institute of Mathematical Geography, Ann Arbor, 1989.

**Griffith, D.** Supercomputer and spatial statistics: a reconnaissance, *Professional Geographer*, Vol. 42, 481-492, 1990a.

**Griffith, D.** A numerical simplification for estimating parameters of spatial autoregressive models, in *Spatial Statistics, Past, Present, and Future*, edited by D. Griffith, Institute of Mathematical Geographers, Ann Arbor, MI, 183-197 1990b.

**Griffith, D., *et al.*** Developing minitab software for spatial statistical analysis: a tool for education and research," *Operational Geographer*, Vol. 8, No. 3., 28-34, 1990c.

**Griffith, D., and C. Amrhein, C.** *Statistical Analysis for Geographers*, Prentice-Hall, Englewood Cliffs, NJ, 1991.

**Griffith, D.** Which spatial statistics techniques should be converted to GIS functions?" in *Geographic Information Systems, Spatial Modelling, and Policy Evaluation*, edited by M. Fisher and P. N. Nijkamp, Springer-Verlag, Berlin, 103-114, 1992a.

**Griffith, D.** Simplifying the normalizing factor in spatial autoregressions for irregular lattices, *Papers in Regional Science*, Vol. 71, 71-86, 1992b.

**Griffith, D., and Sone, A.** *Some Trade-offs Associated with Computational Simplifications for Estimating Spatial Statistical/Econometric Models: Preliminary Results*, Discussion Paper No. 103, Department of Geography, Syracuse University, Syracuse, NY, 1993.

**Haining, R.** *Spatial Data Analysis in the Social and Environmental Sciences*, Cambridge University Press, Cambridge, 1990.

**Haining, R., and Wise, S. M.** *GIS and Spatial Data Analysis: Report on the Sheffield Workshop*, Regional Research Laboratory Initiative Discussion Paper Number 11, University of Sheffield, UK, 1991.

**Hillis, D.** *The Connection Machine*, MIT Press, Cambridge, MA, 1985.

**Hillis, D., and Steele, G. L. Jr.** Data parallel algorithms, *Communications of ACM*, Vol. 29, pp. 1170-1183, 1986.

**Hillis, D.** The connection machine, *Scientific American*, Vol. 256, No. 6, 108-115, 1987.

**IBM.** *Engineering and Scientific Subroutine Library Guide and Reference*, 1992.

**Li, B.** Opportunities and challenges of parallel processing in spatial data analysis: initial experiments with data parallel map analysis, *GIS/LIS '92*, 445-458, 1992a.

**Li, B.** Prospects of parallel processing in geographic data analysis, in *Development and Potentials of GIS: Inside and Outside China*, Hui Lin (eds), Science Press, Beijing, China, 76-89, 1992b.

**Li, B.** Suitability of topological data structures for data parallel operations in computer cartography, *Auto-Carto 11*, 434-443, 1993a.

**Li, B.** Developing network-oriented GIS software for parallel computing, *GIS/LIS'93*, 403-413, 1993b.

**Maspar Computer Corporation.** *Maspar Parallel Application Language (MPL) User Guide*, Sunnyvale, CA, 1991a.

**Maspar Computer Corporation.** *Maspar FORTRAN User Guide*, Sunnyvale, CA, 1991b.

**McCauley, J., and Engel, B.** Spatial statistics and interpolation procedures for GRASS, paper presented to *The 8th Annual GRASS Users' Conference and Exhibition*, March 14-16, 1993, Reston, Virginia, 1993.

**Ord, J. K.** Estimation methods for models of spatial interaction, *Journal of the American Statistical Association*, Vol. 70, 120-126, 1975.

**Press, W., *et al.*** *Numerical Recipes in C, the Art of Scientific Computing*, Cambridge University Press, Cambridge, 1988.

**Puppo, E., L. Davis, D. DeMenthon, and Y. A. Teng**, Parallel Terrain Triangulation, *Proceedings of the 5th International Symposium on Spatial Data Handling*, Charleston, SC, 632-641, 1992.

**Quinn, M. J.** Data-parallel programming on multicomputers, *IEEE Software*, Sept. 1990, 69-76, 1990.

**Rose, J.R. and Steele, G. L. Jr.** C*: An Extended C Language for Data Parallel Programming, *Technical Report PL 87-5*, Thinking Machines Corp., Cambridge, MA, 1986.

**Sandhu, J. S. and Marble, D.** An investigation into the utility of the Cray X-MP supercomputer for handling spatial data, *Proceedings, Third International Symposium on Spatial Data Handling*, Sydney, Australia, 253-267, 1988.

**Smith, J. R.** *The Design and Analysis of Parallel Algorithms*, Oxford University Press, New York and Oxford, 1993.

**Thinking Machines Corporation.** *The Connection Machine CM-5 Technical Summary*, Cambridge, MA, 1991a.

**Thinking Machines Corporation.** *Programming in FORTRAN*, Cambridge, MA, 1991b.

**Thinking Machines Corporation.** *Programming in C\**, Cambridge, MA, 1991c.

**Thinking Machines Corporation.** *CM FORTRAN Reference Manual*, Cambridge, MA, 1992.

**Thinking Machines Corporation.** *CM FORTRAN Utility Library Reference Manual*, Cambridge, MA, 1993a.

**Thinking Machines Corporation.** *CMSSL for CM Fortran: CM-5 Edition*, Cambridge, MA, 1993b.

**Trew, A., and Wilson, G.** *Past, Present, Parallel: A Survey of Available Parallel Computer Systems*, Springer-Verlag, London, 1992.

**Upton, G., and Fingleton, B.** *Spatial Data Analysis by Example*, John Wiley & Sons, New York, 1989.

## Chapter 7

**Anselin, L.** *SpaceStat: A program for the analysis of spatial data*. Department of Geography, University of California, Santa Barbara, CA, 1991.

**Anselin, L., Dodson, R. F., and Hudak, S.** Linking GIS and spatial data analysis in practice. *Geographical Systems*, 1, 2-23, 1993a.

**Anselin, L. and Getis, A.** Spatial statistical analysis and geographic information systems. *Annals of Regional Science*, 26, 19-33, 1992.

**Anselin, L. and Griffith, D. A.** Do spatial effects really matter n regression analysis? *Papers of the Regional Science Association*, 65, 11-34, 1988.

**Anselin, L., Hudak, S., and Dodson, R. F.** Spatial Data Analysis and GIS: Interfacing GIS and econometric software. *NCGIA Technical Paper 93-7*. University of California, Santa Barbara, CA, 1993b.

**Ben Chaabane, H., Cyiza, P., and Rushengura, A.** *L'Indice des Priorités Communales: Un Indice Synthétique du Taux de Développement des Communes Rwandaises*. Kigali, Rwanda: MINIPLAN, Rép. Rwandaise, 1991.

**Ben Chaabane, H., and Cyiza, P.** *Méthodologie d'Elaboration de Données sur le Revenu Rural des Communes Rwandaises*. Kigali, Rwanda: MINIPLAN Direction de la Planification, Division des Stratéties de Développement Communal et Régional, 1992.

**Berry, L., Olson, J. M., Campbell, D. J., and Brown, D. G.** The Rwanda Society-Environment Project: a pilot study to link socioeconomic and physical data within an environmental information system. An interim report. *Proceedings of the Second Annual Meeting of the CIESIN User's Group*, Atlanta, GA, 1994.

**Brown, D. G., Bian, L., and Walsh, S. J.** Response of a distributed watershed erosion model to variations in input data aggregation levels. *Computers and Geosciences*. 19(4), 499-509, 1993.

**Brown, D. G. and Bara, T. J.** Recognition and reduction of systematic error in elevation and derivative surfaces from 7-1/2 minute DEMs. *Photogrammetric Engineering and Remote Sensing*. 60(2), 189-194, 1994.

**Campbell, D. J.** Environmental stress in Rwanda No. 1: Preliminary analysis. *Rwanda Society-Environment Project Working Paper*, 4, Department of Geography, Michigan State University, East Lansing, MI, 1994.

**Campbell, D. J., Olson, J. M., and Berry, L.** Population pressure, agricultural productivity and land degradation in Rwanda: an agenda for collaborative training, research and analysis.

*Rwanda Society-Environment Project Working Papers,* 1, Department of Geography, Michigan State University, East Lansing, MI, 1993.

**Can, A.** Residential quality assessment: alternative approaches using GIS. *Annals of Regional Science,* 26, 97-110, 1992.

**CPR (Carte Pedologique du Rwanda).** *Carte d'Aptitude des Sols du Rwanda.* Kigali, Rwanda: MINAGRI and Cooperation Technique Belge, 1993.

**Cordy, C. and Griffith, D.** Efficiency of least squares estimators in the presence of spatial autocorrelation. *Communications in Statistics* B, 22, 1161-1179, 1993.

**Delepierre, G.** Les Régions Agricoles du Rwanda. *Bulletin Agricole du Rwanda,* 8(4), 216-225, 1975.

**DSA (Division des Statistiques Agricoles).** *Enquête Nationale Agricole 1989: Production, Superficie, Rendement, Elevage et Leur Evolution 1984-1989.* Kigali, Rwanda: MINAGRI, Rép. Rwandaise, 1991.

**Eastman, J. R.** *IDRISI: A Grid-Based Geographic Analysis System,* version 3.2. Clark University School of Geography, Worcester, MA, 1990..

**ESRI.** *Arc/Info, User's Guide.* Environmental Systems Research Institute, Redlands, CA, 1992..

**Flowerdew, R., and Green, M.** Statistical methods for inference between incompatible zonal systems. In M. F. Goodchild and S. Gopal, Eds., *Accuracy of Spatial Databases.* Taylor and Francis, London, 239-247, 1989.

**Griffith, D. A.** A spatially adjusted ANOVA model. *Geographical Analysis,* 10, 296-301, 1978.

**Griffith, D. A.** Geometry and spatial interaction. *Annals of the Asociation of American Geographers,* 72(3), 332-346, 1982.

**Griffith, D.A.** Estimating spatial autoregresive model parameters with commercial statistical packages. *Geographical Analysis,* 20, 176-186, 1988.

**Griffith, D. A.** A spatial adjusted N-way ANOVA model. *Regional Science and Urban Economics,* 22, 347-369, 1992.

**Griffith, D. A., Lewis, R., Li, B., Vasiliev, I., Knight, S., Yang, X.** 1990. Developing Minitab Software for spatial statistical analysis: a tool for education and research. *Operational Geographer,* 8, 28-33, 1990.

**Haining, R.** *Spatial Data Analysis in the Social and Environmental Sciences.* Cambridge University Press, Cambridge, 1990.

**Haslett, J, Bradley, R., Craig, P., Unwin, A., and Wills, G.** Dynamic graphics for exploring spatial data with application to locating global and local anomalies. *American Statistician,* 45(3), 234-242, 1991.

**Lowell, K.** Utilizing discriminant function analysis with a geographical information system to model ecological succession. *International Journal of Geographical Information Systems,* 5(2), 175-191, 1991.

**Martin, D. and Bracken, I.** Techniques for modelling population-related raster databases. *Environment and Planning A,* 23, 1069-1075, 1991.

**MINAGRI.** *Production Agricole en 1987: Bilan d'Autosufficance Alimentaire par Commune et par Habitant.* Kigali, Rwanda: MINAGRI, Rép. Rwandaise, 1989.

**MINIPLAN.** *Syntése des principaux résultats du recensement général de la population et de l'Habitat 1978.* Kigali, Rwanda: Rép. Rwandaise, 1982.

**MINIPLAN.** *Recensement Général de la Population et de l'Habitat au 15 aout 1991: Résultats Préliminaires Echantillon au 10e.* Kigali, Rwanda: MINIPLAN, Rép. Rwandaise, 1992.

**MINITRAP.** *Bulletin Climatologique Année 1991.* Kigali, Rwanda: MINITRAP, Rép Rwandaise, 1991.

**Nzisabira, J.** Accumulation du Peuplement Rural et Ajustements Structurels du Système d'Utilisation du Sol au Rwanda Depuis 1945. *Bulletin Agricole du Rwanda,* 2, 117-127, 1989.

**Oliver, M., Webster, R., and Gerrard, J.** Geostatistics in physical geography. Part 2: Applications. *Transactions, Institute of British Geographers,* 14, 270-286, 1989.

**Olson, Jennifer M.** *The Impact of Socioeconomic Factors on Migration Patterns in Rwanda.* Unpublished M.A. Thesis. Department of Geography, Michigan State University, 1990.

**Olson, Jennifer M.** Farming systems of Rwanda: echoes of historic divisions reflected in current land use. *Rwanda Society-Environment Project Working Paper*, 2, Department of Geography, Michigan State University, East Lansing, MI, 1994a..

**Olson, Jennifer M.** Demographic responses to resource constraints. *Rwanda Society-Environment Project Working Paper*, 7, Department of Geography, Michigan State University, East Lansing, MI, 1994b.

**Owen, D.** The Starship. *Communications in Statistics*, 17, 315-341, 1988.

**Prioul, C. and Sirven, P.** *Atlas du Rwanda*. Kigali, Rwanda: Imprimérie Moderne Nantasie Coueron, 1981.

**Semple, R. K. and Green, M. B.** Classification in human geography. . In G. L. Gaile and C. J. Willmott, Eds., *Spatial Statistics and Spatial Models*. Dordrecht: Reidel, 133-145.

**Shryock, H. S. and Seigel, J. S.** *The Methods and Materials of Demography*, Academic Press, Inc., New York, 1976.

**Tobler, W.** Smooth pycnophylactic interpolation for geographical regions. *Journal of the American Statistical Association*, 74(367), 519-536, 1979.

**Upton, G. J. G.** Information from regional data. In, D. A. Griffith, Ed., *Spatial Statistics: Past, Present, and Future*, Monograph Series, Institute of Mathematical Geography, Ann Arbor, MI, Monograph #12, 315-360, 1990.

# Chapter 8

**Becker, G. S.** An economic analysis of fertility, in A. J.Coale, (ed.), *Demographic and Economic Change in Developed Countries*, Princeton University Press, Princeton, NJ, 1960.

**Becker, G. S.** A theory of the time allocation. *Economic Journal*, 75, 493-517, 1965.

**Becker, G. S.** A theory of social interactions. *Journal of Political Economy*, .82, 1963-93, 1974.

**Bulatao, R. A. and Lee, R. D.** (eds), *Determinants of Fertility in Developing Countries*, two volumes, Academic Press, New York, 1983.

**Cleland, J. and Wilson, C.** Demand theories of the fertility transition: An iconoclastic view. *Population Studies*, 41, 5-30, 1987.

**Cliff, A. and K. Ord.** *Spatial Processes*, Pion, London, 1981.

**Coale, A. and Watkins, S.** (eds) *The Decline of Fertility in Europe*, Princeton University Press, Princeton, NJ, 1986.

**Daniel, C., and Wood, F. S.** *Fitting Equations to Data: ComputerAnalysis of Multifactor Data*, John Wiley and Sons, New York, 1983.

**Davis, K.** The theory of changes and response in modern demographic history. *Population Index*, 29, 345-66, 1963.

**Eastline, R. A.** The economics and sociology of fertility: A synthesis, in C. Tilly (ed.), *Historical Studies of Changing Fertility*, Princeton University Press, Princeton, NJ, 1978.

**Feeney, G.; Wang, F.; Zhou, M.; and, Xiao, B.** Recent fertility dynamics in China: results from the 1987 one percent population survey. *Population and Development Review*, 15, no.1, 297-322, 1989.

**Feng, H. M.** The Contextual Determinants of Contemporary Chinese Fertility Transition, unpublished manuscript, Department of Geography, Syracuse University, Syracuse, NY, 1993.

**Freedman, R.** Theories of fertility decline: A reappraisal. *Social Forces*, 48, 1-17, 1979.

**Gasim, A.** First-order autoregressive models: a method for obtaining eigenvalues for weighting matrices. *Journal of Statistical Planning and Inference*, 18, 391-98, 1988.

**Goldscheider, C.** *Population, Modernization, and Social Structure*, Little Brown, Boston, 1971.

**Greenhalgh, S.** Toward a political economy of fertility: Anthropological contributions. *Population and Development Review*, 16, no.1, 85-106, 1990.

**Greenhalgh, S.; Zhu, C.; and, Li, N.** Restraining population growth in three Chinese villages, 1988-93. *Population and Development Review*, 20, no.2, 365-95, 1994.

**Griffith, D. A. and Amrhein, C.** *Statistical Analysis for Geographers*, Prentice Hall, Englewood Cliffs, NJ, 1991.

**Griffith, D. A.** *Spatial Autocorrelation: A Primer*, Association of American   Geographers, Washington, D.C., 1987.

**Griffith, D. A.** Estimating Spatial Autoregressive Model Parameters with Commercial Statistical Packages. *Geographical Analysis*, 20, No.2, 176-86, 1988.

**Griffith, D. A.** Simplifying the normalizing factor in spatial autoregressions for irregular lattices. *Papers in Regional Science*, 71, 71-86, 1992.

**Griffith, D.A.** *Spatial Regression Analysis on the PC: Spatial Statistics Using SAS*, Association of American Geographers, Washington, D.C., 1993.

**Haining, R.** The use of added variable plots in regression   modeling with spatial data. *The Professional Geographer*, 42, 336-44, 1990.

**Haining, R.** *Spatial Data Analysis for Social and Environmental Sciences*, Oxford University Press, London, 1991.

**He-Nan Province Census Bureau**, *He-Nan Province Population Census Statistics, 1990*, China Statistics Publishing Company, Beijing, China, 1992.

**Johnsson, S.R.** Implicit policy and fertility during development.  Population and Development Review, 17, no.3, 377-414, 1991.

**Knodel, J. and van de Walle, E.** Lessons from the past: policy implications of historical fertility studies. *Development Review*, .5, no.2, 217-45, 1979.

**Lesthaeghe, R. and  Surkyn, J.** Cultural dynamics and economic theories of fertility change. *Population and Development Review*, 14, no.1, 1-45, 1988.

**Luther, N.Y. Jr.; Feeney, G.; and, Zhang, W.** One-child families or a baby boom? Evidence from China's 1987 one-per-hundred survey. *Population Studies*, 44, no.2,  341-57, 1990.

**Nambookiri, N.K.** Some observations on the economic framework for fertility analysis. *Population Studies*, 26, 185-206, 1972.

**Notestein, F.** Economic problems of population change, in *Proceedings of the Eighth International Conference of Agri-Economics*, Oxford University Press, London,      1953.

**Notestein, F.** Population - the long view, in T.W.Schultz (ed.), *Food for the World*, University of Chicago Press, Chicago, 1954.

**Ord, K.** Estimation methods for models of spatial interaction. *Journal of American Statistical Association*, 70, 120-26, 1975.

**Peng, Peiyuan**, Accomplishments of China's family planning program: a statement by a Chinese official. *Population and Development Review*, 19, no.2, 399-403, 1993.

**Peng, X.** Major determinants of China's fertility transition. *The China Quarterly*, No.117 (March), 1-37, 1989.

**Pye, L.** *The Dynamics of Chinese Politics*, Delgeschlager, Gunn & Hain, Publishers, Cambirdge, MA, 1981.

**Ripley, B**. *Statistical Inference for Spatial Processes*,  Cambridge University Press, Cambridge, 1988.

**SAS Institute Inc.** *SAS/STAT User's Guide* (two volumes), SAS Institute Inc., Cary, NC, 1989.

**Schultz, T. P.** *Economics of Population*, Addison-Wesley, MA, 1981.

**Shryock, H.S., and Siegel, J. S.** *et al.*  *The Methods and Materials of Demography*, U.S.Bureau of Census, Washington, D.C., Vol.2, 1980.

**Smith, H.** Integrating theory and research on the institutional determinants of fertility. *Demography*, 26, 171-184, 1989.

**Willis, R. J.** A new approach to the economic theory of fertility behavior. *Journal of Political Economy*, 81, 14-69, 1973.

**Wolf, A. P.** The preeminent role of government intervention in China's family   revolution. *Population and Development Review*, 12, 106-16, 1986.

**Yi, Z.** Is the Chinese family planning program 'tightening up'? *Population and Development Review*, 15, No.2 (June), 1989.

# Chapter 9

**Anselin, L.** *SPACESTAT TUTORIAL: A Workbook for Using SpaceStat in the Analysis of Spatial Data*. Technical Software Series S-92-1, NCGIA, University of California, Santa Barbara,1992.

**Anselin, L., and Can, A.** Model comparison and model validation issues in empirical work on urban density functions. *Geographical Analysis*, 18: 179-197, 1986.

**Batty, M., and Kim, K.** Form follows function: reformulating urban population density functions. *Urban Studies*, 29: 1043-1070, 1992.

**Baumont, C.** Preferences spatiales et éspaces urbaines multicentriques. *Document de Travail No. 9308*, Laboratoire d'Analyse et de Techniques Economiques, Université de Bourgogne, Dijon, France, 1993.

**Can, A., and Griffith, D.** Spatial dependence in urban density functions. Paper presented at the 40th North American meetings of RSAI, Houston, TX, November 12-14, 1993.

**Griffith, D.** Modelling urban population density in a multi-centered city. *Journal of Urban Economics*, 9: 298-310, 1981.

**Griffith, D.** Estimating spatial autoregressive model parameters with commercial statistical packages. *Geographical Analysis*, 20: 176-186, 1988.

**Griffith, D.** Simplifying the normalizing factor in spatial autoregressions for irregular lattices, *Papers in Regional Science*, 71: 71-86, 1992.

**Griffith, D.** *Spatial Regression Analysis on the PC: Spatial Statistics Using SAS*. The Association of American Geographers, Washington D.C., 1993.

**Griffith, D.; Haining, R.; and Arbia, G.** Heterogeneity of attribute sampling error in spatial data sets. *Geographical Analysis*, 26: 300-320, 1994.

**Hill, F.** Spatio-temporal trends in urban population density: a trend surface analysis, in *The Form of Cities in Central Canada*, edited by L. Bourne; MacKinnon, R.; and Simmons, J. University of Toronto Press, Toronto, pp. 103-119, 1973.

**Kau, J., C. Lee and Sirmans, C.** *Urban Econometrics*. Vol. 6, Research in Urban Economics Series, JAI Press, Greenwich, CT, 1986.

**Thrall, G.** Statistical and theoretical issues in verifying the population density function, *Urban Geography*, 9: 518-537, 1988.

# Chapter 10

**Barlett, M.S.** Nearest neighbour models in the analysis of field experiments (with discussion). *J. Roy. Statist. Soc. Ser. B*. 40:147-174, 1978.

**Besag, J.E. and Kempton, R** Statistical analysis of field experiments using neighbouring plots. *Biometrics* 42:231-251, 1986.

**Brownie, C.; Bowman, D. T.; and Burton, J. W..** Estimating spatial variation in analysis of data from field trials: A comparison of methods. *Agron. J.* 85:1244-1253, 1993.

**Cliff, A.D. and Ord, J. K.** The comparison of means when samples consist of spatially autocorrelated observations. *Env. Plan. A.* 7:725-734, 1975.

**Cressie, N.** 1991. *Statistics for Spatial Data*. Wiley, New York. p. 900, 1991.

**Glass, G.V.; Peckham, P.D.; and Sanders, J. R.** Consequences of failure to meet assumptions underlying the fixed effects analysis of variance and covariance. *Rev. Educ. Res.* 42(3):237-288, 1972.

**Green, P.J.; Jennison, C.; and Seheult, A. H..** Analysis of field experiments by least squares smoothing. *J. Roy. Statist. Soc. Ser. B.* 47(2):299-315, 1984.

**Griffith, D.A.** Estimating spatial autoregressive model parameters with commercial statistical packages. *Geogr. Anal.* 20(2):176-186, 1988b.

**Griffith, D.A.** A numerical simplification for estimating parameters of spatial autoregressive models. p. 185-195. *In* D.A. Griffith (ed.) *Spatial Statistics: Past, Present and Future*. Institute of Mathematical Geography, Ann Arbor, MI, 1990.

**Griffith, D.A.** What is spatial autocorrelation? Reflections on the past 25 years of spatial statistics. *L'Espace geographique* 3:265-280, 1992.

**Griffith, D.A.** *Spatial Regression Analysis on the PC: Spatial Statistics using SAS.* Assoc. Am. Geog. Washingtion D.C. p. 130, 1993.

**Grondona, M.O. and Cressie, N.** Using spatial considerations in the analysis of experiments. *Technomet.* 33(4):381-392, 1991.

**Haining, R.P.** A spatial model for high plains agriculture. *Ann. Ass. Am. Geog.* 68:(4):493-504, 1978.

**Haining, R.P.** 1980. Spatial autocorrelation problems. *In* D.T. Herbert and R.J. Johnson (eds.), *Geography and the urban environment, progress in research and applications.* 3:1-44, Wiley, New York, 1980.

**Legendre, P.; Oden, N. L.; Sokal, R. R.; Vaudor, A.; and Kim J.** Approximate analysis of variance of spatially autocorrelated regional data. *J. Class.* 7:53-75, 1990.

**Miron, J.** Spatial autocorrelation in regression analysis: a beginners guide. *In* G.L. Gaile and C.J. Willmot (ed.) *Spatial Statistics and Models.* Dordrectht. p. 201-222, 1984.

**Moran, P.** The interpretation of statistical maps. *J. Roy. Statist. Soc. Ser. B* 37:243-251, 1948.

**Mulla, D.J.; Bhatti, A. J.; and Kunkel, R.** Methods for removing spatial variability from field research trials. *Advances in Soil Science.* Springer-Verlag New York Inc., 13:201-213, 1990.

**Olson, K.R.; Carmer, S. G.; and Olson, G. W.** *Assessment of effects of soil variability on maximum alfalfa yields,* GEDMAB, New York, 36:1-14, 1985.

**Petersen, R.G.** *Design and Analysis of Experiments.* Marcel Dekker, Inc. p. 429, 1985.

**SAS Institute Inc.** *SAS Procedures Guide,* release 6.03 edition. Cary, NC. 210 p., 1988.

**Tukey, J.W.** Comparing individual means in the analysis of variance. *BIOKA.* 5:99-114, 1949.

**van Es, H.M. and van Es, C. L.** The spatial nature of randomization and its effect on the outcome of field experiments. *Agron. J.* 85:420-428, 1993.

**Zimmerman, D.L. and Harville, D. A.** A random field approach to the analysis of field-plot experiments and other spatial experiments. *Biometrics,* 47:223-239, 1991.

# References, Alphabetically

The following listing offers student and researcher alike the opportunity to gain an understanding, at a glance, of the handbook's emphasis.

## A

**Adams, J., et al.**. *Fortran90 Handbook, Complete ANSI/ISO Reference*, McGraw-Hill Book Company, New York, 1992.

**Ahuja, N., and Schachter, B.** *Pattern Models.* Wiley, New York,1983.

**Amrhein, C. G.** Searching for the elusive aggregation effect: evidence from statistical simulation. *Environment and Planning A* (forthcoming), 1994.

**Amrhein, C. G. and Flowerdew, R.** The effect of data aggregation on a Poisson regression model of Canadian migration. *Environment and Planning A* 24, 1381-1391, 1992.

**Anselin, L.** *Spatial Econometrics: Methods and Models*, Kluwer Academic Publisher, Dordrecht, The Netherlands, 1988.

**Anselin, L.** *SpaceStat: A program for the analysis of spatial data.* Department of Geography, University of California, Santa Barbara, CA, 1991.

**Anselin, L.** *SPACESTAT TUTORIAL: A Workbook for Using SpaceStat in the Analysis of Spatial Data.* Technical Software Series S-92-1, NCGIA, University of California, Santa Barbara,1992.

**Anselin, L., and Can, A.** Model comparison and model validation issues in empirical work on urban density functions. *Geographical Analysis*, 18: 179-197, 1986.

**Anselin, L., Dodson, R. F., and Hudak, S.** Linking GIS and spatial data analysis in

**Anselin, L. and Getis, A.** Spatial statistical analysis and geographic information systems. *Annals of Regional Science*, 26, 19-33, 1992.

**Anselin, L., and Griffith, D.** Do spatial effects really matter in regression analysis? *Papers of the Regional Science Association*, 65, 11-34, 1987.

**Anselin, L., and Hudak.** Spatial econometrics in practice: a review of software options. *Regional Science and Urban Economics*, 22, 509-536, 1992.

**Anselin, L., Hudak, S., and Dodson, R. F.** Spatial Data Analysis and GIS: Interfacing GIS and econometric software. *NCGIA Technical Paper 93-7.* University of California, Santa Barbara, CA, 1993b.

**Arbia, G.** *Spatial Data Configuration in Statistical Analysis of Regional Economic and Related Problems.* Kluwer Academic Publishers, Dordrecht, 1989.

**Arbia, G.** The use of GIS in spatial statistical surveys. *International Statistical Review*, 61, 339-359, 1993.

## B

**Barlett, M.S.** Nearest neighbour models in the analysis of field experiments (with discussion). *J. Roy. Statist. Soc. Ser. B.* 40:147-174, 1978.

**Bartlett, M.** *The Statistical Analysis of Spatial Pattern.* Chapman and Hall, London, 1975.

**Batty, M., and Kim, K.** Form follows function: reformulating urban population density functions. *Urban Studies*, 29: 1043-1070, 1992.

**Baumont, C.** Preferences spatiales et éspaces urbaines multicentriques. *Document de Travail No. 9308*, Laboratoire d'Analyse et de Techniques Economiques, Université de Bourgogne, Dijon, France, 1993.

**Becker, G. S.** An economic analysis of fertility, in A. J.Coale, (ed.), *Demographic and Economic Change in Developed Countries*, Princeton University Press, Princeton, NJ, 1960.

**Becker, G. S.** A theory of the time allocation. *Economic Journal*, 75, 493-517, 1965.

**Becker, G. S.** A theory of social interactions. *Journal of Political Economy*, .82, 1963-93, 1974.

**Bellhouse, D. R.** Some optimal designs for sampling in two dimensions. *Biometrika*, 64, 605-611, 1977.

**Bellhouse, D. R.** Spatial sampling in the presence of a trend. *Journal of Statistical Planning and Inference*, 5, 365-375, 1981.

**Bellhouse, D. R.** Systematic sampling of periodic functions. *Canadian Journal of Statistics*, 13, 17-28, 1985.

**Bellhouse, D. R.** *Systematic sampling. Handbook of Statistics, Vol. 6* (P. R. Krishnaiah and C. R. Rao, Eds.). Elsevier Science Publishers, Amsterdam, 1988.

**Ben Chaabane, H., Cyïza, P., and Rushengura, A.** *L'Indice des Priorités Communales: Un Indice Synthétique du Taux de Développement des Communes Rwandaises.* Kigali, Rwanda: MINIPLAN, Rép. Rwandaise, 1991.

**Ben Chaabane, H., and Cyïza, P.** *Méthodologie d'Elaboration de Données sur le Revenu Rural des Communes Rwandaises.* Kigali, Rwanda: MINIPLAN Direction de la Planification, Division des Stratéties de Développement Communal et Régional, 1992.

**Berry, L., Olson, J. M., Campbell, D. J., and Brown, D. G.** The Rwanda Society-Environment Project: a pilot study to link socioeconomic and physical data within an environmental information system. An interim report. *Proceedings of the Second Annual Meeting of the CIESIN User's Group*, Atlanta, GA, 1994.

**Besag, J.E. and Kempton, R** Statistical analysis of field experiments using neighbouring plots. *Biometrics* 42:231-251, 1986.

**Bian, L. and Walsh, S.** Scale dependencies of vegetation and topography in a mountainous environment of Montana. *The Professional Geographer* 45, 1-11, 1993.

**Blair, P. and Miller, R. E.** Spatial aggregation in multiregional input-output models. *Environment and Planning A* 15, 187-206, 1983.

**Blalock, H. M.** *Causal Inferences in Nonexperiental Research.* University of North Carolina Press, Chapel Hill, 1964.

**Blelloch, G.** *Vector Models for Data-Parallel Computing*, MIT Press, Cambridge, MA, 1991.

**Bookstein, Fred L.** *The Measurement of Biological Shape and Shape Change.* Berlin, Springer-Verlag, 1978.

**Boots, B., and Dufouraud, D.** A programming approach to minimizing and maximizing spatial autocorrelation statistics. *Geographical Analysis* 26(1):54-66, 1994.

**Brown, D. G. and Bara, T. J.** Recognition and reduction of systematic error in elevation and derivative surfaces from 7-1/2 minute DEMs. *Photogrammetric Engineering and Remote Sensing.* 60(2), 189-194, 1994.

**Brown, D. G., Bian, L., and Walsh, S. J.** Response of a distributed watershed erosion model to variations in input data aggregation levels. *Computers and Geosciences.* 19(4), 499-509, 1993.

**Brown, D. and Walsh, S.** Spatial autocorrelation in remotely sensed and GIS data. *Proceedings* of the ACSM/ASPRS Annual Convention, New Orleans, LA, Vol. 3, 13-39, 1993.

**Brownie, C.; Bowman, D. T.; and Burton, J. W..** Estimating spatial variation in analysis of data from field trials: A comparison of methods. *Agron. J.* 85:1244-1253, 1993.

**Brus, D.J., and De Gruijter, J.J.** Design-based versus model-based estimates of spatial means: Theory and applications in environmental soil science. *Environmetrics*, 4, 123-152, 1993.

**Buckland, S.T., Anderson, D.R., Burnham, K.P., and Laake, J.L.** *Distance Sampling: Estimating Abundance of Biological Populations.* Chapman & Hall, New York, 1993.

**Bulatao, R. A. and Lee, R. D.** (eds), *Determinants of Fertility in Developing Countries*, two volumes, Academic Press, New York, 1983.

**Bureau of the Census.** *Census of Population and Housing, 1990: Summary Tape File 1 on CD-ROM (Connecticut).* The Bureau, Washington, 1992a.

**Bureau of the Census.** *Census of Population and Housing, 1990: Summary Tape File 3 on CD-ROM (Connecticut).* The Bureau, Washington, 1992b.

# C

**Campbell, D. J.** Environmental stress in Rwanda No. 1: Preliminary analysis. *Rwanda Society-Environment Project Working Paper*, 4, Department of Geography, Michigan State University, East Lansing, MI, 1994.

**Campbell, D. J., Olson, J. M., and Berry, L.** Population pressure, agricultural productivity and land degradation in Rwanda: an agenda for collaborative training, research and analysis. *Rwanda Society-Environment Project Working Papers*, 1, Department of Geography, Michigan State University, East Lansing, MI, 1993.

**Can, A.** Residential quality assessment: alternative approaches using GIS. *Annals of Regional Science*, 26, 97-110, 1992.

**Can, A., and Griffith, D.** Spatial dependence in urban density functions. Paper presented at the 40th North American meetings of RSAI, Houston, TX, November 12-14, 1993.

**Chorley, R.** [chairman]. *Handling Geographic Information.* Her Majesty's Stationary Office, Report to the Select Committee on GIS, London, 1987.

**Cleland, J. and Wilson, C.** Demand theories of the fertility transition: An iconoclastic view. *Population Studies*, 41, 5-30, 1987.

**Cliff, A.D. and Ord, J. K.** *Spatial Autocorrelation.* Pion, London, 1973.

**Cliff, A. D., and Ord, J. K.** *Spatial Processes, Models & Applications*, Pion, London, 1981.

**Cliff, A.D. and Ord, J. K.** The comparison of means when samples consist of spatially autocorrelated observations. *Env. Plan. A.* 7:725-734, 1975.

**Coale, A. and Watkins, S.** (eds) *The Decline of Fertility in Europe*, Princeton University Press, Princeton, NJ, 1986.

**Cochran, W.G.** Relative accuracy of systematic and stratified random samples for a certain class of population. *Annals of Mathematical Statistics*, 17, 164-177, 1946.

**Cochran, W.G.** Sampling Methods (3$^{rd}$ ed.). Wiley, New York, 1977.

**Cordy, C.B.** An extension of the Horvitz-Thompson theorem to point sampling from a continuous universe. *Statistics & Probability Letters*, 18, 353-362, 1993.

**Cordy, C.B., and Thompson, C.M.** An application of the deterministic variogram to design-based variance estimation. *Mathematical Geology*, 27, 173-205, 1995.

**Cordy, C., and Griffith, D.** Efficiency of least squares estimators in the presence of spatial autocorrelation. *Communications in Statistics--Simulation and Computation*, 22, 1161-1179, 1993.

**CPR (Carte Pedologique du Rwanda).** *Carte d'Aptitude des Sols du Rwanda.* Kigali, Rwanda: MINAGRI and Cooperation Technique Belge, 1993.

**Cressie, N. A. C.** *Statistics for Spatial Data.* Wiley, New York, 1991.

**Current, J. R. and Schilling, D. A.** Analysis of errors due to demand data aggregation in the set covering and maximal covering location problems. *Geographical Analysis* 22, 116-26, 1990.

# D

**Dacey, Michael F.** Analysis of central place and point patterns by a nearest neighbour method. Lund Studies in Geography, B, Human Geography, 24, 55-75, 1962.

**Dacey, Michael F.** Imperfections in the uniform plane. *Michigan Inter-University Community of Mathematical Geographers*, John D. Nystuen, Editor, May, 1964. Reprinted, *Solstice: An Electronic Journal of Geography and Mathematics*, Summer, 1994.

**Daniel, C., and Wood, F. S.** *Fitting Equations to Data: Computer Analysis of Multifactor Data*, John Wiley and Sons, New York, 1983.

**Das, A.C.** Two-dimensional systematic sampling and the associated stratified and random sampling. *Sankhya*, 10, 95-108, 1950.

**Davis, K.** The theory of changes and response in modern demographic history. *Population Index*, 29, 345-66, 1963.

**De Gruijter, J.J., and Ter Braak, C.J.F.** Model-free estimation from spatial samples: A reappraisal of classical sampling theory. *Mathematical Geology*, 22, 407-415, 1990.

**Delepierre, G.** Les Régions Agricoles du Rwanda. *Bulletin Agricole du Rwanda*, 8(4), 216-225, 1975.

**Deming, W.E.** *Some Theory of Sampling.* Wiley, New York, 1950.

**De Vries, P.G.** Sampling Theory for Forest Inventory. Springer-Verlag, New York, 1986.

**DSA (Division des Statistiques Agricoles).** *Enquête Nationale Agricole 1989: Production, Superficie, Rendement, Elevage et Leur Evolution 1984-1989.* Kigali, Rwanda: MINAGRI, Rép. Rwandaise, 1991.

**Dunn, R., and Harrison, A.R.** Two-dimensional systematic sampling of land use. *Applied Statistics,* 42, 585-601, 1993.

**Durrett, R.** Stochastic spatial models, *Forefronts* (newsletter of the Cornell Theory Center), 9 (#4, Spring), 4-6, 1994.

**Dykes, J.** Area-value data: new visual emphases and representations, in *Visualization in Geographical Information Systems,* edited by H. Hearnshaw and D. Unwin. Wiley, New York, pp. 103-114, 1994.

# E

**Eastline, R. A.** The economics and sociology of fertility: A synthesis, in C. Tilly (ed.), *Historical Studies of Changing Fertility,* Princeton University Press, Princeton, NJ, 1978.

**Eastman, J. R.** *IDRISI: A Grid-Based Geographic Analysis System,* version 3.2. Clark University School of Geography, Worcester, MA, 1990..

**ElGindy, H.** Optimal parallel algorithms for updating planar triangulations, *Proceedings of the Fourth International Symposiums on Spatial Data Handling,* Zurich, Switzerland, 200-208, 1990.

**ESRI.** *ARC Command References,* Redlands, CA, 1991.

**ESRI.** *Arc/Info, User's Guide.* Environmental Systems Research Institute, Redlands, CA, 1992..

# F

**Fang, T. P., and Piegl, L.** Delaunay Triangulation Using a Uniform Grid, *IEEE Computer Graphics and Applications,* Vol. 13, 36-47, 1993.

**Feeney, G.; Wang, F.; Zhou, M.; and, Xiao, B.** Recent fertility dynamics in China: results from the 1987 one percent population survey. *Population and Development Review,* 15, no.1, 297-322, 1989.

**Feng, H. M.** The Contextual Determinants of Contemporary Chinese Fertility Transition, unpublished manuscript, Department of Geography, Syracuse University, Syracuse, NY, 1993.

**Florax, R., and Rey, S.** The impacts of misspecified spatial interaction in linear regression models, in *New Directions in Spatial Econometrics,* edited by L. Anselin and R. Florax, Springer-Verlag, Berlin, pp. 111-135, 1995.

**Flowerdew, R., and Green, M.** Statistical methods for inference between incompatible zonal systems. In M. F. Goodchild and S. Gopal, Eds., *Accuracy of Spatial Databases.* Taylor and

**Flynn, M. J.** Very high-speed computing systems, *Proceedings of the IEEE,* Vol. 54, 1901-1909, 1966.

**Fotheringham, A. S. and Wong, D. W. S.** The Modifiable Areal Unit Problem in multivariate statistical analysis. *Environment and Planning A* 23, 1025-44, 1991.

**Fotheringham, A. S.** Scale-independent spatial analysis. In *Accuracy of Spatial Databases,* edited by M. F. Goodchild and S. Gopal, pp.221-28, and Francis, London, 1991.

**Fox, G.** *Parallel Computers and Complex Systems,* SCCS-370, Syracuse University, Syracuse, NY, 1992.

**Freedman, R.** Theories of fertility decline: A reappraisal. *Social Forces,* 48, 1-17, 1979.

# G

**Gasim, A.** First-order autoregressive models: a method for obtaining eigenvalues for weighting matrices. *Journal of Statistical Planning and Inference,* 18, 391-98, 1988.

**Gatrell, Anthony C.** Complexity and redundancy in binary maps. *Geographical Analysis,* Vol. IX, No. 1, 29-41, 1977.

**Gehlke, C. E. and Biehl, K.** Certain effects of grouping upon the size of the correlation coefficient in Census tract material. *Journal of the American Statistical Association Supplement* 29, 169-70, 1934.

**Getis, A. and Boots, B.** *Models of Spatial Processes: An Approach to the Study of Point, Line, and Area Patterns.* Cambridge University Press, Cambridge, 1978.

**Gilbert, R.O.** *Statistical Methods for Environmental Pollution Monitoring.* Van Nostrand Reinhold, New York, 1987.

**Glass, G.V.; Peckham, P.D.; and Sanders, J. R.** Consequences of failure to meet assumptions underlying the fixed effects analysis of variance and covariance. *Rev. Educ. Res.* 42(3):237-288, 1972.

**Goldscheider, C.** *Population, Modernization, and Social Structure,* Little Brown, Boston, 1971.

**Goodchild, M.** *Spatial Autocorrelation.* CATMOG, Norwich, England, 1986.

**Goodchild, M., Haining, R., and Wise, S.** Integrating GIS and spatial data analysis: problems and possibilities. *International Journal of Geographical Information Systems,* 6, 407-423, 1992.

**Green, M.** Ecological fallacies and the modifiable areal unit problem. *Research Report No. 27,* North West Regional Research Laboratory, Lancaster University, UK, 1993.

**Green, P.J.; Jennison, C.; and Seheult, A. H..** Analysis of field experiments by least squares smoothing. *J. Roy. Statist. Soc. Ser. B.* 47(2):299-315, 1984.

**Greenhalgh, S.** Toward a political economy of fertility: Anthropological contributions. *Population and Development Review,* 16, no.1, 85-106, 1990.

**Greenhalgh, S.; Zhu, C.; and, Li, N.** Restraining population growth in three Chinese villages, 1988-93. *Population and Development Review,* 20, no.2, 365-95, 1994.

**Griffith, D.** A numerical simplification for estimating parameters of spatial autoregressive models, in *Spatial Statistics, Past, Present, and Future,* edited by D. Griffith, Institute of Mathematical Geographers, Ann Arbor, MI, 183-197 1990b.

**Griffith, D. A.** A spatially adjusted ANOVA model. *Geographical Analysis,* 10, 296-301, 1978.

**Griffith, D.** *Advanced Spatial Statistics, Advanced Studies in Theoretical and Applied Econometrics,* Kluwer Academic Publishers, Dordrecht, The Netherlands, 1988a.

**Griffith, D.** Advanced spatial statistics for analysing and visualizing geo-referenced data. *International Journal of Geographical Information Systems,* 7, 107-123, 1993b.

**Griffith, D.** Estimating spatial autoregressive model parameters with commercial statistical packages, *Geographical Analysis,* Vol. 20, 176-186, 1988b.

**Griffith, D. A.** Geometry and spatial interaction. *Annals of the Asociation of American Geographers,* 72(3), 332-346, 1982.

**Griffith, D.** Modelling urban population density in a multi-centered city. *Journal of Urban Economics,* 9: 298-310, 1981.

**Griffith, D.** Simplifying the normalizing factor in spatial autoregressions for irregular lattices, *Papers in Regional Science,* Vol. 71, 71-86, 1992b.

**Griffith, D.** *Spatial Autocorrelation, A Primer,* Resource Publications in Geography, American Association of American Geographers, Washington, DC, 1987.

**Griffith, D.** Spatial regression Analysis on the PC: spatial statistics using MINITAB. *Discussion Paper #1,* Institute of Mathematical Geography, Ann Arbor, MI, 1989.

**Griffith, D.** *Spatial Regression Analysis on the PC: Spatial Statistics Using SAS.* Association of American Geographers, Washington, D.C., 1993c.

**Griffith, D.** (ed.). *Spatial Statistics: Past, Present, and Future.* Institute of Mathematical Geography, Ann Arbor, MI, 1990.

**Griffith, D.** Supercomputer and spatial statistics: a reconnaissance, *Professional Geographer,* Vol. 42, 481-492, 1990a.

**Griffith, D.A.** What is spatial autocorrelation? Reflections on the past 25 years of spatial statistics. *L'Espace geographique* 3:265-280, 1992.

**Griffith, D.** Which spatial statistics techniques should be converted to GIS functions? in *Geographic Information Systems, Spatial Modelling and Policy Evaluation,* edited by M. Fischer and P. Nijkamp. Springer-Verlag, 103-114, Berlin, 1993a.

**Griffith, D., and C. Amrhein, C.** *Statistical Analysis for Geographers*, Prentice-Hall, Englewood Cliffs, NJ, 1991.

**Griffith, D., and Csillag, F.** Exploring relationships between semi-variogram and spatial autoregressive models. *Papers in Regional Science*, 72, 283-295, 1993.

**Griffith, D.; Haining, R.; and Arbia, G.** Heterogeneity of attribute sampling error in spatial data sets. *Geographical Analysis*, 26: 300-320, 1994.

**Griffith, D. A., Lewis, R., Li, B., Vasiliev, I., Knight, S., Yang, X.** Developing Minitab Software for spatial statistical analysis: a tool for education and research. *Operational Geographer*, 8, 28-33, 1990.

**Griffith, D., and Sone, A.** Trade-offs associated with computational simplifications for estimating spatial statistical models. *Working Paper*, l'Institut de Mathématiques Economiques, Université de Bourgogne, Dijon, France (with French Resumé), 1992.

**Griffith, D., and Sone, A.** Some trade-offs associated with computational simplifications for estimating spatial statistical/econometric models: preliminary results. *Discussion Paper* No. 103, Department of Geography, Syracuse University, 1993.

**Griffith, D., et al.** Developing minitab software for spatial statistical analysis: a tool for education and research," *Operational Geographer*, Vol. 8, No. 3., 28-34, 1990c.

**Grondona, M., and Cressie, N.** Using spatial considerations in the analysis of experiments. *Technometrics*, 33, 381-392, 1991.

# H

**Haggett, Peter; Cliff, Andrew D.; and Frey, Allan.** *Locational Analysis in Human Geography* (in two volumes), 2nd edition. Wiley, New York, 1977.

**Haining, R.P.** A spatial model for high plains agriculture. *Ann. Ass. Am. Geog.* 68:(4):493-504, 1978.

**Haining, R.P.** 1980. Spatial autocorrelation problems. *In* D.T. Herbert and R.J. Johnson (eds.), *Geography and the urban environment, progress in research and applications.* 3:1-44, Wiley, New York, 1980.

**Haining, R.** *Spatial Data Analysis for Social and Environmental Sciences*, Oxford University Press, London, 1991.

**Haining, R.** The use of added variable plots in regression modeling with spatial data. *The Professional Geographer*, 42, 336-44, 1990.

**Haining, R., and Wise, S. M.** *GIS and Spatial Data Analysis: Report on the Sheffield Workshop*, Regional Research Laboratory Initiative Discussion Paper Number 11, University of Sheffield, UK, 1991.

**Haslett, J, Bradley, R., Craig, P., Unwin, A., and Wills, G.** Dynamic graphics for exploring spatial data with application to locating global and local anomalies. *American Statistician*, 45(3), 234-242, 1991.

**He-Nan Province Census Bureau,** *He-Nan Province Population Census Statistics, 1990*, China Statistics Publishing Company, Beijing, China, 1992.

**Hill, F.** Spatio-temporal trends in urban population density: a trend surface analysis, in *The Form of Cities in Central Canada*, edited by L. Bourne; MacKinnon, R.; and Simmons, J. University of Toronto Press, Toronto, pp. 103-119, 1973.

**Hillis, D.** *The Connection Machine*, MIT Press, Cambridge, MA, 1985.

**Hillis, D.** The connection machine, *Scientific American*, Vol. 256, No. 6, 108-115, 1987.

**Hillis, D., and Steele, G. L. Jr.** Data parallel algorithms, *Communications of ACM*, Vol. 29, pp. 1170-1183, 1986.

**Hordijk, L.** Problems in estimating econometric relations in space. *Papers of the Regional Science Association*, 42: 99-115, 1979.

**Hubert, L. J.; Golledge, R. G.; and, Costanzo, C. M.** Generalized procedures for evaluating spatial autocorrelation. *Geographical Analysis* 13, 224-33, 1981.

# I

**IBM.** *Engineering and Scientific Suboutine Library Guide and Reference*, 1992.

**IBM** Exploring new worlds with GIS. *Directions*, Summer/Fall, 12-19, 1991.

**Isaaks, E. and Srivastava, R.** *An Introduction to Applied Geostatistics.* Oxford University Press, Oxford, England, 1989.

## J

**Jager, H.I., and Overton, W.S.** Explanatory models for ecological response surfaces. In *Environmental Modeling with GIS,* M.R. Goodchild, B.O. Parks, and L.T. Steyaert (ed.). Oxford University Press, pp. 422-431, 1993.
**Jessen, R.J.** *Statistical Survey Techniques.* Wiley, New York, 1978.
**Johnsson, S.R.** Implicit policy and fertility during development. Population and Development Review, 17, no.3, 377-414, 1991.

## K

**Kau, J., C. Lee and Sirmans, C.** *Urban Econometrics.* Vol. 6, Research in Urban Economics Series, JAI Press, Greenwich, CT, 1986.
**Kish, L.** *Survey Sampling.* Wiley, New York, 1965.
**Kish, L.** *Statistical Design for Research.* Wiley, New York, 1987.
**Knodel, J. and van de Walle, E.** Lessons from the past: policy implications of historical fertility studies. *Development Review,* .5, no.2, 217-45, 1979.

## L

**Legendre, P.; Oden, N. L.; Sokal, R. R.; Vaudor, A.; and Kim J.** Approximate analysis of variance of spatially autocorrelated regional data. *J. Class.* 7:53-75, 1990.
**Lesthaeghe, R. and Surkyn, J.** Cultural dynamics and economic theories of fertility change. *Population and Development Review,* 14, no.1, 1-45, 1988.
**Li, B.** Opportunities and challenges of parallel processing in spatial data analysis: initial experiments with data parallel map analysis, *GIS/LIS '92,* 445-458, 1992a.
**Li, B.** Prospects of parallel processing in geographic data analysis, in *Development and Potentials of GIS: Inside and Outside China,* Hui Lin (eds), Science Press, Beijing, China, 76-89, 1992b.
**Li, B.** Suitability of topological data structures for data parallel operations in computer cartography, *Auto-Carto 11,* 434-443, 1993a.
**Li, B.** Developing network-oriented GIS software for parallel computing, *GIS/LIS'93,* 403-413, 1993b.
**Lindgren, B.** *Statistical Theory,* 3rd ed. Macmillan, New York,1976.
**Linthurst, R.A., Landers, D.H., Eilers, J.M., Brakke, D.F., Overton, W.S., Meier, E.P., and Crowe, R.E.** Characteristics of lakes in the eastern United States. Volume I: Population descriptions and physico-chemical relationships. U.S. Environmental Protection Agency, 401 M Street SW, Washington, DC 20460. (EPA-600/4-86/007a), 1986.
**Lowell, K.** Utilizing discriminant function analysis with a geographical information system to model ecological succession. *International Journal of Geographical Information Systems,* 5(2), 175-191, 1991.
**Luther, N.Y. Jr.; Feeney, G.; and, Zhang, W.** One-child families or a baby boom? Evidence from China's 1987 one-per-hundred survey. *Population Studies,* 44, no.2, 341-57, 1990.

## M

**Mandallaz, D.** (1993). *Geostatistical Methods for Double Sampling Schemes: Application to Combined Forest Inventories. Chair of Forest Inventory and Planning,* Department of Forest and Wood Sciences, ETH-Zentrum, CH-8092, Zürich, 1993.
**Mardia, K., and Marshall, R.** Maximum likelihood estimation of models for residual covariance in spatial regression. *Biometrika,* 71, 135-146, 1984.
**Martin, D. and Bracken, I.** Techniques for modelling population-related raster databases. *Environment and Planning A,* 23, 1069-1075, 1991.
**Maspar Computer Corporation.** *Maspar Parallel Application Language (MPL) User Guide,* Sunnyvale, CA, 1991a.

**Maspar Computer Corporation.** *Maspar FORTRAN User Guide*, Sunnyvale, CA, 1991b.

**Matérn, B.** *Spatial Variation*, 2nd ed. Springer-Verlag, Berlin,1986.

**McCauley, J., and Engel, B.** Spatial statistics and interpolation procedures for GRASS, paper presented to *The 8th Annual GRASS Users' Conference and Exhibition*, March 14-16, 1993, Reston, Virginia, 1993.

**MINAGRI.** *Production Agricole en 1987: Bilan d'Autosufficance Alimentaire par Commune et par Habitant*. Kigali, Rwanda: MINAGRI, Rép. Rwandaise, 1989.

**MINIPLAN.** *Syntése des principaux résultats du recensement général de la population et de l'Habitat 1978*. Kigali, Rwanda: Rép. Rwandaise, 1982.

**MINIPLAN.** *Recensement Général de la Population et de l'Habitat au 15 aout 1991: Résultats Préliminaires Echantillon au 10e*. Kigali, Rwanda: MINIPLAN, Rép. Rwandaise, 1992.

**MINITRAP.** *Bulletin Climatologique Année 1991*. Kigali, Rwanda: MINITRAP, Rép Rwandaise, 1991.

**Miron, J.** Spatial autocorrelation in regression analysis: a beginners guide. *In* G.L. Gaile and C.J. Willmot (ed.) *Spatial Statistics and Models*. Dordrectht. p. 201-222, 1984.

**Moellering, Harold.** Real maps, virtual maps and interactive cartography, pp. 109- 132 in Gaile and Willmott (ed.), *Spatial Statistics and Models*, Reidel, 1984.

**Moellering, H. and Tobler, W.** Geographical variances. *Geographical Analysis* 4, 34-50, 1972.

**Moran, P.** The interpretation of statistical maps. *J. Roy. Statist. Soc. Ser. B* 37:243-251, 1948.

**Mulla, D.J.; Bhatti, A. J.; and Kunkel, R.** Methods for removing spatial variability from field research trials. *Advances in Soil Science*. Springer-Verlag New York Inc., 13:201-213, 1990.

## N

**Nambookiri, N.K.** Some observations on the economic framework for fertility analysis. *Population Studies*, 26, 185-206, 1972.

**National Research Council (Mapping Science Committee; Commission on Physical Sciences, Mathematics, and Resources).** *Spatial Data Needs: The Future of the National Mapping Program*, National Academy Press, Washington, D.C., 1990a.

**National Research Council (Board on Mathematical Sciences).** *Renewing U.S. Mathematics: A Plan for the 1990s*. National Academy Press, Washington, D.C., 1990b.

**National Research Council (Panel on Spatial Statistics and Image Processing).** *Spatial Statistics and Digital Image Analysis*. National Academy Press, Washington, D.C., 1991.

**National Science Foundation.** *Solicitation: National Center for Geographic Information and Analysis*. Biological, Behavioral, and Social Sciences Directorate, Washington, D.C., 1987.

**Ness, Gayl D.; Drake, William D.; Brechin, Steven R., eds.** *Population-Environment Dynamics*. Ann Arbor, University of Michigan Press, 1993.

**Notestein, F.** Economic problems of population change, in *Proceedings of the Eighth International Conference of Agri-Economics*, Oxford University Press, London, 1953.

**Notestein, F.** Population - the long view, in T.W.Schultz (ed.), *Food for the World*, University of Chicago Press, Chicago, 1954.

**Nystuen, Jeffrey A.; McGlothin, Charles C.; and, Cook, Michael S.** The underwater sound generated by heavy rainfall. *Journal of the Acoustical Society of America*, 93 (6), 3169-3177, 1993.

**Nzisabira, J.** Accumulation du Peuplement Rural et Ajustements Structurels du Système d'Utilisation du Sol au Rwanda Depuis 1945. *Bulletin Agricole du Rwanda*, 2, 117-127, 1989.

## O

**Odland, J.** *Spatial Autocorrelation*. Sage, Beverly Hills, CA, 1988.

**Okabe, A., Boots, B., and Sugihara, K.** *Spatial Tessellations: Concepts and Applications of Voronoi Diagrams.* Wiley, New York, 1992.

**Olea, R.A.** Sampling design optimization for spatial functions. *Mathematical Geology*, 16, 369-392, 1984.

**Oliver, M., Webster, R., and Gerrard, J.** Geostatistics in physical geography. Part 2: Applications. *Transactions, Institute of British Geographers*, 14, 270-286, 1989.

**Olson, Jennifer M.** *The Impact of Socioeconomic Factors on Migration Patterns in Rwanda.* Unpublished M.A. Thesis. Department of Geography, Michigan State University, 1990.

**Olson, Jennifer M.** Farming systems of Rwanda: echoes of historic divisions reflected in current land use. *Rwanda Society-Environment Project Working Paper*, 2, Department of Geography, Michigan State University, East Lansing, MI, 1994a..

**Olson, Jennifer M.** Demographic responses to resource constraints. *Rwanda Society-Environment Project Working Paper*, 7, Department of Geography, Michigan State University, East Lansing, MI, 1994b.

**Olson, Judith M.** Autocorrelation and visual map complexity. *Annals of the Association of American Geographers* 65(2,189 - 204, 1975.

**Olson, K.R.; Carmer, S. G.; and Olson, G. W.** *Assessment of effects of soil variability on maximum alfalfa yields*, GEDMAB, New York, 36:1-14, 1985.

**Openshaw, S.** A geographical solution to scale and aggregation problems in region-building, partitioning and spatial modelling. *Transactions of Institute of British Geographers* 2, 459-72, 1977b.

**Openshaw, S.** *Concepts and Techniques in Modern Geography, Number 38. The Modifiable Areal Unit Problem.* Geo Books, Norwich, 1984.

**Openshaw, S.** Optimal zoning systems for spatial interaction models. *Environment and Planning A* 9, 169-84, 1977a.

**Openshaw, S. and Taylor, P. J.** A million or so correlation coefficients: three experiments on the modifiable areal unit problem. Pages 127-144 in N. Wrigley (ed.) *Statistical Applications in the Spatial Sciences*, Pion Limited, London, 1979.

**Ord, J. K.** Estimation methods for models of spatial interaction, *Journal of the American Statistical Association*, Vol. 70, 120-126, 1975.

**Overton, W.S., and Stehman, S.V.** Properties of designs for sampling continuous spatial resources from a triangular grid, *Communications in Statistics - Theory and Methods*, 21, 2641-2660, 1993.

**Overton S., and Stehman, S.** Statistical properties of designs for sampling continuous functions in two dimensions using a triangular grid. *Technical Report* No. 143, Department of Statistics, Oregon State University, 1990.

**Overton, W.S., and Stehman, S.V.** The Horvitz-Thompson theorem as a unifying perspective for probability sampling: with examples from natural resource sampling. *American Statistician* (forthcoming), 1995.

**Overton, W.S., and Stehman, S.V.** Variance estimation in the EMAP strategy for sampling discrete ecological resources, *Environmental and Ecological Statistics*, 1, 133-152, 1994.

**Overton, W.S., Kanciruk, P., Hook, L.A., Ellers, J.M., Landers, D.H., Brakke, D.F., Blick, D.J., Linthurst, R.A., DeHaan, M.D.** *Characteristics of Lakes in the Eastern United States. Volume II: Lakes Sampled and Descriptive Statistics for Physical and Chemical Variables*, EPA600/486007b, Washington, DC, U. S. Environmental Protection Agency, 1986.

**Overton, W.S., White, D., and Stevens, D.L.** *Design Report for EMAP: Environmental Monitoring and Assessment Program.* Washington, DC, U. S. Environmental Protection Agency (EPA/600/3-91/053), 1991.

**Owen, D.** The Starship. *Communications in Statistics*, 17, 315-341, 1988.

# P

**Paelinck, J. and Klaassen, L.** *Spatial Econometrics.* Saxon House, Farnborough, England, 1979.

**Peng, Peiyuan**, Accomplishments of China's family planning program: a statement by a Chinese official. *Population and Development Review*, 19, no.2, 399-403, 1993.

**Peng, X.** Major determinants of China's fertility transition. *The China Quarterly*, No.117 (March), 1-37, 1989.

**Petersen, R.G.** *Design and Analysis of Experiments*. Marcel Dekker, Inc. p. 429, 1985.

**Plane, David A. and Rogerson, Peter A.** *The Geographical Analysis of Population*. Wiley, New York, 1994.

**Press, W., *et al.*** *Numerical Recipes in C, the Art of Scientific Computing*, Cambridge University Press, Cambridge, 1988.

**Prioul, C. and Sirven, P.** *Atlas du Rwanda*. Kigali, Rwanda: Imprimérie Moderne Nantasie Coueron, 1981.

**Puppo, E., L. Davis, D. DeMenthon, and Y. A. Teng,** Parallel Terrain Triangulation, *Proceedings of the 5th International Symposium on Spatial Data Handling*, Charleston, SC, 632-641, 1992.

**Putman, S. H. and Chung, S-H.** Effects of spatial systems design on spatial interaction models. 1: the spatial definition problem. *Environment and Planning A* 21, 27-46, 1989.

**Pye, L.** *The Dynamics of Chinese Politics*, Delgeschlager, Gunn & Hain, Publishers, Cambirdge, MA, 1981.

# Q

**Quenouille, M.H.** Problems in plane sampling. *Annals of Mathematical Statistics*, 20, 355-375, 1949.

**Quinn, M. J.** Data-parallel programming on multicomputers, *IEEE Software*, Sept. 1990, 69-76, 1990.

# R

**Ripley, B.** *Statistical Inference for Spatial Processes*. Cambridge University Press, Cambridge, England, 1988.

**Ripley, B.D.** *Spatial Statistics*. Wiley, New York, 1981.

**Robinson, A. H.** The necessity of weighing values in correlation analysis of areal data. *Annals, Association of American Geographers* 46: 233-236, 1956.

**Robinson, W. S.** Ecological correlations and the behavior of individuals. *American Sociological Review* 15: 351-357, 1950.

**Rose, J.R. and Steele, G. L. Jr.** C*: An Extended C Language for Data Parallel Programming, *Technical Report PL 87-5*, Thinking Machines Corp., Cambridge, MA, 1986.

**Royall, R. M.** On finite population sampling theory under certain linear regression models. *Biometrika*, 57, 377-387, 1970.

# S

**Sadowski, Frank G. and Covington, Stephen J.** *Processing and Analysis of Commercial Satellite Image Data of the Nuclear Accident near Chernobyl, U.S.S.R.* U. S. Geological Survey, Bulletin 1785. U.S. Government Printing Office, Washington, D. C., 1987.

**Sandhu, J. S. and Marble, D.** An investigation into the utility of the Cray X-MP supercomputer for handling spatial data, *Proceedings, Third International Symposium on Spatial Data Handling*, Sydney, Australia, 253-267, 1988.

**Särndal, C.E., Swensson, B., and Wretman, J.** (1992). *Model Assisted Survey Sampling*. Springer-Verlag, New York,1992.

**SAS Institute Inc.** *SAS Procedures Guide*, release 6.03 edition. Cary, NC. 210 p., 1988.

**SAS Institute Inc.** *SAS/STAT User's Guide* (two volumes), SAS Institute Inc., Cary, NC, 1989.

**Schultz, T. P.** *Economics of Population*, Addison-Wesley, MA, 1981.

**Smith, H.** Integrating theory and research on the institutional determinants of fertility. *Demography*, 26, 171-184, 1989.

**Semple, R. K. and Green, M. B.** Classification in human geography. . In G. L. Gaile and C. J. Willmott, Eds., *Spatial Statistics and Spatial Models*. Dordrecht: Reidel, 133-145.

**Shryock, H. S. and Seigel, J. S.** *The Methods and Materials of Demography*, Academic Press, Inc., New York, 1976.
**Shryock, H.S., and Siegel, J. S.** *et al.* *The Methods and Materials of Demography*, U.S.Bureau of Census, Washington, D.C., Vol.2, 1980.
**Smith, J. R.** *The Design and Analysis of Parallel Algorithms*, Oxford University Press, New York and Oxford, 1993.
**Steel, D.G.; Holt, D.; and, Tranmer, M.** Modelling and adjusting aggregation effects. Paper presented at the Annual Conference of the US Bureau of the Census, 1993.
**Stehman, S.V., and Overton, W.S.** Comparison of variance estimators of the Horvitz-Thompson estimator for randomized variable probability systematic sampling. *Journal of the American Statistical Association*, 89, 30-43, 1994.
**Stehman, S., and Overton, W.** Variance estimation for fixed-configuration, systematic sampling. *Technical Report* No. 134, Department of Statistics, Oregon State University, 1989.
**Stetzer, F.** Specifying weights in spatial forecasting models: the results of some experiments. *Environment and Planning A*, 14, 571-584, 1982.
**Stuart, A.** *The Basic Ideas of Scientific Sampling* (1984 ed.), Charles Griffin and Company, London, 1962.

## T

**Thinking Machines Corporation.** *CM FORTRAN Reference Manual*, Cambridge, MA, 1992.
**Thinking Machines Corporation.** *CM FORTRAN Utility Library Reference Manual*, Cambridge, MA, 1993a.
**Thinking Machines Corporation.** *CMSSL for CM Fortran: CM-5 Edition*, Cambridge, MA, 1993b.
**Thinking Machines Corporation.** *Programming in C\**, Cambridge, MA, 1991c.
**Thinking Machines Corporation.** *Programming in FORTRAN*, Cambridge, MA, 1991b.
**Thinking Machines Corporation.** *The Connection Machine CM-5 Technical Summary*, Cambridge, MA, 1991a.
**Thompson, D'Arcy W.** *On Growth and Form*. Cambridge University Press, 1917.
**Thrall, G.** Statistical and theoretical issues in verifying the population density function, *Urban Geography*, 9: 518-537, 1988.
**Tobler, W.** Cellular geography. In *Philosophy in Geography*, edited by S. Gale and G. Olsson, pp. 379-86, Reidel, Dordrecht, 1979.
**Tobler, W.** Frame independent spatial analysis. In *Accuracy of Spatial Databases*, edited by M. F. Goodchild and S. Gopal, pp. 115-22. Taylor and Francis, London, 1991.
**Tobler, Waldo R.** *Map Transformations of Geographic Space*. Ph.D. Thesis, University of Washington, 1961.
**Tobler, Waldo R.** Preliminary representation of world population by spherical harmonics. *Proceedings of the National Academy of Sciences of the United States of America*, 89:6262, 4 Jul. 15 1992
**Tobler, W.** Smooth pycnophylactic interpolation for geographical regions. *Journal of the American Statistical Association*, 74(367), 519-536, 1979.
**Trew, A., and Wilson, G.** *Past, Present, Parallel: A Survey of Available Parallel Computer Systems*, Springer-Verlag, London, 1992.
**Tukey, J.W.** Comparing individual means in the analysis of variance. *BIOKA*. 5:99-114, 1949.

## U

**Upton, G.** Information from regional data, in *Spatial Statistics: Past, Present, and Future*, edited by D. Griffith, Institute of Mathematical Geography, Ann Arbor, pp. 315-359, 1990.

**Upton, G.J.G., and Fingleton, B.** *Spatial Data Analysis by Example, Volume I: Point Pattern and Quantitative Data.* Wiley, New York, 1985.
**Upton, G. J. G., and Fingleton, B.** *Spatial Data Analysis by Example, Volume II: Categorical and Directional Data.* Wiley, New York, 1989.

# V

**van Es, H.M. and van Es, C. L.** The spatial nature of randomization and its effect on the outcome of field experiments. *Agron. J.* 85:420-428, 1993.

# W

**Warnecke, L.** GIS in the states: applications abound. *GIS World*, 3, (# 3), 54-58, 1990.
**Warnecke, L.** *State Geographic Information Activities Compendium.* Council of State Governments, Lexington, KY, 1991.
**Warntz, William.** *Macrogeography and Income Fronts.* Philadelphia, Regional Science Research Institute, Monograph #3, 1965.
**Wartenberg, D.** Multivariate spatial correlation: a method for exploratory geographical analysis *Geographical Analysis* 17, 263-83, 1985.
**Webster, R., and Oliver, M.A.** *Statistical Methods in Soil and Land Resource Survey.* Oxford University Press, New York, 1990.
**Willis, R. J.** A new approach to the economic theory of fertility behavior. *Journal of Political Economy*, 81, 14-69, 1973.
**Wolf, A. P.** The preeminent role of government intervention in China's family revolution. *Population and Development Review*, 12, 106-16, 1986.
**Wolter, K.** Introduction to Variance Estimation. Springer-Verlag, New York, 1985.
**Wong, D.W.S.** Spatial dependency of segregation index. Paper presented at 41st North American Meeting of Regional Science Association International, Niagara Falls, Ontario, Canada, Nov.17-20, 1994.
**Wrigley, N.** Revisiting the Modifiable Areal Unit Problem and the ecological fallacy. In *Festschrift for Peter Haggett*, edited by Cliff, A.D., Gould, P.R., Hoare, A.G. and Thrift, N.J., Blackwell, Oxford, 1994.

# X

# Y

**Yates, F.** *Sampling Methods for Censuses and Surveys* (4th ed.). Charles Griffin and Company, London, 1981.
**Yi, Z.** Is the Chinese family planning program 'tightening up'? *Population and Development Review*, 15, No.2 (June), 1989.

# Z

**Zimmerman, D.L. and Harville, D. A.** A random field approach to the analysis of field-plot experiments and other spatial experiments. *Biometrics*, 47:223-239, 1991.
**Zubrzycki, S.** Remarks on random, stratified and systematic sampling in a plane. *Colloquium Mathematicum*, 6, 251-264, 1958.

When a term appears once in a chapter, that is generally a good indication that there is much material concerning that term within that chapter. An index is a statistical tool by which one can gauge the content of a book; in the spirit of this particular handbook, we cast this traditional statistical tool in a spatial format--so that the reader can appreciate the relative locations of topics, from an alphabetical perspective, within the handbook. The matrix below shows a set of terms and the chapters in which they appear. A darkened square indicates that the term on the left appears in the chapter; a blank indicates that it does not.

| CHAPTER NUMBER | 1 | 2 | 3 | 4 | 5 | 6 | 7 | 8 | 9 | 10 |
|---|---|---|---|---|---|---|---|---|---|---|
| adjacency | | ■ | | ■ | | | | | | |
| aggregation | ■ | | | ■ | | ■ | | | | |
| aggregation effect | ■ | | | ■ | | | | | | |
| analysis of variance | | | | | | ■ | ■ | | | ■ |
| ANOVA | | | ■ | | | ■ | ■ | | | ■ |
| area | ■ | ■ | ■ | | ■ | ■ | ■ | ■ | ■ | |
| areal | ■ | ■ | ■ | ■ | ■ | ■ | ■ | ■ | | ■ |
| areal sampling | | | ■ | | | | | | | |
| autocorrelation | ■ | ■ | ■ | ■ | ■ | ■ | ■ | ■ | ■ | ■ |
| autoregression | | | | | | ■ | ■ | ■ | | |
| autoregressive | ■ | | | ■ | | ■ | ■ | ■ | | ■ |
| autoregressive response model | | | | | | ■ | | | ■ | ■ |
| bias | ■ | | ■ | ■ | ■ | | ■ | ■ | ■ | |
| bivariate | | | | ■ | ■ | | | | | |
| complexity | ■ | ■ | | | ■ | ■ | | | | |
| conditional autoregressive model | | | | | | ■ | | | | |
| connectivity | | | | ■ | | ■ | ■ | ■ | ■ | ■ |
| contiguity | | | | ■ | ■ | | ■ | ■ | | |
| correlogram | ■ | | ■ | | | | | | | |
| density | ■ | | ■ | | | ■ | ■ | ■ | ■ | ■ |
| distribution | ■ | ■ | ■ | | ■ | ■ | ■ | ■ | ■ | ■ |
| eigenvalue | | | | ■ | | ■ | | ■ | ■ | ■ |
| error | ■ | | ■ | ■ | ■ | ■ | ■ | ■ | ■ | ■ |
| error modeling | | | | ■ | | | | | | |
| estimator | ■ | | ■ | ■ | | ■ | ■ | | ■ | |
| FORTRAN | | | | | ■ | | | | | |

| CHAPTER NUMBERS | 1 | 2 | 3 | 4 | 5 | 6 | 7 | 8 | 9 | 10 |
|---|---|---|---|---|---|---|---|---|---|---|
| simultaneous autoregressive model | | | | | | ■ | | | | |
| spatial autocorrelation | ■ | ■ | ■ | ■ | ■ | ■ | ■ | ■ | ■ | ■ |
| spatial autoregression | | | | | | ■ | ■ | ■ | ■ | |
| spatial dependency | | | | ■ | ■ | ■ | ■ | | | |
| spatial econometrics | ■ | | | ■ | ■ | ■ | | | ■ | |
| spatial mean | | ■ | ■ | | | | | | | |
| spatial regression | ■ | ■ | | | ■ | ■ | | ■ | ■ | ■ |
| spatial sampling | ■ | ■ | ■ | | | | | | | |
| spatial statistics | ■ | ■ | ■ | ■ | ■ | ■ | ■ | ■ | ■ | ■ |
| spatially adjusted | | | ■ | | | | ■ | | | |
| standard error | ■ | | ■ | ■ | | | ■ | ■ | ■ | |
| stochastic | ■ | | ■ | | ■ | | | ■ | | ■ |
| strata | | | ■ | | | | | | | |
| supercomputer | ■ | | | | | ■ | | | | |
| tessellation | ■ | | ■ | ■ | | | | | | |
| Thiessen | | | ■ | | | ■ | | | | |
| time series | ■ | | | | | | | | | ■ |
| transition | ■ | | | | | | | ■ | | |
| translation | | | ■ | | | | | | | |
| unbiased | ■ | | ■ | ■ | ■ | | | ■ | ■ | |
| universe | | | ■ | | | | | | | |
| variance | ■ | | ■ | ■ | ■ | | | ■ | ■ | ■ |
| variogram | ■ | | ■ | | | | ■ | | | |
| visualization | ■ | ■ | | | | ■ | ■ | | | |
| Voronoi | ■ | | | | | ■ | | | | |
| weight | ■ | ■ | | ■ | ■ | ■ | ■ | ■ | ■ | ■ |

Milton Keynes UK
Ingram Content Group UK Ltd.
UKHW021623071024
449327UK00020BA/1160